Smart Technologies for Sustainable Development Goals: Zero Hunger

Editors

A. Jose Anand

Professor, Department of Electronics & Communication Engineering, KCG College of Technology, Chennai, India

Saravanan Krishnan

Associate Professor, Department of Computer Science & Engineering, College of Engineering, Anna University, Chennai, India

CRC Press
Taylor & Francis Group
Boca Raton London New York

CRC Press is an imprint of the
Taylor & Francis Group, an **informa** business
A SCIENCE PUBLISHERS BOOK

First edition published 2025
by CRC Press
2385 NW Executive Center Drive, Suite 320, Boca Raton FL 33431

and by CRC Press
4 Park Square, Milton Park, Abingdon, Oxon, OX14 4RN

CRC Press is an imprint of Taylor & Francis Group, LLC

Library of Congress Cataloging-in-Publication Data (applied for)

ISBN: 978-1-032-88267-3 (hbk)
ISBN: 978-1-032-94242-1 (pbk)
ISBN: 978-1-003-56966-4 (ebk)

DOI: 10.1201/9781003569664

Typeset in Times New Roman
by Prime Publishing Services

Preface

In recent years, the global community has made remarkable strides toward addressing poverty, hunger, and inequality through the United Nations Sustainable Development Goals (SDGs). Among these, **SDG 2: Zero Hunger** stands as a fundamental commitment to eradicate hunger and ensure food security for all by 2030. However, achieving this goal requires innovative approaches and the integration of advanced technologies to tackle multifaceted challenges such as food production, distribution, and equitable access. This book, *Smart Technologies for Sustainable Development Goals: Zero Hunger*, explores the transformative role of smart technologies in achieving this ambitious objective.

This edited volume brings together contributions from experts and researchers across diverse disciplines, offering comprehensive insights into how emerging technologies can revolutionize hunger alleviation strategies. The chapters collectively emphasize that while hunger is a complex global challenge, smart solutions—ranging from precision agriculture to blockchain-based supply chain transparency—can create a sustainable path toward food security.

The book is organized into the following chapters:

1. **Introduction to Smart Technologies to Ensure Zero Hunger: SDG2** lays the groundwork, outlining how technological advancements intersect with the goal of eliminating hunger.
2. **Understanding Sustainable Development Goal 2: Zero Hunger** provides an in-depth analysis of the key principles and targets under SDG 2.
3. **Amalgamation of Technological Innovations in Hunger Alleviation** explores how various technologies can synergize to address different aspects of food insecurity.
4. **Nourishing Nations: Harnessing Nutritional Food Security for Zero Hunger** discusses the importance of focusing on nutritional value alongside food production.
5. **Precision Farming in Urban Environments to Ensure Zero Hunger** examines how technology-driven urban agriculture can contribute to food security in densely populated areas.
6. **Blockchain for Transparent Food Supply Chains** highlights the potential of blockchain technology to enhance traceability and trust in global food supply chains.
7. **Mobile Solutions for Food Access and Distribution** investigates how

mobile technologies can bridge gaps in food access, especially in remote and underserved regions.

8. **Climate Smart Agriculture (CSA) and Food Security in Rural Communities of Developing Countries** assesses the qualitative impact of CSA practices in mitigating food insecurity.
9. **Smart Agriculture: Cultivating Sustainable Solutions for a Green Future** focuses on technology-driven sustainable agricultural practices that foster environmental stewardship.
10. **Community Empowerment through Technology** underscores the role of technology in enabling communities to take ownership of local food security initiatives.
11. **Leveraging Machine Learning for Paddy Disease Identification in Sustainable Agriculture** demonstrates the practical applications of machine learning in improving crop yields and resilience.
12. **Innovative Technologies for Alleviating Hunger: Challenges and Opportunities** explores both the promise and hurdles associated with implementing advanced technologies in hunger alleviation.
13. **Multidimensional Aspects of Zero Hunger in Sub-Saharan Africa** offers a regional perspective on the diverse factors influencing hunger and potential interventions.
14. **Smart Aqua Culture Cultivation to Achieve Zero Hunger in Indonesia** examines how technological advancements in aquaculture can address food insecurity in coastal regions.
15. **Leveraging Information and Communication Technology (ICT) Solutions for Food Security in Nigeria** highlights how ICT can transform food systems in resource-constrained settings.
16. **Augmented Reality-based Food Wastage Reductions in Oman Food Restaurants** showcases a novel approach to reducing food waste through augmented reality.

This book aims to inspire policymakers, researchers, practitioners, and students to harness smart technologies for creating innovative, sustainable, and context-specific solutions to combat hunger. By presenting a wide range of perspectives and case studies from across the globe, it underscores the critical role of technological integration in shaping a future where no one goes hungry. As we navigate the path toward 2030, let us recognize that the journey to Zero Hunger is not just a technological challenge but a collective responsibility demanding collaboration, creativity, and unwavering commitment.

A. Jose Anand
Saravanan Krishnan

Contents

1 | Introduction to Smart Technologies to Ensure Zero Hunger: SDG2

Amal Ray[1*], Ashutosh Kumar Srivastava[1], Sukhbir Singh[2] and Mahadev Bar[3]

[1] Lovely Professional University, Phagwara, Punjab, India
[2] Faculty of Fashion and Design, SGT University, Gurugram Haryana, India
[3] Production Engineering Laboratory, University of Technology Tarbes Occitanie Pyrénées (UTTOP), Tarbes, France

Over the past few decades, unpredictable weather conditions and insects considerably hamper the quality and quantity of crops. Farmers face many problems, especially limited resources like water, soil, workforce, crop failure, poor quality of the crop, crop loss due to insects and disease attacks. Food security remains a pressing issue in countries like India for multiple reasons, including the shortage of agricultural land, harsh weather conditions, and geographical location. The constraints on cultivated land, as well as resources, may lead to an acute problem related to the food crisis in third-world countries such as India. Thus, there is a need to take necessary steps to improve crop productivity to achieve zero hunger with the help of the technological approach to overcome the impact of the food crisis.

In this chapter are discussed the many facets of agriculture, including new technology, challenges, advantages, and potential contributions to sustainable agricultural practices and global food security for the ever-growing world population. The greenhouse cultivation process evolved from being covered with shade material to an advanced, sophisticated, controlled, or semi-controlled environment. In this scenario, the properties of covering materials play an essential role.

1. Introduction

Even though Agriculture occupies 44% of the world's habitable land, 48

*Corresponding author: textile.amal@gmail.com

million square kilometres make up the entire region (UN Food and Agriculture Organization, 2019), there are still challenges with food availability because of factors like population growth, poverty, climate change, insufficient food distribution, biofuels, corruption, inadequate storage facilities, a lack of awareness, proper knowledge, and education [1]. Complicating matters is the significant proportion of individuals living below the poverty line (BPL). Due to droughts and floods brought on by climate change, livestock, forestry, fisheries, and aquaculture have all been negatively impacted by farming as well as agricultural activities [2]. To overcome those critical challenges, the introduction of closely supervised nutrition programmes needs to be prioritized. For most of our civilization, humans survived by gathering edible plants and killing animals in the vast grasslands, woods, and other untamed environments. Indeed, a sophisticated system of hunting and gathering underpins much of our evolutionary progress, providing diverse and healthy food [3]. All this change occurred roughly 10,000 years ago, when agriculture came at the same time in many civilizations on Earth, resulting in a food-related system that is still quite prevalent today. The agriculture sector dominates the worldwide landscape. The agricultural industry dominates the global environment. As mentioned earlier, a substantial portion of the world's land area is under agriculture, and more food is produced than ever before. Much of this development has come at the expense of large amounts of farmed land. Other environmental constraints are being tested as a result of agricultural expansion [3]. Around half of the freshwater of the whole world is used for irrigating cultivated land, soil erosion has surpassed soil formation, and chemical herbicides and pesticides have caused widespread contamination. However, the agriculture sector accounts for approximately up to one-third of greenhouse gas (GHG) emissions. While this growth has significant environmental consequences, it has not always resulted in improved dietary and nutrient intakes [4].

Global food security is currently a serious concern, exacerbated by the world's ever-increasing population, climate change, and rising wealth, all of which contribute to increasingly resource-intensive diets. According to the United Nations Food and Agriculture Organization (NUFAO), worldwide food production must grow by around 70% to feed roughly 10 billion people by 2050 [5–6]. It has created an alarming situation of supply and demand into balance by using cutting-edge technology to increase agricultural productivity. It is anticipated that the global population will be around 9.55 billion population in 2055 [7].

Over-population will stress on the agricultural resources, for instance, cultivated land and water. Thus, farmers are encouraged by new farming techniques such as greenhouse cultivation methods, mulching techniques, genetic modification crops, and cultivated climate-sustainable crops to improve the use of agricultural resources [8]. The adaptation of new innovative technology can be a viable solution towards meeting the global food demand. The problems are further aggravated due to the reduction of workforce in the agriculture sector as people move to urban areas to earn their livelihoods.

This chapter aims to give a vision and framework for ways to accomplish the Sustainable Development Goal for Zero Hunger (SDG2), which focuses on hunger reduction, food security, nutrition improvement, and sustainable agriculture. To increase the agricultural productive capacity of developing nations, SDG2 focuses on building up rural infrastructure, promoting agricultural research, encouraging technological development, and producing crops [9]. Therefore, accomplishing the SDG2 needs new technology as well as climate-smart agriculture (CSA), which adjusts food production to a changing environment while reducing its negative consequences [10]. CSA efforts are to enhance agricultural productivity, improve climate resilience, and minimize GHG emissions. It focuses on reducing agricultural emissions through strategies such as integrated pest management, the usage of renewable energy, and conservation agriculture [11].

2. Role of Food Security in Achieving Zero Hunger

Hunger is the term used to describe the physical discomfort brought on by not consuming enough food. The FAO adopted the broad resolution of food security at the World Food Summit in Rome in 1996 [12]. It claims that food security occurs when everyone has constant physical and economic access to sufficient, safe, and nutritious food to suit their dietary needs and food choices for an active and healthy lifestyle [13]. This definition is used by many analysts. The definition underlines the importance of food safety, sufficiency, and nutrition. It addresses consumer behaviour by their preference and test, which are influenced, among other things, by the cultural characteristics of the communities to be applicable in practice and move beyond prescriptive biological concerns. Food uses are often described as the body's optimal usage of various nutrients included in the diet. Individuals' appropriate energy and nutritional intake are the result of effective care and feeding practices, food preparation, dietary diversity, and family-wide food distribution [14]. Food security for zero hunger requires a complex strategy that includes social protection, providing safe and nutritious food, particularly for children, and changing food systems to build a more inclusive and sustainable society [15]. As a result, both urban and rural communities will require investments, as well as social protection. Consequently, underprivileged individuals will have access to food and may better their living conditions.

2.1. Availability

In this segment, availability is the first element, which takes into account the issue of food security. It is demonstrated by total food production that is sufficient to satisfy domestic demand [16].

2.2. Access

The requirement is shown at the home level by the access pillar. According to Sen et al. (1981), the characteristic of starvation is that some people do not have enough food to consume. It is not the fact that there is little food available, although the latter may contribute to the former as only one of several potential reasons. Because of this well-known quote, demand-side policy making has gained international attention by taking into account factors like pricing, purchasing power, and the distribution of goods both within and between households. The demand side of this process connects concerns of income poverty and distribution to food security [17].

2.3. Application

The appropriate use of food expenditures by households on affordable, nutritionally necessary items is referred to as the usage elements. It also takes into account factors like how well people absorb the food they eat and related concerns like safe preservation, appropriate cooking, and hygienic surroundings [18].

2.4. Stability

Finally, the stability element takes into account how households may be more susceptible to food insecurity as a result of intertemporal variations in access or availability [19].

3. Concept of Poverty

Poverty and the income-based poverty line are closely connected concepts. People who are BPL with a benchmark income or consumption of food and other resources are considered impoverished. This benchmark can be the lowest acceptable income or level of consumption that a community can afford to provide a satisfactory standard of living. The percentage of the population that lives in poverty is known as the extent of poverty, often known as the headcount ratio of poverty [20]. The different dimensions of sustainable agriculture are shown in Figure 1.

The link between income and nutritional intake is frequently used to determine the poverty line on a normative basis. An individual's calorie requirement usually determines how much food they require. The average per capita energy use increases and tends to plan areas at a high level of income if the data arrange households by per capita income level and progress from low-income to high-income groups [21]. This relation suggests that the level of consumer spending at which energy intake is only enough to fulfill society's average calorie requirement level may be used to determine the poverty line. However, in this approach, the particular families whose income is below the poverty line are known as poor.

Individual calorie requirements typically dictate how much food they need, but additional nutrients enhance meal quality and help create a balanced diet. Nutritionists advise a balanced diet of foods that match the needs for main nutrients while considering social and economic factors for different age, sex, and activity

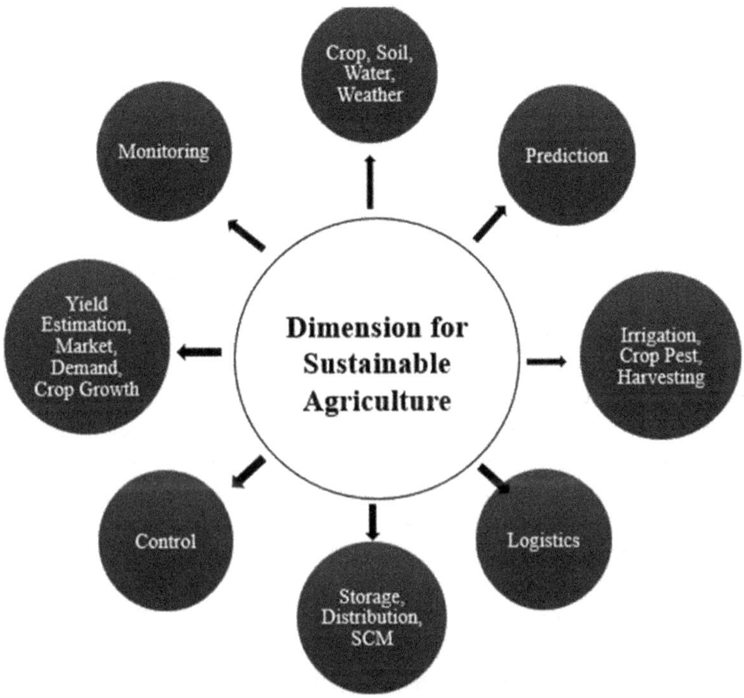

Figure 1: Different dimensions of sustainable agriculture.

groups [22]. This is similar to how they recommend specialized nutrient diets. When a certain allowance is given for fundamental essential non-food goods, the enlarged food may be considered as a basic need food. The population-weighted average of the suggested dietary tolerances for different categories may be used to determine the average food needs of a society. Therefore, an alternative normative poverty level may be estimated using the cost of essential consumer demands. It's significant to keep in mind that poverty is determined by the comparison of the price of the food entirely with income or total consumption expenditure, not by comparing the cost of the food item in detail. Even though a household's actual use of a given item may surpass or fall short of the norm used in the basic needs of food, it is still considered impoverished if its income is insufficient to cover the entire cost of the food [23]. However, a specific family's income equal to the poverty line may or may not have consumption equal to the calorie norm even under the calorie-based approach to setting the line. The equivalence applies to an average level of family income [24].

4. Strategic Plan for Sustainable Agriculture

Sustainable agriculture is very crucial to any family and society in a broad way as it remains the source of living, business, reducing poverty, employment opportunities, and also the sourcing of raw materials [25]. The importance of agriculture was fascinating from the aspect of national growth as a means of producing wealth and a source of income for farmers by exhibiting them to external trading such as the export of products. Thus, agriculture plays an important role in both human civilization and building economic growth for the nation [26].

Moreover, this is fascinating to see how the concept of the smart farming process is getting more attention from farmers, agronomists, and agricultural researchers alike. One of the prominent approaches to the smart farming process is the use of greenhouse technology [27]. It is a closed environment cultivation process that leverages selective information and communication technology to improve the quality of crops with minimal intervention from humans [28]. The advancement of Internet of Things (IoT) technology has a vast potential for innovative methods and smart solution development that can revolutionize the agriculture sector. Thus, integrating the IoT with a greenhouse can transform it into a smart and automated greenhouse, one of the possible solutions for sustainability is considered to be one of the ideal solutions to hanger issues [29]. However, sustainability is the capacity to continue at present levels of activity without irreversibly depleting natural resources. The sustainability of the ecosystem depends on the protection of natural resources [30–31]. According to the UN, sustainable development is the growth that makes sure present demands are satisfied without the ominous capacity of the next generations to meet their own, as stated in the Brundtland Report [32]. Deeply, the concept is predicated on the notion that natural resources are finite and must be utilized with caution to preserve the current quality of living standard without sacrificing that of the coming generations. A sustainable civilization is contingent upon the dynamic equilibrium between natural and human systems. There are three components to the idea of sustainability, which are shown in Figure 2.

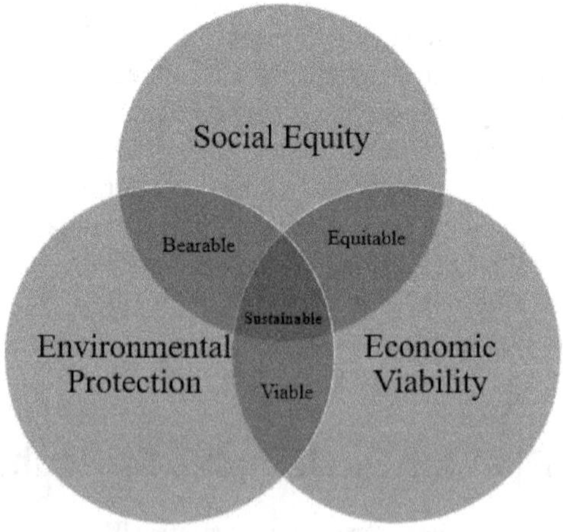

Figure 2: The three main pillars of sustainability [33].

Environmentally sustainable is the direction in which natural resources can be harvested without wiping out non-renewable resources. It will become not sustainable if it cannot be engaged in advance for an indefinite period. Due to increasing pollution levels, the death of seas resulting in more and more harsh climatic conditions, it is crucial to prioritize the sustainability precept of defending the environment [34]. Finally, a sustainable economic process concerns the capacity of a country to hold a consistent level of economic production throughout time. To achieve public support for environmental preservation, businesses must reduce the negative environmental consequences of their production practices. However, the debate will focus primarily on the environmental pillar of sustainability.

Agriculture and its related sectors are important to the accomplishment of the 17 SDGs acquired by 193 nations in 2015 and determine aims and adopted as scheduled in 2030 [35]. The agriculture sector is the biggest employer in the world and the main source of food and the economy, especially for the poor family, who live under the poverty line. A prominent majority of the total population in developing nations is dominating the global landscape, specifically, Africa, Latin America, and Asia depend on agriculture for their livelihood [36]. More than 40% of the world's land surface is under agriculture, and more food is produced than at any time in the history of humanity [37]. The meaning of sustainable agriculture belief means different things to different individuals. According to the definition, sustainable refers to something that may be maintained or continued throughout time. Sustainable agriculture is not supposed to harm people or the environment. It is even recognized for its ability to mitigate the negative effects of traditional approaches, such as reducing soil erosion, revitalizing the earth's biodiversity, and enhancing food nutrition. Furthermore, sustainable agriculture

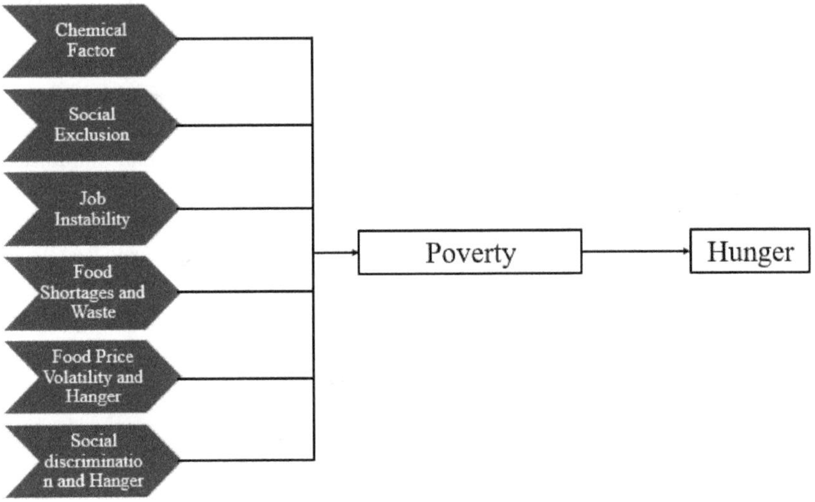

Figure 3: Different factors affecting poverty and hunger.

seeks to improve not just the environment for future generations, but also the growth of the economy and society [38]. Also, sustainable agriculture is the management of agricultural resources in order to meet the changing human needs while conserving the environment and enhancing productivity. The factors that are affecting the poverty level and hunger are shown in Figure 3.

To eliminate hunger by 2030, urgent coordinated policy measures must be implemented to address entrenched inequities, promote sustainable farming practices, reform food systems, and mitigate the impact of war and epidemics on global nutrition and food security. Thus, it can be considered that sustainability in the agro-industry is very important to maintain the balance of the environment.

5. Challenges for Food Security

Food security challenges are a worldwide problem, ranging from invertebrates to top predators such as humans. Every living species on the inhabitable Earth is concerned about food security to overcome hunger under harsh climatic circumstances [39].

5.1. Degradation of the Land and Soil

Excessive use or misuse of agrochemicals, deforestation, and natural disasters pose major challenges to the production of food in a sustainable manner. Currently, almost one-third of the soil on Earth has degraded.

5.2. Climate Change

About 70% of India's yearly rainfall and 60% of its net land under cultivation are irrigated by the monsoon. India's agricultural output is already declining due to altering precipitation forms and a rise in the frequency and severity of extreme weather consequences like heat waves, dryness, heavy rainfalls and floods, which pose a severe threat to food security. Over the last 15 years, India has had over 10 significant outbreaks of invasive pests and weeds, such as the fall armyworm in 2018 that nearly destroyed all maize harvests and the locust attacks that have been seen mainly in Rajasthan and Gujarat in 2020. In addition, high seed costs, which cause manual labor and low production, are another issue facing the agriculture sector. In small and marginal produce particularly, the majority of agricultural operations are still done by hand using basic resources, despite mechanization in some areas [40]. Farmers frequently take loans at excessive interest rates from traders, commission agents, and moneylenders to buy agricultural products at lower prices. The aforementioned elements and many more create challenges to India's food security.

6. Government Initiative for Food Security

The First Five-Year Plan, initiated by the government in 1951, distinguished that nutrition was a crucial aspect in maintaining people's health and had an impact on an individual's productive potential. It emphasized the prevalence of undernutrition in the country, as well as a dearth of preventive foods to supplement primary grains, which contributes to malnutrition. The Second Five-Year Plan admitted that it was not possible to offer appropriate nutrition to the whole population throughout the Plan period [41]. Mentioning that damage to proper development and growth at an early stage can't be kept still by allowing for enough nutrition. However, this started certain strategies focused on nutritional improvement for the susceptible groups of the population e.g., nursing mothers, infants, and children [42].

On the other hand, low crop production and rising population are two major factors for restructuring the existing policy to increase productivity to meet domestic food demand. A large-scale dependency on imports could not be a viable solution for the large country since world trade in the food grain market is small, particularly for rice where the volume is only around 10% of consumption. On the other hand, international trade volume is not thin for wheat, and economic processes in the food market are widely predominant [43]. It was clear to government policymakers that the strategy of food security had to mostly depend on domestic food production. However, the scheme did not rule out international trade agreements as the need to add on domestic production with some imports [44]. There are mainly four dimensions of food security that are discussed below:

6.1. Physical Availability of Food

Food availability refers to the supply side of food security and it is controlled by the level of food production, stock levels, and food grain export-import.

6.2. An Economic and Physical Approach to Food

Individual family's food security is not guaranteed by an adequate supply of food on a national or international scale. Concerns about limited food access have resulted in a more comprehensive strategy that focuses on incomes, expenditures, markets, and pricing to achieve food security objectives.

6.3. Food Use

Food uses are often defined as the process by which the human body produces the most diversified nutrients in food. Adequate energy for individual and nutritional intake is achieved by correct care and feeding methods, food preparation, various dietary options, and food distribution within the family. This defines an individual's nutritional condition, as well as the correct biological metabolism of the food consumed.

6.4. Stability of the Other Three Dimensions Over Time

Food insecurity persists, and individuals do not have enough access to food, putting their nutritional worth in danger. Adverse climate conditions, political instability, or economic problems such as unemployment, inflation, and rising food costs can all influence food security.

7. Environment Factor on Agriculture and Food Security

The effectiveness of agricultural production depends upon the different environmental factors. These include elements that influence photosynthesis, such as light intensity, atmospheric CO_2 concentration, water supply, and mineral content. Micro-climatic conditions, particularly temperature and humidity are essential factors. In addition, the quality of the soil, including the presence of minerals, water circulation, and oxygenation of roots, has a substantial impact on agricultural growth. Even the presence of a deficiency in one of these factors can limit crop production, irrespective of variations in other factors. This leads to considerations about the feasibility of controlling environmental factors.

8. Benefits of Sustainable Agriculture for SDG2

Organic farming, crop rotation methods, and urban rooftop cultivation for vegetable crops, aquaponic and vertical farming are the available sustainable agriculture methods that can improve soil health as well as boost crop production. Some of the most commonly presented ideas for increasing agricultural climate resilience are energy-saving irrigation, controlled environment farming, preservation agriculture, energy-efficient cold storage, and animal grazing management [45]. Besides that, diverse cropping patterns can enable farms to better endure harsh weather situations such as droughts and floods. Sustainable agriculture prioritizes biodiversity and the protection of natural resources. Agroecological and no-till farming approaches both assist in avoiding soil erosion and protect water quality [46]. Maintaining natural ecosystems on or near farms can help with biodiversity, crop pollination, and organic pest management. Sustainable agriculture may enhance the livelihoods of farmers and rural communities by offering them more stable earnings. It reduces production costs by reducing the dependency on expensive inputs such as synthetic fertilizers and pesticides. Sustainable methods usually prioritize local markets and communities, which benefits rural economies. Sustainable agricultural methods, such as soil CO_2 absorption and less chemical usage can help to reduce GHG emissions [43]. Sustainable agriculture indirectly helps to ensure long-term food security by reducing climate change and preserving consistent environmental conditions for batter crop production [48]. Overall, sustainable agriculture provides a complete food production approach that addresses both long-term environmental and social sustainability as well as present food security concerns.

9. Different Approaches to Sustainable Agriculture

9.1. Sustainable Farming Practices

9.1.1. Organic Farming

Natural alternatives and soil health get priority above synthetic pesticides and fertilizers in organic farming. In addition to generating organic food, it promotes biodiversity and reduces chemical pollution.

9.1.2. Crop Rotation and Diversification

To improve soil health and interrupt insect life cycles, crops are rotated in a field throughout the season. Diversification is the practice of adding many crops or livestock to a farm to enhance resilience and dependency on a single crop [49].

9.1.3. Conservation Tillage

Leaving agricultural waste on the field lowers soil disturbance. Crop production and the environment benefit because of its capacity to prevent soil erosion, retain moisture, and trap carbon in the soil [50].

9.1.4. Agroforestry

Growing crops or animals along bushes or trees on the same area of land benefits farmers by generating a variety of items such as fruits and wood while also increasing biodiversity. It promotes soil fertility and carbon sequestration.

9.1.5. IPM (Integrated Pest Management)

This process uses many pest-control strategies that avoid the overuse of chemicals. This includes pest-resistant varieties, crop monitoring, and biological treatments, all of which reduce environmental harm and promote long-term pest management during crop production.

9.2. Sustainable Livestock Farming for Animals

Animal welfare, economic viability, and ecologically friendly techniques are the first prioritized in sustainable crop production. It works to maintain resources, reduce GHG emissions, and promote ethical animal welfare [51]. Rotational grazing, smart feed management, and reduced antibiotic usage are examples of sustainable methods that strike a balance between meeting global protein demands and defending the environment for future generations [52].

9.3. Empowering Smallholder Farmers

Small farmers play an important part in world food production. They contribute significantly to the world's food supply and employ a sizable fraction of the agricultural labour force, particularly in developing nations. Their diversified, localized farming practices promote crop diversity, contribute to food security, and generally emphasize long-standing, climate-tolerant approaches. To accomplish global food security and sustainable agriculture, smallholders must be supported with resources, technology, and expertise [53].

Many initiatives and legislation improve small-scale farmers' access to critical resources, empowering them. Extension services provide valuable agricultural education and information exchange, whereas microfinance programmes provide loans and financial services. Subsidies for inputs like seeds and fertilizer help to lower production costs. Fair-trade initiatives increase market access, which ensures better prices for their products [54]. Insurance for agriculture covers agricultural losses caused by natural disasters, but land tenure rights provide ownership of land and access. The adoption of economical and sustainable agricultural technology, as well as aid from regional farmer cooperatives. It helps small-scale farmers enhance productivity, income, and resilience.

10. Renewable Energy for Sustainable Development on Agriculture

Promoting renewable resources such as solar, wind, biomass, tidal, geothermal, small-scale hydro, biofuels, and wave-generated power can assist in mitigating negative environmental consequences. These renewable resources have significant promise in the agriculture industry. Subsidies should be provided to farmers to encourage the usage of renewable energy technologies. The idea of sustainable agriculture is focused on achieving a delicate balance between improving agricultural yields and preserving economic stability while using fewer natural resources and minimizing negative environmental repercussions. This reduces dependency on fossil fuels, lowers energy prices, and eliminates environmental effects. Mainly four different factors affecting renewable energy resources are shown in Figure 4.

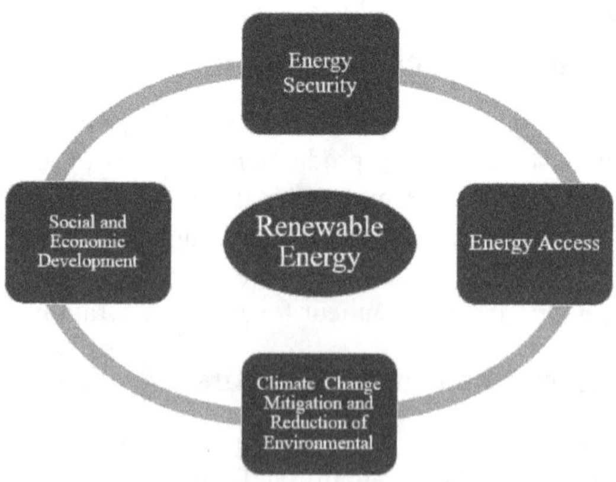

Figure 4: Factors affecting renewable energy [55].

11. Technology and Innovation in Sustainable Agriculture

Sustainable agricultural practices have changed as a result of technology, especially precision agriculture and digital technologies. Precision agriculture utilizes information from GPS, sensors, and drones to ensure the appropriate use of resources [56]. Farmers can accurately administer water, fertilizer, energy-efficient irrigation, and pesticides, reducing crop waste while also benefiting the environment. Farmers may make educated decisions with the help of digital tools and applications that provide real-time weather, crop monitoring, and market current data. Furthermore, drought-resistant, pest-tolerant, and nutrient-rich crops may be developed using genetic engineering and biotechnology, enhancing production and sustainability. These advancements not only boost output but also promote ecologically friendly farming practices, which are critical for a more resilient and sustainable agricultural success. Agriculture is evolving as a consequence of inventions such as vertical farming and aquaponics [57]. Vertical farming may assist urban areas by stacking crops in regulated circumstances while using fewer space and resources. Aquaponics, which combines aquaculture with hydroponics, uses fish excrement to fertilize plant growth [58]. Both approaches contribute to more sustainable and effective food production by reducing water consumption, pesticide use, and food miles. However, vertical farming and aquaponics boost year-round agricultural productivity in cities with limited resources, hence improving food security. Weather-related agricultural failures are a common cause of crop failures across the world [59]. However, urban residents may produce locally through rooftop and vertical farming. For example, fresh vegetables and seafood reduce dependency on long food supply chains and ensure a more reliable and accessible food source for communities, thus improving food security.

12. Sustainable Agriculture and Climate Resilience

Sustainable agricultural strategies boost resistance to global warming by increasing soil health, conserving water, diversifying crops, and extending crop cycles. Agroforestry and crop cycle rotation are both less susceptible to harsh weather conditions [60]. Soil resistance to droughts and floods is enhanced by reduced chemical usage and carbon sequestration. Regardless of changing climate conditions, these strategies promote long-term agricultural sustainability. Achieving the SDG-2 requires smart agriculture, which adapts food production to a changing environment while mitigating its adverse effects [61]. It increases agricultural sustainability by ensuring consistent food production in the face of altering weather and development conditions. CSA promotes food security, sustainable development, and the broader objective of eradicating hunger by reducing GHG emissions and conserving natural resources [62].

13. Recycling

Organic wastes are a rich supply of biological material and important plant nutrients, therefore agricultural recycling of these wastes looks to be a potential alternative method for enabling value addition and innovative applications. Application of organic waste in cultivated land stabilized by procedures such as composting, vermicomposting, and anaerobic digestion produces excellent organic fertilizer, such as compost, which improves soil fertility and crop production [63]. Furthermore, the method includes indirect environmental benefits such as lower GHG emissions, land conservation as a result of less waste landfilling, and chemical fertilizer alternatives. The waste hierarchy refers to the '3Rs' (reduce, reuse, and recycle), which are illustrated in Figure 5 and define waste management solutions based on their desirability. Economically, agricultural use of organic waste saves the cost of landfilling, trash transfer, imports, and chemical fertilizer manufacture, while also creating opportunities for rural employment. However, the successful use of organic wastes for agricultural applications necessitates a thorough and stringent risk assessment to avoid the negative impacts of contaminants such as heavy metals and persistent organic pollutants, ensuring agro-environmental sustainability [64]. The main purpose of this study is to describe the advantages and disadvantages of this practice, allowing us to design a method or plan that maximizes environmental and economic benefits.

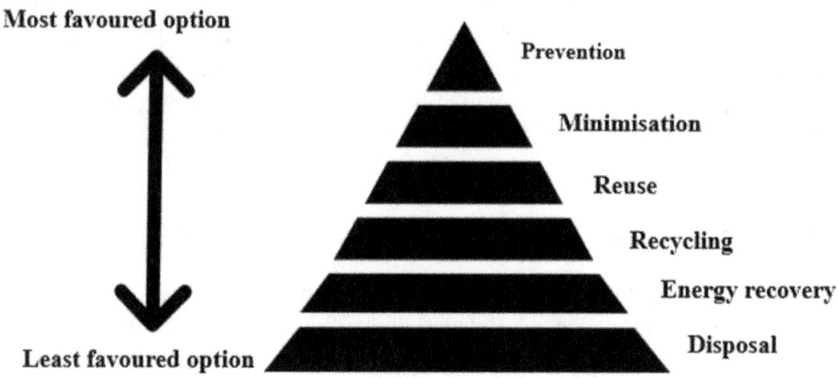

Figure 5: 3 R's hierarchy for agriculture wastes' management [65].

14. Conclusion

A consolidated idea of most of the important characteristics of sustainable way food production, and food security to accomplishing zero hunger is provided in this chapter. However, achieving the Zero SDG2 is a difficult but vital undertaking that requires a multifaceted strategy. Organic farming, crop diversity, conservation tillage, agroforestry, and integrated pest management are all examples of sustainable agricultural methods that are critical for increasing food security, eliminating hunger, and promoting sustainability. We may strive toward a civilization in which no one goes hungry while conserving the planet's scarce resources by prioritizing approaches and equipping farmers with new developing technologies. Adopting smart agriculture techniques ensures a more resilient and secure future for everyone and the next generation, which helps to fight hunger and also unpredictable weather conditions.

References

1. Kaviyarasan, G., Rahale, S. and Subramaniam, T.. (2024, Jan.). Revolutionizing agriculture: A deep dive into modern fertilizer application strategies. *In*: *Recent Trends in Soil Science*, SSSA, 119–137.
2. Sundari, M.S., Anand, A.J. and Nagarajan, T.S. (2022). Smart aerial imaging solution in precision agriculture. *In*: *International Conference on Data Science, Agents, and Artificial Intelligence* (*ICDSAAI*), Chennai, India, 1–3.
3. Behera, B.K., Rout, P.K. and Behera, S. (2019). *Move Towards Zzero Hunger*. doi: 10.1007/978-981-32-9800-2.
4. Lile, R., Ocnean, M. and Balan, I.M. (2023, Dec.). Challenges for zero hunger (SDG2): Links with other SDGs. *In*: *2 Zero Hunger*, 1–59. doi: 10.3390/books978-3-03897-863-3-2.
5. Bruinsma, J. (2009). *Expert Meeting on How to Feed the World in 2050*. FAO, Rome.
6. Otekunrin, O.A. (2024, Dec.). Countdown to the 2030 global goals: A bibliometric analysis of research trends on SDG 2 – Zero Hunger. *Curr. Res. Nutr. Food Sci.*, 11: 1338–1362. doi: 10.12944/CRN FSJ.11.3.34.
7. Rahman, F. and Abhijeet Madhukar Haval. (2024). Application of machine learning techniques and the Internet of Things for smart, sustainable agriculture. *BIO Web Conf.*, 05021(82): 1–11.
8. Alazzai, W.K., Abood, B.S.Z., Al-Jawahry, H.M. and Obaid, M.K. (2024). Precision farming: The power of AI and IoT technologies. *E3S Web Conf.*, 491: 0–5. doi: 10.1051/e3sconf/202449104006.
9. Sivakumar, S. (2024, Jan.). Precision agriculture technologies. *In*: *Sustainable Crop Management in Modern Agronomy*, 117–145. ND Global Publication.
10. Sarker, M.N.I., Hossain, B., Shi, G. and Firdaus, R.B.R. (2023). Promoting net-zero economy through climate-smart agriculture: Transition towards sustainability. *Sustain. Sci.*, 18(5): 2107–2119. doi: 10.1007/s11625-023-01379-0.
11. Md. Shariful Islam, A.K.M.K.P. and Md. Shahriar Kabir. (2023). Climate-smart agriculture promoting sustainable agriculture in Bangladesh. *J. Agricultural Rural Res.*, 7(1): 1–7.
12. United Nations. (2018). Goal 2: Zero Hunger – United Nations Sustainable Development. *In*: *Goal 2: Zero Hunger*, 1–5 [Online]. Available: https://www.un.org/sustainabledevelopment/hunger/.
13. Haldar, N. and Rinitta Pearlin A. (2023). Sustainable and climate smart agriculture for food security: A review. *J. Exp. Agric. Int.*, 45(11): 229–239. doi: 10.9734/jeai/2023/v45i112253.

14. Bin Alam, M.F., Tushar, S.R., Zaman, S.M., Gonzalez, E.D.R.S., Bari, A.B.M.M. and Karmaker, C.L. (2023). Analysis of the drivers of Agriculture 4.0 implementation in the emerging economies: Implications towards sustainability and food security. *Green Technol. Sustain.*, 1(2): 100021. doi: 10.1016/j.grets.2023.100021.
15. Khan, M.R. et al. (2022, Jul.). *Sustainable Agriculture Leads to Zero Hunger.* [Online]. Available: https://iksadyayinevi.com/wp-content/uploads/2022/07/SUSTAINABLE-AGRICULTURE-LEADS-TO-ZERO-HUNGER-.pdf%0Ahttps://iksadyayinevi.com/home/sustainable-agriculture-leads-to-zero-hunger/.
16. Wijerathna-Yapa, A. and Pathirana, R. (2022). Sustainable agro-food systems for addressing climate change and food security. *Agric.*, 12(10): 1–26. doi: 10.3390/agriculture12101554.
17. Tilman, D., Balzer, C., Hill, J. and Befort, B.L. (2011). Global food demand and the sustainable intensification of agriculture. *Proc. Natl. Acad. Sci. U.S.A.*, 108(50): 20260–20264. doi: 10.1073/pnas.1116437108.
18. Lee, M., Rivard, C., Wang, W., Pliakoni, E., Gude, K. and Rajashekar, C.B. (2021). Spectral blocking of solar radiation in high tunnels by poly covers: Its impact on nutritional quality regarding essential nutrients and health-promoting phytochemicals in lettuce and tomato. *Horticulturae*, 7(12): 524. doi: 10.3390/horticulturae7120524.
19. Kotykova, O., Babych, A. Iagodzinska, and A. Tabatskova. (2022). Global initiatives on implementation of zero hunger policy. *Agric. Resour. Econ.*, 2: 30–49. doi: 10.51599/are.2022.08.02.02.
20. Bhavani, R. and Rampal, P. (2020). Harnessing agriculture for achieving the SDGs on poverty and zero hunger. *ORF Issue Br.*, 407: 16.
21. Singh, K.P., Singh, S.K. and Singh, S.K. (2013). An economic analysis of milk production in Sultanpur District of Uttar Pradesh. *Econ. Aff.*, 58(2), 83–87. [Online]. Available: http://search.ebscohost.com/login.aspx?direct=true&db=ecn&AN=1442035&site=ehost-live%5Cnhttp://www.ndpublisher.in/ndpjournal.php?j=EA
22. Çakmakçı, R., Salık, M.A. and Çakmakçı, S. (2023). Assessment and principles of environmentally sustainable food and agriculture systems. *Agric.*, 13(5). doi: 10.3390/agriculture13051073.
23. Food Security Information Network. (2020). Global report on food crises 2020. *Fao. Org.*, 1–240. [Online]. Available: https://www.wfp.org/publications/2020-global-report-food-crises
24. Shanthakumar, S.D.S. (2023). *Achieving Zero Hunger in India: Challenges and Policies*. Springer Link.
25. Akella, G.K., Wibowo, S., Grandhi, S. and Mubarak, S. (2023). A systematic review of blockchain technology adoption barriers and enablers for smart and sustainable agriculture. *Big Data Cogn. Comput.*, 7(2) doi: 10.3390/bdcc7020086.
26. Bathaei, A. and Štreimikienė, D. (2023). A systematic review of agricultural sustainability indicators. *Agric.*, 13(2) doi: 10.3390/agriculture13020241.
27. Javaid, M., Haleem, A., Singh, R.P. and Suman, R. (2022, July). Enhancing smart farming through the applications of Agriculture 4.0 technologies. *Int. J. Intell. Networks*, 3: 150–164. doi: 10.1016/j.ijin.2022.09.004.
28. Balatsouras, C., Karras, A., Karras, C., Karydis, I. and Sioutas, S. (2023). Ecosystem for smart agriculture applications. *Sensors*, 23(9486): 1–34.
29. Dhanaraju, M., Chenniappan, P., Ramalingam, K., Pazhanivelan, S. and Kaliaperumal, R. (2022). Smart farming: Internet of Things (IoT)-based sustainable agriculture. *Agric.*, 12(10): 1–26 doi: 10.3390/agriculture12101745.

30. Maraveas, C., Piromalis, D., Arvanitis, K.G., Bartzanas, T. and Loukatos, D. (2022). Applications of IoT for optimized greenhouse environment and resources management. *Comput. Electron. Agric.*, 198(April): 106993 doi: 10.1016/j.compag.2022.106993.
31. Marouani, R., Mahamat, C., Adnen, C. and Fst, A. (2024, Jan.). Design and Implementation of a Smart Photovoltaic Hydroponic Greenhouse for Sustainable Agriculture in Tunisia Design and Implementation of a Smart Photovoltaic Hydroponic Greenhouse for Sustainable Agriculture in Tunisia. *Preprints*, 3–26.doi: 10.20944/preprints202401.0643.v1.
32. Sunderland, T.C.H., O'Connor, A., Muir, G., Nerfa, L., Nodari, G.R., Widmark, C. et al. (2019). SDG 2: Zero hunger-challenging the hegemony of monoculture agriculture for forests and people. *Sustain. Dev. Goals Their Impacts for People*, (Harari 2014), 48–71. doi: 10.1017/9781108765015.004.
33. Maraveas, C. and Bartzanas, T. (2021). Application of Internet of Things (IoT) for optimized greenhouse environments. *AgriEngineering*, 3(4): 954–970 doi: 10.3390/agriengineering3040060.
34. Grigorieva, E., Livenets, A. and Stelmakh, E. (2023). Adaptation of agriculture to climate change: A scoping review. *Climate*, 11(10). doi: 10.3390/cli11100202.
35. Rijah, M. (2023). Developing strategies for "Zero Hunger" – Angel Wings. *J. Inf. Syst. Inf. Technol.*, 7(August), 1–18.
36. Musa, S.F.P.D., Basir, K.H. and Luah, E. (2021). The role of smart farming in sustainable development. *Int. J. Asian Bus. Inf. Manag.*, 13(2): 1–12. doi: 10.4018/ijabim.20220701.oa5.
37. Karunathilake, E.M.B.M., Le, A.T., Heo, S., Chung, Y.S. and Mansoor, S. (2023). The path to smart farming: Innovations and opportunities in precision agriculture. *Agric.*, 13(8). doi: 10.3390/agriculture13081593.
38. Velazquez, J.B. and Rulka, A. (2023). Welcome to the inaugural issue of sustainable food technology. *Sustain. Food Technol.*, 1(1), 7–8. doi: 10.1039/d2fb90001g.
39. Miguel-Rojas, C. and Pérez-de-Luque, A. (2023). Nanobiosensors and nanoformulations in agriculture: New advances and challenges for sustainable agriculture. *Emerg. Top. LifeSci.*, 7(2): 229–238. doi: 10.1042/ETLS20230070.
40. Ammann, J., Umstätter, C. and El Benni, N. (2022). The adoption of precision agriculture enabling technologies in Swiss outdoor vegetable production: A Delphi study. *Precis. Agric.*, 23(4): 1354–1374. doi: 10.1007/s11119-022-09889-0.
41. Mahdi, S.S. (2018, Jan.). *Climate Change and Agriculture in India: Impact and Adaptation*. Springer Link. doi: 10.1007/978-3-319-90086-5.
42. Gummagolmath, K.C. (2021). Domestic and export competitiveness of major agricultural commodities in India with special reference to Telangana. 1(1), 1–149.
43. Singh, S.P. (2017). An Economic Analysis of Shade net Cultivation in Chhattisgarh Plain. Semantic Scholar.
44. Balusamy, S.R., Joshi, A.S., Perumalsamy, H., Mijakovic, I. and Singh, P. (2023). Advancing sustainable agriculture: A critical review of smart and eco-friendly nanomaterial applications. *J. Nanobiotechnology*, 21(1). doi: 10.1186/s12951-023-02135-3.
45. Lalramhlimi, B., Mukherjee, D., Chakraborty, I., Ghosh, N., Chattopadhyay, A. and Dey, Ramesh C. (2022). Fruit and vegetable wastes as livestock feeds. *In*: *Fruits and Vegetable Wastes Valorization to Bioproducts*, pp. 139–168. May 2023. Research Gate. doi: 10.1007/978-981-16-9527-8.

46. Dhanalakshmi, R., Jose Anand, Arun Kumar Sivaraman and Sita Rani. (2022). IoT-based water quality monitoring system using cloud for agriculture use. *In*: *Cloud and Fog Computing Platforms for Internet of Things*, 187–196. Routledge.
47. Plastras, S., Polimeni, S., Skoutas, D.N., Kormentzas, G. and Skianis, C. (2023). Sustainable networking solutions in remote IoT environments: Use cases, challenges, and solutions for smart agriculture. *Key Challenges Geogr.*, Part F2252, 307–325. doi: 10.1007/978-3-031-40747-5_17.
48. Gabriel, S.E. (2023). Appraising the obligations of the Nigerian State in achieving zero hunger. *ABUAD Law J.*, 11(1): 45–77.
49. Xie, Q., Ni, J.Q., Bao, J. and Su, Z. (2019, Aug.). A thermal environmental model for indoor air temperature prediction and energy consumption in pig building. *Build. Environ.*, 161. doi: 10.1016/j.buildenv.2019.106238.
50. Chinasa, O.B., Ukonze, A.P.J. and Okad, D.A.O. (2022). Sustainable skill acquisition in agriculture as a panacea for economic growth and achieving zero hunger in Nigeria. *Int. J. Sci. Manag. Res.*, 05(04): 226–236. doi: 10.37502/ijsmr.2022.5417.
51. Michelle, V. and Gunawan, J. (2023). Support to sustainable development goal no. 2 – Zero Hunger: A study from animal feed sub sector industries. *ARRUS J. Soc. Sci. Humanit.*, 3(4): 557–564. doi: 10.35877/soshum1991.
52. Bhatt, P., Rani, A., Gahlot, M. and Bhandari, V. (2019, Jun.). Recycling and reuse of agrotextiles used in agricultural farms. *Int. J. Curr. Microbiol. Appl. Sci.*, 8(06): 1566–1576. doi: 10.20546/ijcmas.2019.806.188.
53. Ramawat, N. and Bhardwaj, V. (2023). *Wild Food Plants for Zero Hunger and Resilient Agriculture*. Springer.
54. Pramanik, K., Mohapatra, P.P., Pradhan, J., Acharya, L.K. and Jena, C. (2020). Factors influencing performance of capsicum under protected cultivation: A review. *Int. J. Environ. Clim. Chang.*, *C* (12): 572–588. doi: 10.9734/ijecc/2020/v10i1230339.
55. Nagar, A.K., Namdeo, S., Verma, B., Shrivastava, S. and Sarkar, R. (2023). Role of climate-smart agriculture towards sustainable food production. *Int. J. Environ. Clim. Chang.*, 13(12): 465–472. doi: 10.9734/ijecc/2023/v13i123703.
56. Duraiarasu, E., Mahalakshmi, S., Jose Anand, A., Manikandan, T. and El Rhalibi, A. (2023). Precision farming for crop prediction. *In*: *Advanced Technologies for Smart Agriculture*, 195–218. Routledge.
57. Dirkes, L., Massanés, J.D., Böttjer, R., Storck, J.L. and Ehrmann, A. (2021). Outdoor vertical farming on textile substrates. *IOP Conf. Ser. Mater. Sci. Eng.*, 1031(1). doi: 10.1088/1757-899X/1031/1/012020.
58. Nath, D. (2023, Sept.). Smart farming: Automation and robotics in agriculture (Chapter 16). *In*: *Recent Trends in Agriculture* (Vol. 5), 281–310. Prasad Central Agricultural University, Pusa , Bihar , India.
59. Ehrmann, A. (2019). On the possible use of textile fabrics for vertical farming. *Tekstilec*, 62(1): 34–41. doi: 10.14502/Tekstilec 2019.62.34-41.
60. Ahmed, A.A., Institutions, D.S. and Patil, V. (2024, Apr.). Smart agriculture: Using IoT to strengthen the agriculture sector and overcoming the challenges of cybersecurity in the context of Agriculture 5.0. *In*: *Green Solutions: A Handbook for Sustainable Business Strategies and Models*, 62–73. New Delhi: Bharti Publications.
61. Roussaki, I., Doolin, K., Skarmeta, A., Routis, G., Claffey, E., Mora, M. et al. (2023). Building an interoperable space for smart agriculture. *Digit. Commun. Networks*, 9(1): 183–193. doi: 10.1016/j.dcan.2022.02.004.
62. Nguyen, V.D., Pham, T.C., Pham, T.T. and Phan, A.N. (2023). Digital and circular technologies for climate-smart and sustainable agriculture: The case of Vietnamese

coffee. *IOP Conf. Ser. Earth Environ. Sci.*, 1278(1): 012003. doi: 10.1088/1755-1315/1278/1/012003.

63. Marinova, V. and Radev, R. (2023). Sustainability of the European textile industry. *E3S Web Conf.*, 452: 1–13. doi: 10.1051/e3sconf/202345205020.
64. Iderawumi, A.M. and Khan, N.F. (2024, Jan.). Precision agriculture: Towards emerging trends in food security and advances in agricultural technology and plant sciences. *Adv. Agric. Technol. Plant Sci.*, 7: 1–4.
65. Nwakaire, J.N. and Obi, O.F. (2016, Oct.). Agricultural waste concept, generation, utilization and management. *Niger. J. Technol.*, 35: 957–964. doi: 10.4314/njt.v35i4.34.

2 | Understanding Sustainable Development Goal 2: Zero Hunger

D.R. Saxena[1], A.D. Saxena[1], N.J. Tupkar[1], F.A. Karim[1*] and Anton L. Irving[2]
[1] KNM, Nagpur, Maharashtra, India
[2] CNASAHN, Northern Caribbean University, Mandeville, Manchester

The Supreme Court of India (2001) directed extensive orders to implement the "Right to Food under the Rights guaranteed by the Indian Constitution/ Fundamentals Rights". It must be called "zero hunger fundamental right for sustenance, children and women, as well as the old age people are the worst sufferers"; the zero hunger food security law must be implemented at the earliest (D.R. Saxena, 2023). The human rights commission observed that "significant human rights problems" currently exist in India (2021). As per the August (2008) World Bank global poverty statistics, approximately 1.4 billion people live at or below the poverty line of $1.25 a day. Advances in agricultural science paved the way for modern farming called conventional farming that gradually replaced the ancient traditional farming leading to imbalances in food production and it is believed that genetically modified crops may restore the yield balance. The green revolution (1960s) was spread to third world countries by the USA, Britain, and Germany to counteract poverty. Natural farming called Fukuoka method or in India known as "Rishi Kheti" means agriculture adopted by the sages that maintains soil fertility and plant yields. This method refrains from applications of tillage, chemical fertilizers, animal waste, pesticides, herbicides, etc. All available local natural materials are applied as nutrients, natural ways of insect pest control, disease control are promoted, etc. Modern intensive agriculture and the international and national policies that support such ventures had impacted the destruction of ecological, hydrological, and other processes. Its single human global objective is to decline and exterminate problems of food shortage, food losses of standing as well as stored food, and hunger through systematic agriculture and business marketing chains, foster sustenance of agriculture through participation for the rural economic development. The

*Corresponding author: firdoskarim6@gmail.com

other face of modern agriculture is organic farming by utilization of crop rotation, economic manuring, farm generated organic resources, biofertilizers to enhance soil quality, control of biological pests and discouraging "Jhum cultivation". Excessive application of chemical fertilizers cause land exhaustion, compaction related decline in soil porosity and burning of vegetation to clear land for growing crops destroy the natural soil microbial biota, earthworms, and soil arthropods that bioengineer the soil structure for aeration, wetting, impart mobility to nutrients, and replenish the groundwater cycle. Sustainable agriculture is concerned with fulfilling the current food demands by preserving the natural resources for the coming generation. It encompasses (a) soil, water, nutrients, etc., management and conservation; (b) employing latest green agriculture technologies like global positioning system to prevent agriculture hazards, using the time-saving agriculture precision equipments, providing cost efficient healthy animal feed, advantageous use of government policies and laws; (c) application of conservation tillage to stop soil erosion, soil pollution, to enhance carbon fixation/sequestration, and stop soil salinity by flood water; (d) agro-ecological principle based integrated crop management practices ensure increased crop yields simultaneously preserving quality of environment for future; (e) avoid continuous monocrop cultures of non-leguminous plants.

The success of sustainable agriculture to ensure zero hunger depends on overcoming problems like policy failures and institutional weaknesses; also in overcoming conflict for sharing of water between urban, rural, interstate, between countries that limit the sustainable water management adoption as envisaged by ICWE (1992). The action plan adoption for right planning-budgeting-decision implementation to optimize the use, distribution of current water supplies will mitigate water scarcity problems. Animal energy and other conventional and renewable non-conventional energy sources, respectively, must be used for irrigation and operating agricultural machine and equipments that are sensor-based and IoT operated (smart agriculture technologies) for stepwise processes that suits a particular situation and farmers economics to achieve various targets: indicators of SDG2 by 2047, that seems impossible by the 2030 agenda of UN due to global spread of Covid-19 Pandemic as well as the prevailing climate shift (D.R. Saxena, 2022, 2023).

1. Introduction

The world population is estimated to elevate to 9 billion by 2050, which is approximately a 25% rise in comparison to the present population (Union, 2050). India, Mexico, and China's contribution in the above escalation of population will be greater (NabbMc, D.E., 2019); there will be more urbanization from the current 49–70% at the global level by 2050 [98, 10]. Currently, 821 million humans globally have no access to food (FAO, 2019) (SDG2: Zero Hunger, A4ID, A Legal Guide). End of poverty and increased prosperity will bring global harmony and conserve the earth. The UN Sustainable Development Goals

(SDGs), also called Agenda (2030), have implemented and addressed issues like inequalities, control and adaptation to climate change, maintain normal economic growth to create suitable jobs, develop cities, manage following significant aspects like the industrialization, oceans, ecosystems, energy use, sustainable production and consumption, peace and justice at the global level, etc. The augment of Covid-19 Pandemic has either deaccelerated or stopped these action-oriented commitments that SDGs may achieve zero-hunger by 2047 (D.R. Saxena, 2023) only by expanding its dimensions. SDGs 17- interdependent goals have 169 prime specific targets whose progress is monitored and assessed employing 232 indicators. The lion's share of agriculture was 18% of GDP and 57% of rural population were in this occupation, but gradually the population from 79.9% in 1951 has declined to a low level of 45.1% in 2011. According to a statistics of Economic Survey (2018) the workers in the agriculture sector will reduce to 25.7% by 2050.

2. Current Use of Smart Technologies Developed in IOT to Improve Agricultural Production and Terminate Misuse of Natural Resources

Smart agriculture practices and smart technologies: Agriculture for food and animal food products roughly commenced 10,000 years ago (Agriculture 1.0) using simple tools, animal energy, and human workforce. The Agriculture 2.0 era in the 19th century benefited by using machines and agrochemicals to enhance yield that created new probabilities related to disturbance in the environment. In the 20th century, the Agriculture 3.0 era documented the fast development of computation and electronic technology that increased food production no doubt, but could partly solve the emerging environmental issues created by the use of new methods. The Agriculture 4.0 era takes advantages of global current technologies like internet of things (IoT), colossal data analysis and interpretation, artificial intelligence (AI), remote sensing (GPS – GIS), etc., to accelerate, conserve, improve benefits per capita food and income through smart agriculture [33]. For timely and effective fruitful real decisions, the extrapolative technique of using big-data in smart agriculture is used by agriculturists [97, 55]. Precision decision-making is being done by application of IoT devices/ sensors, robotics, etc., developed in the field of artificial intelligence (AI); now also used in the agriculture sector to make farming workforce more efficient in raising food production [63]. Accurate competent and rapid monitoring have become possible, reliable, and predictable through modern ICTs (information and communication technologies). Mention must be made of IoT, GPS sensors, drones, robotics, precision equipment, actuators, and data analysis. These can optimize every need from labour, etc., to grow quality products [71, 63].

Target-oriented progress with reference to the SDGs is published annually by the UN Security General, but at the core is the body of a high level political forum to review this global performance. SDG 2 focuses on ending hunger, achieving food security means' availability (overall production), based on economic status and physical location access to food, and its utilization in raw or cooked state, and continuous access to improved nutrition through sustainable agriculture (FAO, 2008). Food sovereignty means freedom or right granted to small-scale and family farmers, respectively, to formulate their own policies regarding crops to be grown, etc., under the UN declaration on the rights of peasants and other people working in rural areas (UNDROP, 2018). Malnutrition is related to hunger and directly to the health of people (WHO, 2020). Elimination of rising wealth inequality in and between national economies globally will not help in wiping issues of hunger. The following challenges to combat zero hunger must be simultaneously functional to make SDG2 successful as mentioned in the draft; this includes the environmental degradation in SDGs 14 and 15, the global climate change in SDG13, poverty in SDG1, armed conflicts in SDG16, unstability or severe inflation in commodity prices in SDG17, absence of appropriate technology, investment, and support from capacity-building in SDG9, better health directly related to sufficient, safe, and nutritious food accessibility according to SDG3, quality education in SDG4, and addressing hunger and food insecurity impacts always tolerated by domestic women, but the nonsteady sources of livelihood have affected the food situation of their children in SDG5 (so the key factor named women empowerment implementation have been envisaged by WHO 2020). Success of SDG2 by 2030 doesn't seem to be attainable because of the Covid-19 Pandemic outbreak, but can be speculated to be possible by 2040 (D.R. Saxena, 2023).

Here, mention of SDG17 for the success of SDG2 must be made regarding the significant roles of multi-stakeholder partnerships, continuous cooperation of concerned governments, the local authorities, civil society, and the private sector. According to the SDGs Report of UN (2017) there is an urgent need to accelerate the actions for fulfillment of SDGs by 2030 to overcome to negative impact of spread of the lethal Covid-19 Pandemic and the five major threats to global food security (World Food Programme, 2020).

SDG 2.1: This mentions total elimination of hunger at the global level and guarantees access to safe, nutritious, and ample quantities of food to all the people, specifically the ultra-poor and vulnerable population; one must keep in mind that infants are the most dominant sufferers. This can be observed from the statistics. Since 2014, total reduction in hunger has occurred but it got nullified by the growing population. Hunger inflicts nearly 11% of the world population, the worst affected is the Sub-Sahara Africans, next in order are the people of South America (UN Report 2019). In the Middle East and North America rise in hunger was induced by human conflicts (TGNRFN, 2019). The progress assessment indicator includes data of per capita dietary energy, current prevailing moderate to severe food insecurity (UN, 2019). The remedy for this is to build resilience in connection with jobs and income opportunities and the level of global food systems that can happen with operations of the adequate safety nets that incorporate food aid, food for work, or cash transfer schemes at the time of crisis. This in turn depends on stable food prices.

SDG 2.2: This targets to end all forms of malnutrition achievable by 2025, the international body agreed targets on growth stunting and body-wasting in children below 5 years of age, and addressing the nutritional needs of adolescent girls, pregnant and lactating women as well as older persons. Worldwide, 45% of deaths occur from hunger, the decrease in chronic malnutrition to 100 million children by 2025 and 83 million by 2030 globally 22%, i.e., 149 million were chronically undernourished and 40% were obese children, respectively, as it appears in the UNICEF, UN Report (2019). Indicators of normal health according to WHO (UN, 2019) criteria are the stunting (height-for-age) and malnutrition conditions (weight-for-height) in children.

SDG Target 2.3: Aims to double the agricultural output and the earnings of small-scale producers, specifically the women, indigenous people, family farmers, pastoralists, and fishermen by 2030. This involves allotment and equal access to land, other productive resources and inputs, knowledge, financial services, markets and opportunities, and non-agriculture jobs or employment. More than 90% of all food cultivation is fine by small-scale producers and small holder farmers. Mostly 75% of the farmers' children are regulators of land under production (Lowder et al., 2016).

SDG 2.4: Producing more but growing more sustainably, SDG Target 2.4 emphasizes ecosystem protection; this imposes constraints or pressures on targets 2.1 and 2.2 which are oriented at provision of more adequate food and nutrition. There are will be encroachment of cropland expansion with the range of ~5–20% and ~10 + 25% for pasture land (Smith et al., 2010; Stehfest et al., 2019), into natural ecosystem forest, biodiversity (Savannahs wetlands), the likelihood to prevent damages due to land-use change may affect many SDG sustainability dimensions: Carbon stocks for SDG13, biodiversity for SDG15, and outbreak of Zoonotic disease (epidemics) affecting SDG3 (good health and well-being), need for clean drinking water (SDG6). Nearly 9.3 Gha land form is used out of the 13.0 Gha land surface, and only one fourth of unused 940 Mha along to unforested land excluding barren, rocks etc. (IPCC, 2009). SDG2: The zero hunger goal and its different targets are very closely connected with which it operates in synergy and for some, even incomplete symbiosis, the SDG needs support for progress aided by key enablers SDGs 5, 8, 10, 11, 16, and 17.

About 25% crop yield decline may occur as a result of increased chonate variability due to climate changes. Meticulous management of food production minimizes GHG emissions, protects environment, preserves natural resources as well as mainstreams the small-scale producers plead urgency. Highly adaptable approach of consumers to adapt to the type of food available in context to change of diet will regulate markets, value chains, and the international trade agreement's or vice-versa. There is no innovative common metric but a "standardized

Universal" metric that calculates sustainability considering environment as the principal factor. These metrics, in combination with a tracker-indicator, assess sustainable total productions in terms of percentage to confirm progress of this target.

SDG 2.5: By 2020, maintain genetic diversity of seed, cultivated plants, farmed domestic animals and their related wild species, including soundly managed and diversified seed and plant banks at national, regional, and international levels and ensure equitable sharing of benefits arising from the urbanization of genetic resources and associated traditional knowledge as internationally agreed, and maintaining and protecting the biodiversity in the global stock of seeds, crops, and livestock which is necessary for food security and resilience. According to an estimation, by 2016, 602 gene banks in 82 countries, 814 regional and international centers had conserved 4.7 million samples of seed and other genetic material; this is a measure for food security and resilience (FAO, 2008). But seed markets and private ownership of genetic patents are hampering small and larger producers' access to biotechnologies advancements (GMO's, gene editing).

SDG 3.0. Internet Technologies in Smart Agriculture: It links remote devices to facilitate smart farming [66] at a specific occasion, this new technology has cast its positive influence on communications, energy, trade , health, etc., to increase efficiency and performances throughout all markets [79, 76, 30]. This technology has its own hurdles in implementation such as poor standards, coverage and interlinking, untrained manpower, anxiety for new technology, expensive investment, and a broken chain market that have slackened the rate of achieving SDG2 targets through smart farming. To preserve, promote food and environment safety, respectively, guidelines and policies for regulating these have been laid down by FAO in association with other government organizations [18, 51]. Robust models, scalability, affordability, and sustainability in smart farming are possible via interactive and easy access to data at the receiver's end (agriculturists, manufacturers, farm engineering companies, etc.) through multiple devices like computers, and smart mobile use. Information on conventional farming methods, techniques, implements, crop pests, diseases, etc., can be retrieved through IoT for sustainable agriculture. Moreover, it is easily accessible, inexpensive, and an interactive tracking platform for information [21].

SDG 4.0. Smart Farming and Technologies: The latitude, longitude, and details about the elevation of a location is accurately recorded by GPS [54] and signals are relayed to ground-based stations to be computed, stored, and retrieved by users. Precise position field data, soil types, weeds, existing pest and physical barriers etc., are available at the farmers disposal to decide the use of seeds, fertilizers, herbicides, etc. [12]. Techniques like electromagnetism, photoelectricity, conductivity, and ultrasound all help in obtaining information on estimates of soil texture and structure, the level of nutrients, humidity, vegetation temperature, etc. Remote sensing (RS) can distinguish crop types, monitor drought, locate stress in the soil and plant conditions, etc. [23]. If the parameters of plant growth cross their threshold levels, the sensors detect it and convey the message to the microcontroller that it is needed until the restoration of parameters to normalcy [80]. Prominent sensors to gauze various parameters depend on characteristics like reliability, memory, portability, durability, coverage, and computational efficiency, that make them suitable for agriculture [60]. The various major sensors have their own purpose-based working and applications (Table 1).

Table 1: Types of sensors and their applications

Sr. No.	Sensors	Applications	Working Procedure
1.	Acoustic sensors	Pest monitoring and detection classifying seed varieties, fruit harvesting (Srivastava et al., 2013).	Measuring the variations in noise level when intermingling with other materials, i.e., soil particles (Kong et al., 2017).
2.	Airflow sensors	Measuring soil air permeability, moisture, and structure in a static position or mobile mode (García-Ramos et al., 2012).	Based on various soil properties, unique identifying signatures (García-Ramos et al., 2012).
3.	Eddy covariance-based sensors	Quantifying exchanges of CO_2, water vapor, methane, or other gases. Measuring surface atmosphere and trace gas fluxes in various agricultural ecosystems (Moureaux et al., 2012).	Measuring continuous flux over large areas (Kumar et al., 2017).
4.	Electrochemical sensors	To analyze soil nutrient levels and pH (Yew et al., 2014).	Nutrients in soil, salinity, and pH are measured using sensors (Cocovi-Solberg et al., 2014)
5.	Electromagnetic sensors	Recording electrical conductivity, electromagnetic responses, residual nitrates, and organic matter in soil (Yunus and Mukhopadhyay, 2011).	Electrical circuits measure the capability of soil particles to conduct or accumulate electrical charge (Millan-Almaraz et al., 2010).

Contd...

6.	Field programmable gate array (FAAA) based sensors	Measuring real-time plant transpiration, irrigation, and humidity (Weiss and Biber, 2011).	Programmable silicon chips and logic blocks are surrounded together by programmable interconnected resources of the digital circuit (Montagnoli et al., 2015).
7.	Light detection and ranging (LIDAR)	Land mapping, soil type determination, farm 3D modeling, erosion monitoring and soil loss, and yield forecasting (Schuster et al., 2017).	Sensors emit pulsed light waves and bounce off when colliding with objects and are returned to the sensor. The time taken for each pulse to return is used for assessment (Schuster et al., 2017).
8.	Mass flow sensors	Yield monitoring based on the amount of grain flow through a combine harvester (Hemmat et al., 2013).	Sensing the mass flow of grain with modules, e.g., grain moisture sensor, data-storage device, and an internal software (Hemmat et al., 2013).
9.	Mechanical sensors	Soil compaction or mechanical resistance	Sensors record the force assessed by strain gauges or load cells (Hemmat et al., 2013).
10.	Optical sensors	Soil organic substances, soil moisture, color, minerals, composition, clay content, etc. Fluorescence-based optical sensors are used to supervise fruit maturation (Murray, 2018). Integrating optical sensors with microwave scattering to characterize orchard canopies (Molina, 2011).	Sensors use light reflectance phenomena to measure changes in wave reflections (Millan-Almaraz et al., 2010).
11.	Optoelectronic sensors	Differentiate plant types to detect weeds in wide-row crops (Andujar et al., 2011).	Sensors differentiate based on reflection spectra (Andujar et al., 2011).
12.	Soft water level-based (SWLB) sensors	Used in catchments to characterize hydrological behaviors (water level	Measuring rainfall, stream flow, and other

Contd...

Contd...

Sr. No.	Sensors	Applications	Working Procedure
		and flow, time-step acquisitions) (Crabit et al., 2011).	water presence options (Crabit et al., 2011).
13.	Telematics sensors	Assessing location, travel routes, and machine and farm operation activities (Mark and Griffin, 2016).	Telecommunication between places (especially inaccessible points) (Mark and Griffin, 2016).
14.	Ultrasonic ranging sensors	Tank monitoring, spray distance measurement, uniform spray coverage, object detection, monitoring crop canopy (Dvorak et al., 2016), and weed detection (Pajares et al., 2013).	An ultrasonic sensor uses a transducer to send and receive ultrasonic pulses that relay information about an object's proximity (Zhmud et al., 2018).
15.	Remote sensing	Crop assessment, yield modeling, forecasting yield date, land cover and degradation mapping, forecasting, the identification of plants and pests, etc. (Yalew et al., 2016).	Satellite-based sensor systems collect, process, and disseminate environmental data from fixed and mobile platforms (Yalew et al., 2016).

SDG 4.1: Grid Soil Sampling and Variable Rate of Technology: The GIS agriculture field map is extrapolated on which at the right place and right time delivery rate of inputs of material is predicted for smart farming [33, 12]. In grid soil sampling, samples are collected from small cells or areas of the subdivision of a field to establish a map for every soil parameter, followed by superimposing the grid lines on the field. The composite samples are the total area of each smaller area called grid point sampling at each intersection of grid lines. Moreover, soil-test values from grid sampling are mapped by interpolation from non-measured locations between sampled points. There is variability for P and K that is specific for each field sampled and improved nutrient management practices for homogenous delivery of fertilizer or manure is the foundation of precision agriculture [58]. These maps are the basis for variable-rate technology (VRT) and are loaded into a variable-rate applicator to improve spatial distribution of nutrients and crop yields, respectively [35], the computer and GPS receiver regulate changes in dispensing the accurate quantity of fertilizers [15, 32].

SDG 4.2: Soil And Plant Sensors: Neo-soil sensors and plant wearables monitor real-time physical and nutrient signals, moisture, pH, temperature as well as pollutants in soil and provide data to manipulate cultivation growth conditions to get optimum yield [2] by lowering stresses related to abiotic and biotic parameters. Subsurface and soil surface spatial variations in nitrogen are detected using NIR – reflectance sensor [83]. Organic nitrogen, phosphorus, and potassium contains are measured using wavelengths of light falling on soil by evaluating soil spectral reflectance in IR (information retrieval) and visible regions respectively to enhance plant growth [27]. Similarly, both N and P are measured by NIRS (near infrared spectrophotometry) technology [52, 57, 56]. Electrical conductivity (ECa) varies in accordance with soil salinity and texture that is measured by specific sensors, optoelectronic, acoustic, impedance sensors, and nanostructured biosensors that are all funded applications to assess soil insects/pests [70].

SDG 4.3: Rate Controllers and Recent Irrigation in Pressurized Systems: They are actually stand-alone systems specifically designed and operated to deliver on a real-time basis in desired field areas materials by determining the speed of vehicles in the farm and accordingly fluctuating the flow-rate of substances [85]. The initiation of use of irrigation machines to control motion, GPS-based controller, sensor technologies, and wireless communication, though commercially not available but already designed, is applied to monitor soil and climatic conditions simultaneously to access flow and pressure of water to prevent wastage of resource [38].

SDG 4.4: Monitoring Crop Yield Sensor Aided: The computer hardware and software, respectively, are integrated to sensors that have been designed to measure crop yield based on force of mass or volume grain flow that is harvested, sorted, processed, etc. Microwave energy emitted and impinged on flowing grains are reflected and that is recorded by mass flow sensor device GPS receivers fitted to yield record monitors, the yield maps of locations and provide the yield data [61, 62].

SDG 4.5: Software Application in Agriculture: Preparing a map, display controller interfacing, data processing, analysis and interpretation, etc., are the multitude of functions executed by the software; they generate overlaying different kinds of maps using advanced geostatistical techniques. Also maps for soil properties, yield maps, variable rate inputs are constructed by the software [34].

SDG 5.0: Currently to enhance agriculture production to achieve SDGs targets of "zero hunger", the conventional farming methods have been supplemented to each step by smart technology using wireless sensors to monitor issues like land suitability, drought, irrigation, pest control, etc. (Table 2).

SDG 5.1: Soil Mapping and Plant Monitoring: The soil and water deficit index estimation is done from satellite data procured through soil moisture maps; later from this information is developed a prediction model taking cognizance of

Table 2: Applications activities and availability devices in smart agriculture practices

Sr. No.	Applications	Facilities	Devices/Sensor
1.	Soil monitoring	Evaluation of soil	Soil electrochemical
2.	Water monitoring	Irrigation	Moisture sensor
3.	Nutrient monitoring	Fertilization	Nutrient sensor
4.	Pest monitoring	Insecticides	Chlorophyll sensors
5.	Environ monitoring	Weedicides	Stem sensor
6.	Farm Implements	Yield monitoring	Temperature sensor
		Storage	Humidity
		Marketing	Fruit size sensor
			Remote sensor
			Robotics
			Communication, mobile, etc.

the physical properties of soil. Analysis of agriculture drought in remote areas is done by frequently collecting soil moisture data exploiting the techniques of remote sensing. To optimize resources usage for cultivation, data on physical, chemical, and biological properties of soil is obtained from soil mapping to decide planting depth and varieties to be planted in specific field conditions of soil like pollutants' drought absorptions rate, soil degradation, etc.; monitoring is done using a broad range of sensors and tools [59, 94]. GPS helps in soil analysis, the data allow critical decisions and implementation at various stages of crop growth in context with irrigation or use of nutrients to restore normal soil fertility [28].

SDG 5.2: The Information from Climate Data: Sensors and satellite imageries are used with IoT techniques. One criterion is the Crop Water Stress Index Based Water Management that calculates water requirements and predicts using the irrigation Index values for every field based on slope or soil differences to enhance water usage efficiently depending on crop type, precipitation, etc. Drip irrigation is an efficient way as 50% world population areas face high water shortages and 168 countries will desertify if water usage is not managed efficiently [74].

SDG 5.3: Site-specific Nutrient Management: The normalized difference vegetation index obtained from satellite images to know the crop nutrient status [13, 69], crop health, vegetation vigor, and density of plant and soil nutrient levels help in correction of soil to improve the fertilization efficiency. Smart fertilization Technologies like GPS [78], Geospatial mapping [84], variable rate technology [19, 25], and autonomous vehicles [47] can be employed for site-specific nutrient management.

SDG 5.4: Agriculture Pest Disease and Their Management: Pests and disease take toll of 20–40% crop yield/year at the global level (FAO, estimation; New lands, 2018); to control losses incurred, pesticides and agrochemicals are applied [48], most of these are deleterious to human and animals eventually leading to environment pollution [20, 72]. Effectiveness of control measures can be enhanced for pest and disease by forecast, modeling, and real-time monitoring at specific locations of the above problems based on IoT devices such as robots, wireless sensors, drones, etc. [49, 95]. During the entire duration of crop growth, data regarding plant health and pest incidence for a specific agriculture field is gathered by field sensors and true images are obtained using RS. Automatic insect traps based on IoT perform trapping, counting, and describing the insect type and outbreak; later this data is uploaded to 'Cloud' for interpretation analysis. IoT management systems employ multispectral image sensing robotic devices that release fine micro amounts of pesticides spray to accurately kill the pests.

SDG 5.5: Yield Monitoring and Forecasting: It predicts the yield before harvest so as to assist the farmer in future planning, decision-making, and re-analysis of yield quality that depends on good pollen [24, 86, 96]. This helps to predict accurate harvesting time of crops for profit. The development and installation of a yield monitor [77] on a harvester machine, linked through a mobile app, gives exact real-time crop harvest as it transmits data automatically to manufacturers web-based platform, to decide the purchase of goods. Desiccating fruit monitoring is done by multiple optical sensors [92]. To know the details of fruit size, color, ripening, etc., multicolor satellite images can be used.

SDG 6: Multiple Applications of IoT in Advanced Farming: Currently various sophisticated sensor technologies and IoT-based technologies have multiple roles in advanced farming methods and in artificial control of environments to produce improved quantitative and qualitative yields.

SDG 6.1: Greenhouse Farming and Protected Cultivation: The ancient most popular method of smart farming under controlled environment started in the 19th century, but accelerated during the 20th century in nations to cope with erratic weather conditions [88]. The effect of environment on plant growth can be surpassed in indoor culture conditions by manipulations of several factors like shed structures and materials to regulate wind, aeration systems, perfection in monitoring parameter decision-support systems, etc., to successfully grow crops that are cultivated traditionally in field [75]. Sensors in indoor cultivation precisely measure parameters like light, temperature, moisture, and atmospheric pressure facilitating control over challenges imposed by local climate conditions and are described as IoT-based greenhouse [5]. Manual inspection is not needed in smart greenhouse farming, and the plants are protected from wind, hailstorms, ultraviolet radiation, and attack by insect pests. Sensors controlling light, humidity, temperature, and required wavelength during the night to grow Hibiscus plants reduces 70–80% of water requirements. Thus, profitable commercial farming in arid areas is possible using IoT [89]; it reduces wastage of resources, stress on environment, pollution, generates disease-free nutritious crops, and the yield is increased manifold (D.R. Saxena 2024). Monitoring and controlling humidity, temperature, moisture, carbon dioxide, intensity of UV light, pH and EC value, water, level of nutrient solution, and quantity of pesticides is carried out by IoT and sensor companionship. Moreover, sensors effectively detect and diagnose and deliver these requirements for growth of rose plants in the greenhouse [89].

SDG 6.2: Vertical Farming, Hydroponics, and Phenotyping: The earlier methods of modern farming have deteriorated soil by erosion, overuse of water has resulted in scarcity in reservoirs, and shrunk arable land. Vertical farming uses a small part of land, provides high yields, and decreases water requirements, etc. [68] (Berkeley and Tomkins 2017). The nondispersive infrared sensor measures and tracks carbon dioxide in vertical farming (VF). The automated smart hydroponics system integrated to local input data, cloud server, and output data can monitor, through the internet, lettuce growth by analyzing pH, solution water nutrients, room temperature, and humidity in real time [53], by placing roots in deep water and employing continuous circulation. The Raspberry pi, and data are processed and monitored automatically to ensure proper water circulation [93]. Phenotyping a crop engineering advanced technique investigates quantitative growth, resistance to various stresses, yield quality, and quantity [90]; image-based sensing and screening of biostimulants help in understanding this technique [67].

SDG 7.0: Smart Farming and Technological Innovations in Engineering: Adopting IoT-based agriculture resolves specific problems in smart farming techniques that need innovation technologies of precision machinery for conventional agriculture mechanization [17], instrumentation, mechatronics, control system, knowledge in artificial-computational intelligence [87], Big Data, satellite, and aerial images strike a harmony between precision agriculture and environment conservation; engineering is a system integrator associating technical experience and potential business skills in public and private sectors, respectively [41], IoT, AI, mobile, remote sensing advanced analytics, the Cloud, RPA and block chain technologies support logistics and maintenance of farm assets [11], machine cum sensor cum other devices generate data like soil monitoring, cereals, fruits, etc. [4], these are used in planning, organizing, and analyzing data, storage in historical archives to be retrieved through systems based on radio-frequency signals [3]. The radio frequency identification or near-field communication and low power wide-area network technologies transmit data to hundreds or thousands of meters away from the source. Foster and Mark [36] believe in adopting technology having heterogeneous factors that are effective. Educational and technical abilities to manage tools enhances farmers' aptitude to process information for making decisions in applications of smart technologies (Feder et al., 1985), use computers [7]; , lack of knowledge and skill and technology awareness affect its use [50] and prevents the adoption of smart farming (Khan et al., 2007) incorporating production, property management, decision-making tools, bridging the existing gap between agriculture and IT may increase user effectiveness by integrating these systems [1].

SDG 8.0: Confronted Current Challenges and Future Expectations: The UN and international communities in their sustainable development goals to end hunger by 2030, anticipate challenges in 2050 to solve problems to achieve "zero hunger" globally. Eight hundred million people globally are facing food scarcity problems [37]. Feeding 10 billion humans, limitations in land expansion, lowering GHGs are challenges that raise questions of water scarcity, climate conditions, rural labor availability, etc. Younger growers must take the responsibility to replace aging old farmers, changing ecological and geographic conditions, and dictating crop selection; 50% population in developing nations are in the agriculture sector and they must use the following for achieving sustainable prospective agriculture: IoT in agriculture depends on connectivity between devices [91], updating information by Wireless sensor networks GPS, upgraded 5G cellular mobile connection availability to farmers will be helpful; by 2022 an estimated 29 billion IoT-based components may operate in agriculture by the creation of 4.1 million data points on-site basis from the farms by 2050 [40]. Similarly, other innovative technologies available to farmers are drones equipped with sensors, 3D cameras for spraying seeds, fertilizers, etc., and unarmed vehicles and robotic application in inclement weather, 3D modeling to monitor height of maize and sorghum, UAV aided [73], leaf area index (LAI) in soybean plants, the measured LAI predicted accuracy corresponding to a hand-held device have been correlated with destructive LAI measurements [22]. UAVs multispectral digital images for evaluating vegetation indices and multitemporal VIs to predict wheat yield [81], evaluation of NDVI, SVI and green-area index for wheat yield [39], monitor plant breeding [29], plant stress to yellow rust disease [82], quantify plant density [42, 43], to reduce wastage of fertilizers and pesticides during spraying an algorithm to self-adjust UAV routes have been developed [31]; these are other management tools and technology. VF and hydroponics are options to shrinking arable land resources, urbanization, and pressures on resources [6], performance analysis by machine learning (ML) in big data systems help to overcome decision-making, crops, animal research, security, etc. [26], ML-based yield prediction based on deep memory model for maize [44], binary classification model with logistic regression technique to assess rainfall intensity [64], short-term memory model to predict soil water content in relation to temperature, rainfall, etc., and time for the next 1, 2, and 7 with greater R^2 compared to artificial herbal networks [98] are technologies enhancing farmers income, healthcare, education as well as nutritious food [65]. Globally, the smart grid technology enables a smooth transition from traditional to smart energy systems, ensuring energy security; alleviating rural electricity problems using smart micrograds linked to the national grid have been possible in Nepal [16]; Dayalbagh in India has a renewable energy smart grid; dairy plant production uses the smart microgrids renewable resources [45] and the mixed integer linear program is a model to systematically and efficiently generate its own energy from a microgrid containing solar panels and wind turbines, forecast wind speed, etc., and sustain the effectiveness and productiveness of energy efficient techniques [8].

The objectives and processes in the design of a model sustainable agro-ecosystem may not be globally or universally applicable for small, medium, and large agriculturists. It includes the following objectives diversified in time and space: dynamically stable, productive and food self-sufficient, conservation and regeneration of natural resources of water, soil, germplasms, nutrients, etc., the economic potential, socially and culturally acceptable/feasible technology, self-promoting and self-help potential. Similarly, the processes that must be used include the soil cover, nutrient recycling, sediment received water harvest and conservation, productive diversity, crop protection, and ecological order. Methods or strategies for a sustainable agro-ecosystem include crop systems (fallow culture, polyculture, no tillage, selective weeding, mulching cover, cropping densities), polyculture (use of rotation with legumes, zonification of production, alleying cropping, use of residues, improved fallow manuring), living and nonliving barriers (no tillage, selective weeding, terracing, zonation, contour planting), regional diversity (forest enrichment, crop zonation, crop mosaics, windbreaks, shelterbelts, genetic diversity (species diversity, cultural and biological control), agro-ecosystem design and reorganization (mimicking natural succession agro-ecosystem analysis methodologies). Diversity within the agro-ecosystem contains polyculture, agroforestry, crop livestock association, and a variety of mixtures.

Some suggestions for Green Sustainable Smart Agriculture are (D.R. Saxena, 2008-2023):

(i) Compulsory electronic SIM card system to access/distribute food, water, clothing, fuel (cooking, vehicle) and electricity, for per capita consumption or per family consumption per year, respectively. Same rules apply for commercial manufactures of the goods and services sector.
(ii) Compulsory manufacture of green innovative non-conventional energy technology for portable cheap solar powered agricultural gadgets and equipped machines for farming, harvesting, post-harvest processing; refrigerated and non-refrigerated storage godowns, as well as transportation facilities to sell products in market to protect losses and scarcity of goods, i.e., insecurity.
(iii) The authors have proposed a prototype portable low cost uni or multifunctional manual or electric/electronic device described as "Nonbiological and biological carbon reducing absorbing innovative smart machine based on redox performance of micronutrient and macronutrient compounds like Ca, KOH, etc. for monitoring greenhouse gasses, oxides of nitrogen and sulfur, hydrogen sulfide, toxic pollutants, etc. to clean air, water, and soil ecosystems; sewerage STP, and industrial effluent ETP systems, etc." Such

an inbuilt mobile "green micro device" can be made. These devices are more useful in fragile Antarctic and Arctic ecosystems for global warming studies and in space research programs, etc.

(iv) Increase global forest land area from existing to 45–50% hectare-wise. Other smart measures such as water harvesting, etc., will be dealt in the chapter.

Conclusion

In India, arable land is shrinking as the farmers are converting their land for non-agricultural residential purposes. Their subsequent generations are searching for better options to earn. Moreover, global warming and El Nino have unpredictable loss incurring outcomes ravaging their socio-economic development by indebting them. The International economy has suffered tremendously from the Covid-19 Pandemic. In India, until January 2024 the cost of consumable and nonconsumable articles have increased manifold and purchasing capacity/power has automatically declined sustainable agriculture development to reduce poverty and achieve zero hunger that seems miles away to be achieved. But, we cannot allow the prevailing conditions to defeat our purpose of SDGs that include the multitude of key issues of technology developed and used for doubling agriculture from the present status. Automation seems to be a partial supportive route to adapt and adopt resource utilization like land, water, fertilizer, labor, pesticides, etc., that is targeted to achieve high yields of food crops to provide quality food under the nutritional food security agenda. Many industries need raw material as well as green energy for steady commercial output; reduction or zero emission of carbon can be supplemented by producing energy from surplus cash crops in some years, but not sustainably for years (D.R. Saxena, 2023).

No Conflict of Interests

All authors have equally contributed in the making of manuscript.

References

1. Abdul Hakkim, V.M., Abhilash Joseph, E., Ajay Gokul, A.J. and Mufeedha, K. (2016). Precision farming: The future of Indian agriculture. *J. Appl. Biol. Biotechnol.*, 4: 68–72. [CrossRef].
2. Adamchuk, V.I., Hummel, J.W., Morgan, M.T. and Upadhyaya, S.K. (2004). On-the-go soil sensors for precision agriculture. *Comput. Electron. Agric.*, 44: 71–91. [CrossRef].
3. Adebayo, S., Ogunti, E.O., Akingbade, F.K. and Oladimeji, O. (2018). A review of decision support systems using mobile applications in the provision of day-to-day information about farm status for improved crop yield. *Period. Eng. Nat. Sci.*, 6: 89–99. [CrossRef].

4. Ahmed, A.T., El Gohary, F., Tzanakakis, V.A. and Angelakis, A.N. (2020). Egyptian and Greek water cultures and hydro-technologies in ancient times. *Sustainability*, 12: 9760. [CrossRef].
5. Akkas,, M.A. and Sokullu, R. (2017). An IoT-based greenhouse monitoring system with Micaz motes. *Procedia Comput. Sci.*, 113: 603–608. [CrossRef].
6. Al-Kodmany, K. (2018). The vertical farm: A review of developments and implications for the vertical city. *Buildings*, 8: 24. [CrossRef].
7. Alvarez, J. and Peter, N. (2006). Adoption of computer based information systems: The case of dairy farmers in Canterbury, NZ and Florida, Uruguay. *Comput. Electron. Agric.*, 50: 48. [CrossRef].
8. Aslam, S., Khalid, A. and Javaid, N. (2020). Towards efficient energy management in smart grids considering microgrids with day-ahead energy forecasting. *Electr. Power Syst. Res.*, 182: 106232. [CrossRef].
9. Augustine C., Balaji K., Dharanikumar S.V. and Jose Anand A. (2023). Urban farming: Case study., *In*: K. Kalaiselvi, Jose Anand A., Poonam Tanwar and Haider Raza (Eds.), *Advanced Technologies for Smart Agriculture*, 321–338. Denmark: River Publishers.
10. Ayaz, M., Ammad-uddin, M., Baig, I. and Aggoune, E.M. (2018). Wireless sensor's civil applications, prototypes, and future integration possibilities: A review. *IEEE Sens. J.*, 18: 4–30. [CrossRef].
11. Bacco, M., Barsocchi, P., Ferro, E., Gotta, A. and Ruggeri, M. (2019). The digitization of agriculture: A survey of research activities on smart farming. *Array*, 3–4: 100009. [CrossRef].
12. Batte, M.T. and VanBuren, F.N. (1999). Precision farming: Factor influencing productivity. *In*: *Proceedings of the Northern Ohio Crops Day Meeting*, Wood County, OH, USA, 21 January.
13. Benincasa, P., Antognelli, S., Brunetti, L., Fabbri, C., Natale, A., Sartoretti, V. and Vizzari, M. 2018. Reliability of NDVI derived by high resolution satellite and UAV compared to in-field methods for the evaluation of early crop N status and grain yield in wheat. *Exp. Agric.*, 54: 604–622. [CrossRef].
14. Benke, K. and Tomkins, B. (2017). Future food-production systems: Vertical farming and controlled-environment agriculture. *Sustain. Sci. Pract. Policy*, 13: 13–26. [CrossRef].
15. Berntsen, J., Thomsen, A., Schelde, K., Hansen, O.M., Knudsen, L., Broge, N., Hougaard, H. and Horfarter, R. (2006). Algorithms for sensor-based redistribution of nitrogen fertilizer in winter wheat. *Precis. Agric.*, 7: 65–83. [CrossRef].
16. Bhattarai, T.N., Ghimire, S., Mainali, B., Gorjian, S., Treiche, H. and Paudel, S.R. (2022). Applications of smart grid technology in Nepal: Status, challenges, and opportunities. *Environ. Sci. Pollut. Res.*, 1–25. [CrossRef] [PubMed].
17. Bochtis, D., Sørensen, C.A.G. and Kateris, D. 2019. *Operations Management in Agriculture*, 1–18. Amsterdam, The Netherlands: Elsevier. [CrossRef].
18. Bonneau, V. and Copigneaux, B. (2017). *Industry 4.0 in Agriculture: Focus on IoT Aspects, European Commission, Digital Transformation Monitor*. Available online: https://ec.europa.eu/growth/tools-databases/dem/monitor/content/industry-40-agriculture- focus-iot-aspects (accessed on 30 December 2020).
19. Bruno, B., Benjamin, D., Davide, C., Andrea, P., Francesco, M. and Luigi, S. (2016). Environmental and economic benefits of variable rate nitrogen fertilization in a nitrate vulnerable zone. *Sci. Total Environ.*, 545–546: 227–235.

20. Carvalho, F.P. (2017). Pesticides, environment, and food safety. *Food Energy Secur.*, 6: 48–60. [CrossRef].
21. Chandhini, K. (2016). A literature study on agricultural production systems using IoT as inclusive technology. *Int. J. Innov. Technol. Res.*, 4: 2727–2731.
22. Chang, A., Jung, J., Maeda, M. and Landivar, J. (2017). Crop height monitoring with digital imagery from unmanned aerial systems (UAS). *Computers and Electronics in Agriculture*, 141: 232–237.
23. Chen, F., Kissel, D.E., West, L.T., Adkin, W., Clark, R., Rickman, D. and Luvall, J.C. (2004). Field scale mapping of surface soil clay concentration. *Precis. Agric.*, 5: 7–26. [CrossRef].
24. Chung, S.O., Choi, M.C., Lee, K.H., Kim, Y.J., Hong, S.J. and Li, M. (2016). Sensing technologies for grain crop yield monitoring systems: A review. *J. Biosyst. Eng.*, 41: 408–417. [CrossRef].
25. Colaço, A.F. and Molin, J.P. (2017). Variable rate fertilization in citrus: A long term study. *Precis. Agric.*, 18: 169–191. [CrossRef].
26. Cravero, A., Pardo, S., Sepúlveda, S. and Muñoz, L. (2022). Challenges to use machine learning in agricultural big data: A systematic literature review. *Agronomy*, 12: 748. [CrossRef].
27. Daniel, K., Tripathi, N.K., Honda, K. and Apisit, E. (2011). Analysis of spectral reflectance and absorption patterns of soil organic matter. *In*: *Proceedings of the 22nd Asian Conference on Remote Sensing*, Singapore, 5–9 November.
28. Dinkins, C.P. and Jones, C. (2013). *Interpretation of Soil Test Results for Agriculture*. MontGuide. Publication no. MT200702AG; Montana State University Extension: Bozeman, MT, USA.
29. Duan, T., Chapman, S., Guo, Y. and Zheng, B. (2017). Dynamic monitoring of NDVI in wheat agronomy and breeding trials using an unmanned aerial vehicle. *Field Crops Res.*, 210: 71–80. [CrossRef].
30. Elijah, O., Rahman, T.A., Orikumhi, I., Leow, C.Y. and Hindia, M.N. (2018). An overview of Internet of Things (IoT) and data analytics in agriculture: Benefits and challenges. *IEEE Internet Things J.*, 5: 3758–3773. [CrossRef].
31. Faial, B.S., Costa, F.G., Pessin, G., Ueyama, J., Freitas, H., Colombo, A., Fini, P.H., Villas, L. et al. (2014). The use of unmanned aerial vehicles and wireless sensor networks for spraying pesticides. *J. Syst. Archit.*, 60: 393–404.
32. Ferguson, R.B., Hergert, G.W, Schepers, J.S., Gotway, C.A., Cahoon, J.E. and Peterson, T.A. (2002). Site-specific nitrogen management of irrigated maize: Yield and soil residual nitrate effects. *Soil Sci. Soc. Am. J.*, 66: 544–553.
33. Ferrandez-Pastor, F.J., Garcia-Chamizo, J.M., Nieto-Hidalgo, M., Mora-Pascual, J. and MoraMartinez, J. (2016). Developing a ubiquitous sensor network platform using Internet of Things: Application in precision agriculture. *Sensors*, 16: 1141. [CrossRef].
34. Ferrández-Pastor, F.J., García-Chamizo, J.M., Nieto-Hidalgo, M. and Mora-Martínez, J. (2018). Precision agriculture design method using a distributed computing architecture on Internet of Things context. *Sensors*, 18: 1731. [CrossRef].
35. Fleming, K.L., Westfall, D.G. and Bausch, W.C. (2000). Evaluating management zone technology and grid soil sampling for variable rate nitrogen application. *In*: *Proceedings of the 5th International Conference on Precision Agriculture*, 1–13. Bloomington, MN, USA, 16–19 July.
36. Foster, A.D. and Mark, R.R. (1995). Microeconomics of technology adoption. *Annu. Rev. Econ.*, 2010, 2: 395–424. [CrossRef].

37. Fróna, D., Szenderák, J. and Rákos, M.H. (2019). The challenge of feeding the world. *Sustainability*, 11: 5816. [CrossRef].
38. Hassan, A., Aitazaz, A.F., Farhat, A., Bishnu, A. and Travis, E. (2020). Precision irrigation strategies for sustainable water budgeting of potato crop in Prince Edward Island. *Sustainability*, 12: 2419. [CrossRef].
39. Hassan, M.A., Yang, M., Rasheed, A., Yang, G., Reynolds, M., Xia, X., Xiao, Y. and He, Z. (2019). A rapid monitoring of NDVI across the wheat growth cycle for grain yield prediction using a multi-spectral UAV platform. *Plant Sci.*, 282: 95–103. [CrossRef]
40. Henriksen, A.V., Edwards, T.C.G., Pesonen, L.A., Green, O. and Sørensen, C.A.G. (2019). Internet of Things in arable farming: Implementation, applications, challenges, and potential. *Biosyst. Eng.*, 191: 60–84. [CrossRef].
41. James, A., Saji, A., Nair, A. and Joseph, D. (2019). CropSense: A smart agricultural system using IoT. *J. Electron. Des. Eng.*, 5: 1–7.
42. Jin, X., Liu, S., Baret, F., Hemerl, M. and Comar, A. (2017). Estimates of plant density of wheat crops at emergence from very low altitude UAV imagery. *Remote Sens. Environ.*, 198: 105–114. [CrossRef].
43. Jose Anand, A., Nirmala Devi, K., Vijayakumar, S. and Shiju C. Chacko. (2023). Decision-making support in smart farming. *In*: K. Kalaiselvi, Jose Anand A., Poonam Tanwar and Haider Raza (Eds.), *Advanced Technologies for Smart Agriculture*, 219–239. Denmark: River Publishers.
44. Kaneko, A., Kennedy, T.W., Mei, L., Sintek, C., Burke, M., Ermon, S. and Lobell, D.B. (2019). Deep learning for crop yield prediction in Africa. *In*: *Proceedings of the International Conference on Machine Learning AI for Social Good Workshop*, LongBeach, CA, USA, 10–15 June.
45. Kedri, J. (2014). *Simulation and Validation of SPV Micro Grid Comprising 518.2 KWp Distributed Solar Power Plants at Dayalbagh Educational Institute*. Ph.D. Thesis, Dayalbagh Educational Institute, Dayalbagh, Agra, India.
46. Khan, A.R., Dubey, M.K., Bisen, P.K. and Saxena, K.K. (2016). Constraints faced by farmers of Narsing Kheda village of Sehore district (Young up to 30 years). *Agricultural and Food Sciences: Economics, Indian Research Journal of Extension Education.* Semantic Scholar.
47. Khan, N., Medlock, G., Graves, S. and Anwar, S. (2018). *GPS Guided Autonomous Navigation of a Small Agricultural Robot with Automated Fertilizing System*; SAE Technical Paper 2018-01-0031. SAE International: Warrendale, PA, USA. [CrossRef].
48. Khattaba, A.S., Habiba, E.D., Ismail, H., Zayanc, S., Fahmya, Y. and Khairya, M.M. (2019). An IoT-based cognitive monitoring system for early plant disease forecasts. *Comput. Electron. Agric.*, 166: 105028. [CrossRef].
49. Kim, S., Lee, M. and Shin, C. (2018). IoT-based strawberry disease prediction system for smart farming. *Sensors*, 18: 4051. [CrossRef] [PubMed].
50. Kimiti, J.M., Odee, D.W. and Vanlauwe, B. (2009). *Area under Grain Legumes Cultivation and Problems Faced by Smallholder Farmers in Legume Production in the Semi-Arid Eastern Kenya*. Cambridge, MA, USA: Academic Press:
51. King, T., Cole, M., Farber, J.M., Eisenbrand, G., Zabaras, D., Fox, E.M. and Hill, J.P. (2017). Food safety for food security: Relationship between global megatrends and developments in food safety. *Trends Food Sci. Technol.*, 68: 160–175. [CrossRef].
52. Kuang, B. and Mouazen, A.M. (2013). Non-biased prediction of soil organic carbon and total nitrogen with vis-NIR spectroscopy, as affected by soil moisture content and texture. *Biosyst. Eng.*, 114: 249–258. [CrossRef].

53. Lakshmanan, R., Djama, M., Selvaperumal, S. and Abdulla, R. (2020). Automated smart hydroponics system using the internet of things. *Int. J. Electr. Comput. Eng.*, 10: 6389–6398. [CrossRef].
54. Lang, L. (1992). GPS + GIS + remote sensing: An overview. *Earth Obs. Mag.*, 1: 23–26.
55. Liakos, K.G., Busato, P., Moshou, D., Pearson, S. and Bochtis, D. (2018). Machine learning in agriculture: A review. *Sensors*, 18: 2674. [CrossRef].
56. Lvova, L. and Nadporozhskaya, M. (2017). Chemical sensors for soil analysis: Principles and applications. *In*: Grumezescu, A.M. (Ed.). *Series Nanotechnology in the Agri-Food Industry*; *New Pesticides and Soil Sensors*, Volume 10: 637–678. Amsterdam, The Netherlands: Elsevier. [CrossRef].
57. Maleki, M.R., Van Holm, L., Ramon, H., Merckx, R., De Baerdemaeker, J. and Mouazen, A.M. (2006). Phosphorus sensing for fresh soils using visible and near infrared spectroscopy. *Biosyst. Eng.*, 95: 425–436. [CrossRef].
58. Mallarino, A.P. and Wittry, D.J. (1997). Use of DGPS, yield monitors, soil testing, and variable rate technology to improve phosphorus and potassium management. *In*: *Proceedings of the Integrated Crop Management Conference*, 267–275. Iowa State University Extension and Outreach: Ames, IA, USA.
59. Martínez-Fernández, J., González-Zamora, A., Sánchez, N., Gumuzzio, A. and Herrero-Jiménez, C.M. (2016). Satellite soil moisture for agricultural drought monitoring: Assessment of the SMOS derived Soil Water Deficit Index. *Remote Sens.*, 177: 277–286. [CrossRef].
60. Muhammad, S.F., Shamyla, R., Adnan, A., Tariq, U. and Yousaf, B.Z. (2020). Role of IoT technology in agriculture: A systematic literature review. *Electronics*, 9: 319. [CrossRef].
61. Naorem, A., Rani, A., Roy, D., Kundu, S., Rao, N.S., Sreekanth, P.D., Kumar, A., Manjaiah, A.M. and Rao, C.S. (2019). Frontier soil technologies for sustainable development goals (SDGs) in India. *In*: Rao, C.S., Senthil, V., Meena, P.C. (Eds.), *Challenges and Emerging Opportunities in Indian Agriculture*, 113–152. Hyderabad, India: National Academy of Agricultural Research Management.
62. Nisha, K., Augustince, C., Jose Anand, A. and Kamesh, M. (2023). Deep learning in smart agriculture. *In*: K. Kalaiselvi, Jose Anand A., Poonam Tanwar, Haider Raza (Eds.), *Advanced Technologies for Smart Agriculture*, 153–175. Denmark: River Publishers.
63. O'Grady, M.J. and O'Hare, G.M.P. (2017). Modeling the smart farm. *Inf. Process. Agric.*, 4: 179–187. [CrossRef].
64. Oswal, N. (2019). Predicting rainfall using Machine Learning Techniques. *arXiv* 2019, arXiv:1910.13827.
65. Ouyang, H., Wei, X. and Wu, Q. (2019). Agricultural commodity futures prices prediction via long- and short-term time series network. *J. Appl. Econ.*, 22: 468–483. [CrossRef].
66. Patil, K.A. and Kale, N.R. (2016). A model for smart agriculture using IoT. *In*: *Proceedings of the 2016 International Conference on Global Trends in Signal Processing, Information Computing and Communication*, 543–545. Jalgaon, India, 22–24 December. IEEE. [CrossRef].
67. Paul, K., Sorrentino, M., Lucini, L., Rouphael, Y., Cardarelli, M., Bonini, P., Reynaud, H., Canaguier, R. et al. (2013). Soil erosion threatens food production. *Agriculture*, 3: 443–463. [CrossRef]

68. Mahalakshmi, S., Anand, A.J. and Sampath, T. (2023). Recommendation of Crops and Fertilizer, Detection of Crop Weed, Pest and Diseases using Machine Learning. *2023 International Conference on Data Science, Agents & Artificial Intelligence* (*ICDSAAI*), 1–6. Chennai, India.
69. Pinheiro Lisboa, I., Melo Damian, J., Roberto Cherubin, M., Silva Barros, P.P., Ricardo Fiorio, P., Cerri, C.C. and Eduardo Pellegrino Cerri, C. (2018). Prediction of sugarcane yield based on NDVI and concentration of leaf tissue nutrients in fields managed with straw removal. *Agronomy*, 8: 196. [CrossRef].
70. Potamitis, I., Rigakis, I., Tatlas, N.A. and Potirakis, S. (2019). *In-Vivo* vibroacoustic surveillance of trees in the context of the IoT. *Sensors*, 19(6): 1366.
71. Raj Kumar, G., Chandra Shekhar, Y., Shweta, V. and Ritesh, R. (2021). Smart agriculture: Urgent need of the day in developing countries. *Sustainable Computing: Informatics and Systems*, 30: 100512.
72. Ramalingam, B., Mohan, R.E., Pookkuttath, S., Gómez, B.F., Sairam Borusu, C.S.C., Wee Teng, T.W. and Tamilselvam, Y.K. (2020). Remote insects trap monitoring system using deep learning framework and IoT. *Sensors*, 20: 5280. [CrossRef].
73. Roth, L., Aasen, H., Walter, A. and Liebisch, F. (2018). Extracting leaf area index using viewing geometry affects new perspectives on high-resolution unmanned aerial system photography. *ISPRS J. Photogramm. Remote Sens.*, 141: 161–175. [CrossRef].
74. Rubio, V.S. and Ma, F.R. (2020). From smart farming towards Agriculture 5.0: A review on crop data management. *Agronomy*, 10: 207. [CrossRef].
75. Shamshiri, R.R., Kalantari, F., Ting, K.C., Thorp, K.R., Hameed, I.A., Weltzien, C., Ahmad, D. and Shad, Z. (2018). Advances in greenhouse automation and controlled environment agriculture: A transition to plant factories and urban agriculture. *Int. J. Agric. Biol. Eng.*, 11: 1–22. [CrossRef].
76. Shi, X., An, X., Zhao, Q., Liu, H., Xia, L., Sun, X. and Guo, Y. (2019). State-of-the-art Internet of Things in protected agriculture. *Sensors*, 19(8).
77. Singh, R. and Singh, G.S. (2017). Traditional agriculture: A climate-smart approach for sustainable food production. *Energy Ecol. Environ.*, 2: 296–316.
78. Sishodia, R.P., Ray, R.L. and Singh, S.K. (2020). Applications of remote sensing in precision agriculture: A review. *Remote Sens.*, 12: 3136. [CrossRef].
79. Sisinni, E., Saifullah, A., Han, S., Jennehag, U. and Gidlund, M. (2018). Industrial Internet of Things: Challenges, opportunities, and directions. *IEEE Trans. Ind. Inform.*, 14: 4724–4734. [CrossRef].
80. Srisruthi, S., Swarna, N., Ros, G.M.S. and Elizabeth, E. (2016). Sustainable agriculture using eco-friendly and energy efficient sensor technology. *In*: *Proceedings of the 2016 IEEE International Conference on Recent Trends in Electronics, Information & Communication Technology* (*RTEICT*), 1442–1446. Bangalore, India, 20–21 May 2016. IEEE. [CrossRef].
81. Stroppiana, D., Migliazzi, M., Chiarabini, V., Crema, A., Musanti, M., Franchino, C. and Villa, P. (2015). Rice yield estimation using multispectral data from UAV: A preliminary experiment in northern Italy. *In*: *Proceedings of the 2015 IEEE International Geoscience and Remote Sensing Symposium* (*IGARSS*), 4664–4667. Milan, Italy, 26–31 July.
82. Su, J., Liu, C., Coombes, M., Hu, X., Wang, C., Xu, Z., Li, Q., Guo, L. and Chen, W.H. (2018). Wheat yellow rust monitoring by learning from multispectral UAV aerial imagery. *Comput. Electron. Agric.*, 155: 157–166. [CrossRef].
83. Sudduth, K.A. and Hummel, J.W. (1993). Soil organic matter, CEC, and moisture

sensing with a portable NIR spectrophotometer. *Trans. ASAE*, 36: 1571–1582. [CrossRef].

84. Suradhaniwar, S., Kar, S., Nandan, R., Raj, R. and Jagarlapudi, A. (2018). Geo-ICDTs: Principles and applications in agriculture. *In*: Obi Reddy, G.P., Singh, S.K. (Eds.), *Geospatial Technologies in Land Resources Mapping, Monitoring, and Management*, Volume 21: 75–99. Geotechnologies and the Environment (Book Series); Cham, Switzerland: Springer. [CrossRef].
85. Sushil, S., Radha Mohan, S., Manhas, S.S. and Shiv Kumar, L. (2014). Potential of Variable Rate Application Technology in India. *AMA Agric. Mech. Asia Afr. Lat. Am.*, 45: 74–89.
86. Talaei, G.H.T.H., Gholami, S., Pishva, Z.K. and Dehaghi, M.A. (2014). Effects of biological and chemical fertilizers nitrogen on yield quality and quantity in cumin (*Cuminum cyminum* L.). *J. Chem. Health Risks*, 4: 55–64.
87. Terence, S. and Purushothaman, G. (2020). Systematic review of Internet of Things in smart farming. *Trans. Emerg. Telecommun. Technol.*, 31: e3958. [CrossRef].
88. Theopoulos, A., Boursianis, A., Koukounaras, A. and Samaras, T. (2018). Prototype wireless sensor network for real-time measurements in hydroponics cultivation. *In*: *Proceedings of the 7th International Conference on Modern Circuits and Systems Technologies (MOCAST)*, Thessaloniki, Greece, 7–9 May. [CrossRef].
89. Tripathy, P.K., Tripathy, A.K., Agarwal, A. and Mohanty, S.P. (2021). MyGreen: An IoT-enabled smart greenhouse for sustainable agriculture. *IEEE Consum. Electron. Mag.*, 10: 57–62. [CrossRef].
90. Tripodi, P., Massa, D., Venezia, A. and Cardi, T. (2018). Sensing technologies for precision phenotyping in vegetable crops: Current status and future challenges. *Agronomy*, 8: 57. [CrossRef].
91. Tzounisa, A., Katsoulasa, N., Bartzanasb, T. and Kittas, C. (2017). Internet of Things in agriculture, recent advances and future challenges. *Biosyst. Eng.*, 164: 31–48. [CrossRef].
92. Udomkun, P., Nagle, M., Argyropoulos, D., Mahayothee, B. and Müller, J. (2016). Multi-sensor approach to improve optical monitoring of papaya shrinkage during drying. *J. Food Eng.*, 189: 82–89. [CrossRef].
93. Usman, N., Arief, P., Gilang, L., Erfan, R. and Hendra, P. (2018). Implementation IoT in system monitoring hydroponic plant water circulation and control. *Int. J. Eng. Technol.*, 7: 122–126.
94. Vågen, T.G., Winowiecki, L.A., Tondoh, J.E., Desta, L.T. and Gumbricht, T. (2016). Mapping of soil properties and land degradation risk in Africa using MODIS reflectance. *Geoderma*, 263: 216–225. [CrossRef].
95. Venkatesan, R., Kathrine, G., Jaspher, W. and Ramalakshmi, K. (2018). Internet of Things based pest management using natural pesticides for small scale organic gardens. *J. Comput. Theor. Nanosci.*, 15: 2742–2747. [CrossRef].
96. Wietzke, A., Westphal, C., Gras, P., Kraft, M., Pfohl, K., Karlovsky, P., Pawelzik, E. Tscharntke, T. and Smit, I. (2018). Insect pollination as a key factor for strawberry physiology and marketable fruit quality. *Agric. Ecosyst. Environ.*, 258: 197–204. [CrossRef].
97. Wolfert, S., Ge, L., Verdouw, C. and Bogaardt, M.J. (2017). Big data in smart farming—A review. *Agric. Syst.*, 153: 69–80. [CrossRef].
98. Zhang, J., Zhu, Y., Zhang, X., Ye, M. and Yang, J. (2018). Developing a long short-term memory (LSTM) based model for predicting water table depth in agricultural areas. *J. Hydrol.*, 561: 918–929. [CrossRef].

3 | Amalgamation of Technological Innovations in Hunger Alleviation

Olawepo, Gabriel Kehinde*, Ajayi-Osho, Folake Abigail, Olukoyejo, Hannah Oluwakemi, Isah, John Ojo and Fatoba, Paul Ojo

Department of Plant Biology, University of Ilorin Nigeria

In the continuous worldwide endeavor to achieve Zero Hunger, a key component of Sustainable Development Goal 2, Sustainable Smart Agriculture emerges as a pivotal and transformative answer. This extensive chapter explores the complex relationship between modern smart technology and the need for sustainable agriculture, highlighting their critical role in resolving the urgent worldwide problem of food insecurity. This investigation provides a thorough and perceptive evaluation of how IoT, AI, Blockchain, and precision farming are transforming sustainable agricultural practices through a careful examination of their underlying ideas and implementations. The chapter thoroughly explains the significant influence of these technologies in reducing hunger while also stressing the obstacles, best practices, and exciting future prospects that lie ahead. It does this by drawing on a multitude of research and empirical evidence. In order to achieve a world free from hunger and food insecurity, this chapter emphasizes the vital importance of fostering collaboration between innovation, policymaking, and academia by making a strong case for the indispensable role of technology in advancing Sustainable Development Goal 2.

1. Introduction

In the agricultural sector, sustainable smart agriculture is a relatively new and innovative field that is quickly gaining recognition and popularity. The idea behind sustainable smart agriculture is to use smart technologies in agriculture to achieve sustainable development goals. The main goal of sustainable smart agriculture is to preserve the planet's capacity to support future generations while also providing the resources required to support the world's current population.

*Corresponding author: olawepo.gk@unilorin.edu.ng

For a very long time, the agricultural sector has been essential to human existence since it produces food and sustains people everywhere. Traditional farming practices, however, have a reputation for being environmentally damaging and unsustainable. A paradigm change in agricultural practices is represented by sustainable smart agriculture, which aims to produce food in an environmentally sustainable, socially responsible, and commercially profitable manner. The ability of sustainable smart agriculture to address the issues of food security, climate change, and environmental degradation is one of its most important advantages. With the world's population predicted to reach 9.7 billion by 2050, there will be a significant increase in the demand for food, which sustainable smart agriculture can help meet by increasing food production while reducing the environmental impact of agriculture [1–3].

Precision agriculture, robotics, and artificial intelligence (AI) are examples of smart technologies that are being utilized to increase the productivity and efficacy of farming techniques. For instance, precision agriculture maximizes crop yields and minimizes waste by utilizing sensors and data analytics. Tasks like planting and harvesting are being automated by robotics to increase efficiency and decrease the need for physical labor. Data analysis using AI is being utilized to provide insights on crop health, weather patterns, and soil quality.

A new field called "sustainable smart agriculture" holds the potential to completely transform the agriculture sector. Sustainable smart agriculture can assist meet food demand while lessening the environmental impact of farming by utilizing smart technologies in agriculture. Assuring that future generations have the resources they require to prosper and accomplishing sustainable development goals are contingent upon this crucial step.

2. Sustainable Smart Agriculture

An agricultural system that employs smart technologies to accomplish sustainable development objectives is known as sustainable smart agriculture. It is a conglomerate of technologies such as AI-powered precision agriculture systems that boost farming productivity and resource efficiency, Blockchain-enabled Internet of Things (IoT) systems that enhance supply chain traceability, alternative proteins that lessen the environmental impact of conventional meat production, and more [4]. By advancing socio-economic fairness, making a profit, and preserving ecosystem health, sustainable smart agriculture aims to address the issues of food security, climate change, and environmental degradation [5].

3. Context of Sustainable Smart Agriculture

Sustainable smart agriculture is essential for achieving Sustainable Development Goal 2 (Zero Hunger). It improves crop yields, product quality, conserves resources, and reduces food waste [5]. It also minimizes environmental impacts by using less water, fertilizer, and reducing greenhouse gas emissions [6].

Adoption challenges include high installation costs, need for specialized staff, and lack of farmer knowledge [7]. Research shows smart farming boosts yields and revenue, reduces food waste, and improves resource efficiency [8–10].

4. Significance of Sustainable Development Goal 2 (Zero Hunger)

The United Nations (UN) approved 17 SDGs in 2015, with "Zero Hunger" being one of them. Ending hunger, achieving food security, enhancing nutrition, and advancing sustainable agriculture are the objectives by 2030 [11]. Achieving SDG2 (Zero Hunger) is important because it tackles the problem of food insecurity and malnutrition. In 2022, the UN reported that 2.4 billion people experienced moderate to severe food insecurity, accounting for around 9.2% of the world's population (735 million people) who were chronically hungry. People who are hungry or malnourished tend to be less productive and more susceptible to illness, which makes it difficult for them to increase their income and improve their standard of living. Attaining the SDG2 (Zero Hunger) has the potential to mitigate poverty, promote health and wellness, increase productivity in the workplace and in schools, and lessen inequality [11].

5. Role of Technology in Advancing Agriculture Sustainability

A major contribution to the advancement of sustainable agriculture is technology. Low-input agriculture can result from farmers using accurate data in their decision-making with the aid of technology. This can be accomplished by lowering greenhouse gas emissions, conserving natural resources, and using fewer or no toxic chemicals. The integration of intelligent technology in sustainable agriculture has the potential to improve sustainability outcomes by raising productivity, quality of products, crop yield, efficiency in the use of resources, food safety, and the standard of living for workers [12, 13]. The potential improvements to sustainability across food systems can be achieved by utilizing a variety of innovative, data-driven technologies for agricultural production and agri-food value chains, as highlighted by the significance of digitalization in reaching sustainability goals in agriculture [12]. The paper also examines high-level legislation and policy in the German and European contexts to emphasize several crucial institutional, societal, and legal prerequisites for utilizing digital agriculture to meet a range of sustainability goals.

6. Global Food Security and Hunger Statistics

The UNs's latest report indicates 828 million people suffered from hunger in 2021, up by 46 million from 2020 and 150 million since Covid-19 began [14].

In 2020, 3.1 billion people couldn't afford a nutritious diet due to rising food costs from the pandemic's financial impact. Wasting affected 45 million children under five, increasing their mortality risk 12-fold. Additionally, 149 million children under five experienced stunted growth due to nutrient deficiencies, and 39 million were overweight [14]. Food and Agriculture Organization (FAO), International Fund for Agricultural Development (IFAD), United Nations Children's Fund (UNICEF), World Food Programme (WFP), and World Health Organization (WHO) produce the State of Food Security and Nutrition in the World report annually, detailing these issues [15].

7. Environmental Impacts of Conventional Agriculture

Conventional agriculture has significant environmental impacts, including soil erosion, water pollution, and soil quality deterioration. Agriculture accounts for 80% of global deforestation and is a major cause of water pollution, soil degradation, and biodiversity loss [16]. The use of synthetic pesticides and fertilizers contaminates groundwater and depletes soil nutrients, reducing land productivity over time. Heavy machinery compacts soil and damages its structure, contributing to greenhouse gas emissions through land clearing and fossil fuel use [17]. Biodiversity loss is also significant, as habitat destruction and monoculture practices reduce plant and animal species diversity [16]. Conversely, sustainable agriculture aims to minimize environmental harm while maintaining food production. Practices like crop rotation, composting, natural pest management, and conservation tillage reduce artificial fertilizer and pesticide use, prevent soil erosion, and promote biodiversity [18].

8. Integration of Smart Technologies in Agriculture

Smart technologies in agriculture use AI, robotics, sensors, drones, and satellite imagery to solve complex issues and enhance sustainability, production, and resource efficiency [19]. These technologies enable precise application of water and fertilizers, improving efficiency and reducing waste. They are crucial for managing post-harvest operations, soil, water, pests, and plants [19]. Smart systems autonomously detect, decide, and act using sensors, global positioning system (GPS), AI, IoT, and robotics to maximize agricultural productivity [19]. Integrating information and communication technology into farming machinery is key to smart farming, utilizing cloud computing and IoT for enhanced practices. Technologies like drones, big data, and satellites make agriculture more efficient [20]. Modern farmers use various technologies: sensors for light, humidity, soil, and temperature control; customized software; connectivity solutions like LoRa and cellular; positioning systems like GPS; and robotics for processing and autonomous operations [21]. These innovations significantly boost agricultural productivity and sustainability.

9. Core Elements of Sustainable Agriculture Practices

The goal of sustainable agriculture is to preserve the planet's capacity to support future generations while producing enough resources to support the world's current population. Its fundamental principles are to maintain the health of the ecosystem, make a profit, and advance socio-economic fairness. Sustainable agriculture aims to address problems such as agricultural diversity loss, animal welfare issues, human health risks, and environmental degradation and the resulting loss of biodiversity [5]. The fundamental components of sustainable agricultural techniques are as follows:

1. **Management of the soil:** Sustainable agriculture places a strong emphasis on the fertility and health of the soil. Crop rotation, cover crops, and decreased tillage are a few techniques that can assist maintain soil productivity and reduce the demand for agricultural chemicals for pest and fertilizer management [22].
2. **Crop management:** Diverse crops, such as heirloom plants, which are frequently adapted to a region's unique environment, should be planted, according to sustainable agriculture. Sustainable agriculture promotes the use of polyculture, in which several crops are produced together, as an alternative to industrial monoculture, which relies on a single crop. In general, polyculture enhances soil quality and can lessen the demand for chemical fertilizers and pesticides [5].
3. **Water management:** Sustainable agriculture places a strong emphasis on the necessity of conserving water and making effective use of available resources. Water conservation and waste reduction can be achieved by implementing techniques like rainwater collection, drip irrigation, and water recycling [5].
4. **Disease/pest management:** Integrated pest management (IPM) approaches, which minimize the use of hazardous pesticides by utilizing a variety of biological, cultural, and chemical methods, are supported by sustainable agriculture [22].
5. **Waste management:** Sustainable agriculture places a strong emphasis on pollution and waste reduction. Reducing waste and pollution can be achieved through actions like recycling, composting, and turning organic waste into energy [22].

10. Relationship between Technology and Sustainability

Technology and sustainability have a complex relationship. While technology can aid sustainability goals, it can also negatively impact society and the environment. A UNs report highlights the importance of technology in achieving SDGs, including environmental preservation, sustainable economic growth, and improved access to healthcare and education [23]. In agriculture, smart technologies like sensors and machine learning optimize resource use, enhancing efficiency and reducing waste. Renewable energy sources like solar and wind power can reduce greenhouse gas emissions and combat global warming [23]. However, technology also poses risks, such as environmental damage from electronic waste and societal issues like job displacement and income inequality from automation [24]. To ensure responsible and sustainable technology use, its social, economic, and environmental impacts must be considered throughout its lifecycle. This includes creating eco-friendly products, promoting sustainable consumption and production, and ensuring equitable access to technology [24].

11. Smart Technologies Driving Sustainable Agriculture

11.1. Internet of Things (IoT) in Agriculture

By facilitating higher resource efficiency, supporting sustainable economic growth, and fostering environmental protection, smart technologies are advancing sustainable agriculture [23]. IoT is one of the main smart technologies enabling sustainable agriculture [21]. Sensors are used by IoT-based agricultural systems to gather data on a range of parameters, including light intensity, temperature, humidity, and soil moisture [25]. The fundamental components of IoT-based smart agriculture are as follows:

1. **Sensor-based data gathering and monitoring:** Sensors are used by IoT-based agricultural systems to gather data on a range of characteristics, including temperature, humidity, light intensity, and soil moisture. After that, machine learning algorithms are used to evaluate this data in order to extract information on crop health, fertilizer needs, and water usage [26].
2. **Precision agriculture techniques:** To maximize the utilization of resources like water, fertilizer, and energy, precision agriculture techniques make use of IoT-based systems. By applying the proper quantity of water and fertilizer to crops based on sensor data, these methods decrease waste and boost productivity. Drones and satellite imagery are also used in precision agricultural techniques to track crop health and pinpoint problem regions [26].

The various uses of IoT technology in smart agriculture are depicted in Figure 1, which includes supply chain optimization, forestry management, precision farming, real-time farm monitoring, data analytics, aquaponics monitoring, and unmanned aerial vehicle (UAV) use.

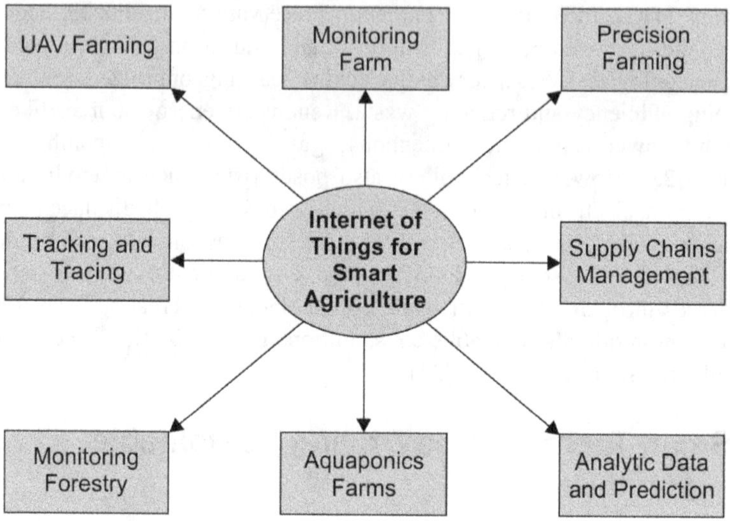

Figure 1: Smart agriculture using the Internet of Things.

CropX

An ag-tech startup called CropX provides soil moisture monitors to farmers so they can maximize irrigation. Farmers can modify their irrigation techniques by using the real-time data on soil moisture provided by these IoT-enabled sensors. By ensuring crops receive the proper amount of water for maximum growth and reducing water waste, this precise management ultimately contributes to higher yields and sustainable agriculture practices.

11.2. Sensor-based Monitoring for Precision Agriculture

The way farmers collect vital information on important agricultural characteristics, such as soil moisture, temperature, humidity, and light intensity, has been completely transformed by sensor-based monitoring technologies. The use of real-time data collecting is crucial to the use of precision agriculture techniques [27]. Farmers are equipped with the knowledge and ability to make well-informed decisions about pest control, fertilization plans, and irrigation schedules by using sensors to continuously monitor these vital aspects. The degree of accuracy and precision with which data is gathered allows farmers to maximize the use of resources, reduce waste, and eventually increase crop yields. In short, sensor-based monitoring is a fundamental component of contemporary farming methods that promote environmentally friendly and sustainable farming practices.

11.3. Improved Resource Management

Farmers utilize IoT sensors to precisely monitor soil moisture levels in their fields. This technology allows real-time data collection, aiding farmers in determining optimal irrigation timing and amounts. Adjusting irrigation techniques based on sensor data helps minimize water waste and ensures crops receive the right amount of water for healthy growth. This control is crucial for reducing water use, mitigating environmental impacts, and improving crop yields and quality. By adopting these sustainable practices, farmers contribute to water conservation and the promotion of eco-friendly farming methods, benefiting future generations [28].

11.4. Reduction in Chemical Use

Sensor-based monitoring is essential to modern agriculture because it gives farmers comprehensive information about the health and growth of their crops. Farmers can target specific problems in their fields, such as pest infestations or nutrient deficiencies, and save time and money by using targeted approaches that reduce the risk of environmental contamination and use less pesticides and fertilizers, allowing the ecosystem to flourish in a healthier and more sustainable way. Precision agriculture has advanced significantly with the use of sensor-based monitoring, which enables farmers to make decisions that are good for the environment and their crops [29].

11.5. Enhanced Crop Health and Yield

IoT-based sensor systems are essential to contemporary agriculture since they allow for real-time crop health monitoring. These cutting-edge technologies are made to identify early symptoms of illness or stress in crops, giving farmers insightful information that enables prompt action. Farmers may proactively address possible problems before they worsen and ultimately reduce crop loss while guaranteeing higher yields by utilizing the data gathered by these sensors. By boosting food production efficiency, this proactive approach benefits individual farmers as well as the global objective of reaching Zero Hunger. Farmers can optimize their resources and make well-informed decisions that result in sustainable agricultural practices by implementing sensor-based monitoring [30]. Sensor-based monitoring helps to achieve Zero Hunger by minimizing crop loss and optimizing output.

11.6. Better Supply Chain Management

Supply chain management systems can now efficiently integrate data from sensors, electronic devices collecting environmental data. This integration enhances the tracking of agricultural products from farm to market, providing stakeholders with real-time visibility into location, condition, and quality. This transparency ensures compliance with food safety regulations, reducing the risk of contamination and spoilage. Additionally, sensor data helps identify issues like temperature changes or transit delays, reducing food waste. By promptly addressing these problems, the supply of wholesome food increases, benefiting producers and consumers. Utilizing sensor technology in supply chain management, as emphasized by [31], enhances overall food security and streamlines distribution processes.

Blue River Technologies' See & Spray system employs AI and computer vision to identify weeds in real time. Targeted herbicide spraying reduces chemical use, promoting sustainable agriculture and contributing to Zero Hunger by minimizing pesticide use.

11.7. Data-driven Decision-making

By utilizing the capabilities of IoT devices like sensors, drones, and GPS technology, precision agriculture approaches have completely changed how farmers approach crop management. Together, these state-of-the-art instruments collect data in real time on critical variables such as crop health, soil moisture content, and meteorological conditions. Farmers are able to make well-informed decisions about planting strategies, irrigation schedules, and the best fertilizer methods by applying this data-driven strategy. Decisions made with this degree of accuracy and precision eventually result in higher crop yields, more efficient use of resources, and less waste. Farmers can increase productivity while reducing their environmental impact by integrating technology and agriculture, which opens the door to more sustainable and effective farming techniques in the future [32].

11.8. Optimized Resource Use

Optimal resource usage is one of the many benefits of precision agriculture that really sticks out. Farmers can precisely gauge the quantity of moisture in the soil and learn important information about how much water is needed for irrigation by using IoT sensors. According to a study by [33], this degree of precision not only helps to minimize water waste but also guarantees that crops receive the hydration required for optimum growth. In addition, the exact application of fertilizers based on soil nutrient levels not only minimizes the environmental effect associated with excessive fertilizer use, but also improves the efficiency of nutrient management procedures. The productivity and sustainability of agricultural experts' farming operations can be greatly increased by implementing these cutting-edge technologies and methods.

11.9. Reduced Chemical Use and Enhanced Crop Yields

Precision agriculture, facilitated by technologies like GPS, drones, and sensors, allows farmers to target specific areas for management, reducing pesticide and herbicide usage. This focused approach not only minimizes environmental contamination but also promotes sustainable farming practices [34]. By producing safer food and preserving ecosystems, precision agriculture contributes to Zero Hunger goals. Monitoring systems help farmers detect early signs of pests, diseases, or nutrient deficiencies, enabling quick action to minimize crop losses and increase yields [35]. Prioritizing early detection and intervention is crucial for optimizing crop output and supporting global food security efforts.

11.10. Improved Food Quality and Traceability

Precision farming methods, based on cutting-edge technologies like IoT-enabled monitoring systems, are critical to improving food product quality and traceability. Farmers can make better decisions and produce higher-quality yields by using real-time data collected from IoT devices to monitor various aspects of crop production. Additionally, the supply chain of agricultural products becomes more transparent and efficient when IoT is integrated with Blockchain technology. With the help of this integration, products can be tracked easily from the farm to the table, giving consumers access to precise information about the source and caliber of the food they eat. By guaranteeing that resources are used more efficiently, this transparency not only raises the bar for food safety requirements but also aids in the reduction of food waste. The way we produce, distribute, and consume food is being revolutionized by the combination of Blockchain technology and IoT-enabled monitoring, which will ultimately result in a more secure and sustainable food system for future generations [36].

12. Technological Applications for Sustainable Agriculture

12.1. AI (Artificial Intelligence) and Machine Learning

1. **Predictive analytics for crop management:** You can anticipate crop yields, identify pests, and improve irrigation plans with the help of AI and machine learning algorithms. For example, V7 Labs has created computer vision technology that uses drone-captured aerial image data to monitor crop and soil conditions. In real time, this technology can assist farmers in determining which regions require pesticide treatment, fertilization, or irrigation, improving crop health and productivity [37].
2. **Resource use optimization:** By forecasting weather patterns, soil moisture content, and other environmental variables, AI and machine learning algorithms can assist farmers in making the best use of their resources. A machine learning system created by the University of Illinois can forecast corn yields based on weather information, soil moisture content, and other variables [38]. When it comes to planting, watering, and harvesting their crops, farmers may use this technology to make more informed decisions.

The applications of AI and robotics in agriculture are shown in Figure 2, where they include supply chain optimization, automating farm inventory management, optimizing AgTech IoT, and encouraging sustainable practices.

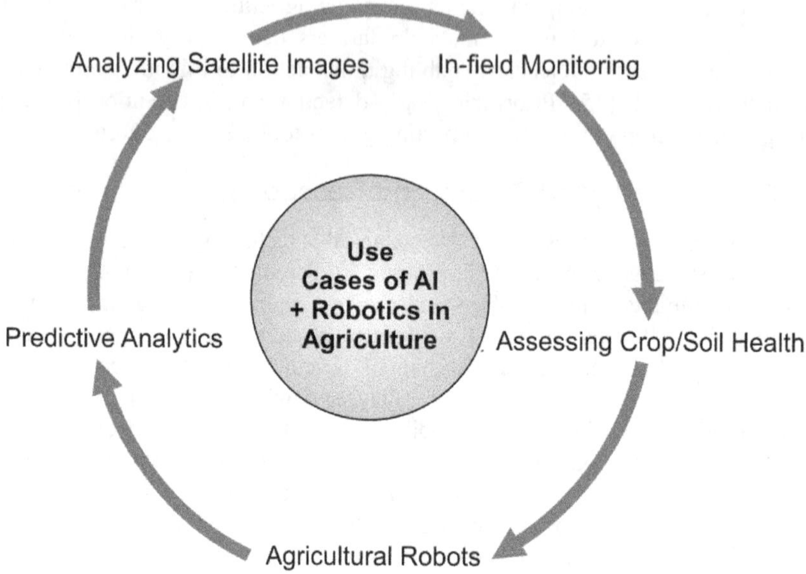

Figure 2: Applications of robotics and AI in agriculture.

12.2. Blockchain and Supply Chain Transparency

1. **Traceability and accountability in agriculture:** A transparent and safe supply chain for agricultural goods can be established using Blockchain technology. A tamper-proof database created by Blockchain technology can record each transaction and help guarantee food goods are traceable from farm to table. Blockchain technology is used by IBM's Food Trust platform to track the origin and path of food goods, giving customers more product transparency and accountability [39].
2. **Ensuring fair trade and quality assurance:** The agricultural sector may also use Blockchain technology to guarantee fair trade and quality assurance. Blockchain can assist both farmers and customers in confirming the legitimacy and caliber of agricultural products by establishing a decentralized trust network. In order to guarantee that coffee growers are paid fairly for their crop, the FairChain Foundation employs blockchain technology to establish an equitable and transparent coffee supply chain [40].

Figure 3 demonstrates how Blockchain technology has the potential to revolutionize the agricultural industry by allowing small farmers to receive safe remittances, updating farm management, guaranteeing corporate accountability, promoting fair pricing, monitoring subsidies, and boosting community agriculture.

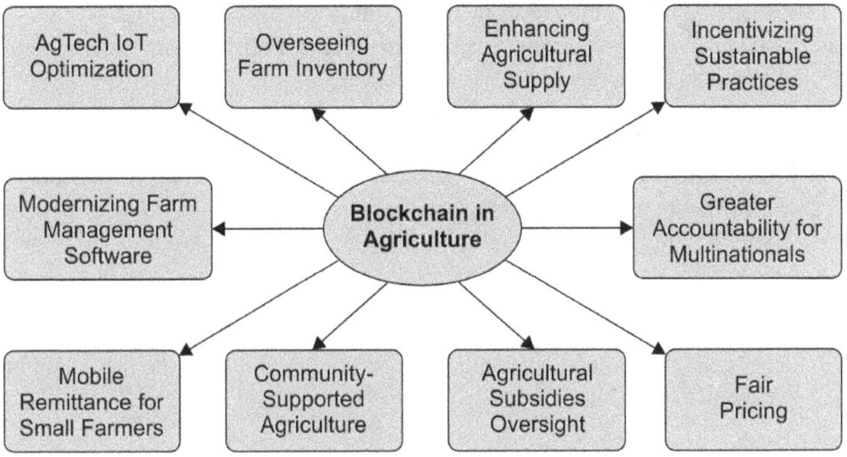

Figure 3: Blockchain in agriculture.

13. Successful Applications of Smart Agriculture Technologies

13.1. Precision Farming in Developing Nations

A good example of this is the Wangree Plant Factory in Thailand, which employs smart technology to improve sustainability performance by raising production productivity, product quality, crop per year, resource use efficiency, and food safety, as well as enhancing the quality of life for its workers [13]. Precision farming is an agricultural approach that maximizes crop yields, reduces waste, and increases productivity. This approach has been successfully implemented in developing nations, where farmers face numerous challenges such as low yields, poor infrastructure, and limited resources.

13.2. IoT integration in Large-scale Farming

Large-scale farming has been completely transformed by the IoT, which gives farmers access to real-time data on the weather, the quality of the soil, the pace of crop development, and the health of the cattle. Farmers may enhance all facets of their business and make well-informed decisions with the use of this data. IoT sensors, for instance, can track the condition of crops, giving farmers precise control over the amount of fertilizer and pesticides needed to achieve maximum efficiency [13, 41].

13.3. Challenges Faced in Implementation

It can be difficult to implement smart agriculture technologies, particularly in underdeveloped countries. Among the primary obstacles are:

i. Inadequate information sharing and coordination between ministries.
ii. Inadequate routes for bottom-up engagement and inadequate representation of non-lead ministries in the writing process.
iii. Financing gaps result in disjointed policies with unclear implementation guidelines, especially at the municipal level.
iv. Insufficient personnel with the necessary skills and training to accurately execute smart agriculture methods.
v. No clear-cut subsidy (liquid price support) for the adoption of smart agriculture products/technology.
vi. A lack of confidence in the potential advantages and risks of smart agriculture in the future.
vii. Uncertainty in policy and regulatory matters pertaining to the use of smart agriculture [13, 42].

14. Overcoming Barriers to Adoption

The following actions are crucial to removing these obstacles:

i. Promote information sharing and collaboration across ministries.
ii. Make sure there are sufficient routes for bottom-up engagement and that non-lead ministries are adequately represented in the writing.
iii. Fund smart agriculture policies and make sure there are clear instructions on how they should be implemented, especially locally.
iv. Provide skilled labor training and development to ensure accurate application of smart agriculture practices.
v. Offer definite subsidies to encourage the use of smart agriculture technologies and goods.
vi. Foster confidence in the potential advantages of smart agriculture and the risk involved.
vii. Assure unambiguous regulation and policy matters for the use of smart agriculture [41].

15. Evaluation of the Impact on SDG2 (Zero Hunger): Achievements and Milestones

Precision farming is one of the most prominent uses of smart agriculture technologies. This creative strategy has been effectively applied in developing countries, where farmers deal with a variety of obstacles like low yields, inadequate infrastructure, and scarce resources. Smart sensors and data analytics are used in precision farming to maximize crop growth, soil health, and irrigation control. With the use of these technologies, farmers are better equipped to manage their crops, increasing yields and enhancing food security. The Wangree Plant Factory in Thailand is one example of a precision farming success story. This cutting-edge facility employs smart technology to improve sustainability performance by raising crop yields and improving worker quality of life by creating a safe and healthy work environment. Production productivity, product quality, crop per year, resource use efficiency, and food safety are all increased with the use of smart agriculture technologies at the Wangree Plant Factory.

16. Addressing Challenges and Unforeseen Consequences

Technologies related to smart agriculture have been hailed as a major answer to the problems confronting the agricultural industry, especially in developing countries. The way food is produced and delivered around the world could be completely changed by these technologies, which maximize agricultural yields and cut waste by utilizing cutting-edge instruments like sensors, drones, and AI.

However, there are a lot of obstacles that must be removed before smart agriculture technology may be implemented in developing countries. The absence of information exchange and collaboration between ministries is one of the main problems. Different government departments and agencies handle different facets of agriculture in many developing countries, and there is sometimes minimal cooperation between them. This may result in a lack of communication and information exchange, which could impede the uptake of smart agriculture technologies.

Financing is still another significant obstacle. In developing countries, a large number of small-scale farmers do not have access to the capital necessary to purchase new technologies. Numerous things, such as excessive interest rates, a deficiency of collateral, and restricted loan availability, may be to blame for this. It might be challenging for farmers to invest in the tools and equipment necessary to apply smart agriculture technology without funding.

In many emerging countries, there is a shortage of professional and trained labor in addition to these difficulties. Because of this, farmers may find it challenging to adopt and maintain smart agriculture technologies if they lack the requisite skills or expertise. It can be challenging to fully profit from these technologies in the absence of competent labor.

17. Emerging Technologies in Sustainable Agriculture

Emerging technologies are becoming essential tools for crop management, quality control, integrated disease management, sustainable production, drone technology, precision agriculture, cloud computing, AI, IoT, machine learning, big data analytics, cloud computing, and genome editing technologies, particularly CRISPR-CAS9-based (clustered regularly interspaced short palindromic repeats-CRISPR-associated protein 9) genetic engineering [43]. By improving agricultural yields, these technologies can lessen the likelihood of global food shortages.

18. Policy Recommendations

By 2050, the global population is projected to reach 9 billion, intensifying food demand. Smart agriculture technologies offer a solution to meet this demand, but their adoption requires collaboration across ministries, funding, and skilled labor. Facilitating collaboration among ministries ensures efficient use of resources and alignment of efforts. Financial support for smart agriculture initiatives mitigates upfront costs and associated risks. Additionally, training programs are essential to equip farmers with the necessary skills for optimal technology utilization. Clear subsidies further incentivize adoption. Establishing farmers' confidence in the benefits and risks of smart agriculture is crucial for widespread adoption. Understanding how these technologies improve yields, reduce waste, and enhance sustainability is key. Farmers must also learn to mitigate potential risks associated with adopting new technologies.

19. Conclusion

Sustainable Smart Agriculture is pivotal for achieving SDG2 by merging agricultural sustainability and technical innovation. AI-driven precision agriculture, Blockchain, and alternative proteins offer revolutionary solutions to food security, climate change, and environmental degradation. Research shows these technologies can enhance crop yields, supply chain traceability, and reduce environmental impact. Through IoT, AI analytics, and Blockchain, Sustainable Smart Agriculture optimizes resource use, crop quality, and environmental stewardship. To achieve Zero Hunger, more efforts are needed to implement these technologies and support smallholder farmers. By embracing technology and sustainable practices, we pave the way for a resilient future without hunger. Sustainable Smart Agriculture serves as both a solution and a catalyst, epitomizing the synergy between innovation and sustainability in shaping a hunger-free world.

References

1. Augustine, C., Balaji, K., Dharani Kkumar, S.V. and Anand, A.J. (2023). Urban farming: Case study. *In*: K. Kalaiselvi, A.J. Anand, P. Tanwar and H. Raza (Eds.), *Advanced Technologies for Smart Agriculture*, 321–338. Denmark: River Publishers.
2. Anand, J.A., Devi, K.N.S., Vijayakumar, S. and Chacko, S.C. (2023). Decision-making support in smart farming. *In*: K. Kalaiselvi, A.J. Anand, P. Tanwar and H. Raza (Eds.), *Advanced Technologies for Smart Agriculture*, 219–239. Denmark: River Publishers.

3. Kamesh, M., Nisha, K., Augustince, C. and Anand, A.J. (2023). Deep learning in smart agriculture. *In*: K. Kalaiselvi, A.J. Anand, P. Tanwar and H. Raza (Eds.), *Advanced Technologies for Smart Agriculture*, 153–175. Denmark: River Publishers.
4. Food and Agriculture Organization (FAO) of the United Nations. (2021). *Smart Agriculture*. http://www.fao.org/smart-agriculture/en/
5. Dubey, A. (2023). Sustainable agriculture. *In*: *Encyclopædia Britannica*. https://www.britannica.com/technology/sustainable-agriculture.
6. Food and Agriculture Organization of the United Nations. (2021). *Climate-Smart Agriculture*. http://www.fao.org/climate-smart-agriculture/en/
7. Sustainable Agriculture Research and Education. (2021). *Smart Farming*. https://www.sare.org/Learning-Center/Topic-Rooms/Sustainable-Agriculture-Education-Directory/Smart-Farming/
8. Kassie, M., Teklewold, H. and Shiferaw, B. (2021). Adoption of smart farming technologies and its impact on smallholder farmers' welfare: Evidence from Ethiopia. *Agricultural Systems*, 194: 103051. https://doi.org/10.1016/j.agsy.2021.103051.
9. Gao, Y., Li, X. and Li, Y. (2021). Blockchain technology in agriculture: A systematic review. *Journal of Cleaner Production*, 315: 128223. https://doi.org/10.1016/j.jclepro.2021.128223.
10. Zhang, Y., Zhang, J. and Li, Y. (2021). Application of Artificial Intelligence in precision agriculture: A review. *Journal of Cleaner Production*, 315: 128142. https://doi.org/10.1016/j.jclepro.2021.128142.
11. United Nations. (2022). *Goal 2: Zero Hunger*. https://www.un.org/sustainabledevelopment/hunger/
12. Santiteerakul, S., Sopadang, A., Tippayawong, K.Y. and Tamvimol, K. (2020). The role of smart technology in sustainable agriculture: A case study of Wangree Plant Factory. *Sustainability*, 12(11): 4640.
13. Santiteerakul, S., Sopadang, A., Tippayawong, K.Y. and Tamvimol, K. (2022). Future agricultural systems and the role of digitalization for achieving sustainability goals: A review. *Agronomy for Sustainable Development*, 42(70). https://doi.org/10.1007/s13593-022-00792-6.
14. United Nations. (2022). *UN Report: Global Hunger Numbers Rose to as Many as 828 Million in 2021*. July 6. https://www.who.int/news/item/06-07-2022-un-report--global-hunger-numbers-rose-to-as-many-as-828-million-in-2021.
15. FAO, WFP, IFAD, UNICEF and WHO. (2021). *2021 State of Food Security and Nutrition in the World – Report and InBrief*. https://www.wfp.org/publications/2021-state-food-security-and-nutrition-world-report-and-inbrief.
16. Union of Concerned Scientists. (2023). *Environmental Impacts of Conventional Agriculture*. https://www.ucsusa.org/resources/environmental-impacts-conventional-agriculture.
17. Galford, G.L., Melillo, J.M., Kicklighter, D.W., Cerri, C.C. and DeFries, R.S. (2010). Greenhouse gas emissions from alternative futures of deforestation and agricultural management in the Southern Amazon. *Proceedings of the National Academy of Sciences* 107(46): 19649–19654. https://doi.org/10.1073/pnas.1000780107.
18. National Institute of Food and Agriculture. (2023). *Sustainable Agriculture*. https://nifa.usda.gov/sustainable-agriculture.
19. Verschae, R. (2023). Smart technologies in agriculture. *In*: *Encyclopedia of Smart Agriculture Technologies*, 1–11. SpringerLink. https://doi.org/10.1007/978-3-030-89123-7_234-1.

20. EOS. (2022). *Smart Farming and Its Technologies Application in Agriculture*. October. https://eos.com/blog/smart-farming/
21. IoT for All. (2020). *What Is Smart Farming? It's the Future of Agriculture*. June 22. https://www.iotforall.com/smart-farming-future-of-agriculture.
22. Agro learner. (2023). *12 Sustainable Agriculture Practices*. https://agrolearner.com/sustainable-agriculture-practices/
23. United Nations. (2021). *The Sustainable Development Goals Report 2021*. https://unstats.un.org/sdgs/report/2021/The-Sustainable-Development-Goals-Report-2021.pdf.
24. United Nations Environment Programme. (2019). *Global Environment Outlook 2019: Healthy Planet, Healthy People*. https://www.unenvironment.org/resources/global-environment-outlook-2019.
25. Dhanaraju, M., Chenniappan, P., Ramalingam, K., Pazhanivelan, S. and Kaliaperumal, R. (2022). Smart farming: Internet of Things (IoT)-based sustainable agriculture. *Agriculture*, 12(10): 1745. https://doi.org/10.3390/agriculture12101745.
26. Singh, A. and Singh, S. (2021). Internet of Things (IoT) in agriculture: A comprehensive review. *In*: *Internet of Things (IoT) in Agriculture*, 1–23. Springer. https://doi.org/10.1007/978-981-16-3379-6_1.
27. Prasad, V. and Kumar, S. (2019). Role of IoT in smart agriculture. *International Journal of Advanced Science and Technology*, 28(16): 77–83. https://doi.org/10.18517/ijaseit.7.1.1771.
28. Dutta, P., Mahapatra, D. and Reddy, V.K. (2020). Application of IoT in agriculture: Opportunities and challenges. *International Journal of Agricultural and Biosystems Engineering*, 14(10): 420–424. https://doi.org/10.5281/zenodo.4314990.
29. Rafique, R., Khan, M.S. and Ali, M. (2020). IoT-based precision agriculture: A systematic review. *International Journal of Computer Applications*, 182(32): 34–40. https://doi.org/10.5120/ijca2020909244.
30. Kumar, P., Gupta, R. and Sharma, M. (2021). Early detection of plant diseases using IoT and sensor-based technology. *Journal of Agricultural Research and Technology*, 6(1): 17–25. https://doi.org/10.19080/jart.2021.06.555676.
31. Amanullah, S. and Farooq, M.U. (2020). IoT-based smart agriculture: Towards sustainable food production and supply chain management. *Journal of Agricultural Informatics*, 10(1): 45–56. https://doi.org/10.17700/jai.2020.10.1.530.
32. Siddharth, R., Gupta, A. and Basu, S. (2018). Precision agriculture using IoT: A survey and perspectives. *International Journal of Computer Applications*, 180(5): 1–6. https://doi.org/10.5120/ijca2018909451.
33. Adel, M.A., El-Maddah, E.A. and Elmahdy, A.H. (2020). Precision agriculture: An emerging technology for sustainable agriculture. *Agricultural Sciences*, 11(5): 329–338. https://doi.org/10.4236/as.2020.115021.
34. Bhatta, G.D. and Joshi, N.K. (2021). Precision agriculture and Internet of Things (IoT) in India: Opportunities, challenges, and implications for achieving zero hunger. *International Journal of Agricultural and Biosystems Engineering*, 15(4): 102–108. https://doi.org/10.5281/zenodo.4325997.
35. Carneiro, R.L., Gontijo, C.A. and Silva, D.F. (2020). Precision agriculture and crop management in IoT. *Journal of Agricultural Engineering and Technology*, 1(1): 21–28. https://doi.org/10.1134/S0024099321040018.
36. Benke, K. and Tomkins, B. (2017). Future food-production systems: Vertical farming and controlled-environment agriculture. *Sustainability*, 9(5): 1–17. https://doi.org/10.3390/su9050866.

37. Rizzoli, A. (2023). *8 Practical Applications of AI in Agriculture*. https://www.v7labs.com/blog/ai-in-agriculture
38. Lenniy, D. (2023). *AI in Agriculture: The Future of Farming*. https://intellias.com/artificial-intelligence-in-agriculture/
39. Gupta, S. (2022). *Top 4 Applications of AI and Machine Learning in Agriculture*. https://techgraph.co/ai/top-4-applications-ai-machine-learning-in-agriculture/
40. Sharma, A., Jain, A., Gupta, P. and Chowdary, V. (2012). Machine learning applications for precision agriculture: A comprehensive review. *IEEE Access*, 9: 4843–4873. https://doi.org/10.1109/ACCESS.2020.3048415.
41. Giles, J., Grosjean, G., Le Coq, J.-F., Huber, B., Bui, V.L. and Läderach, P. (2021). Barriers to implementing climate policies in agriculture: A case study from Viet Nam. *Frontiers in Sustainable Food Systems*, 5: 439881.
42. Eastern Peak. (2023). *IoT in Agriculture: 9 Technology Use Cases for Smart Farming (and Challenges to Consider)*. March 1. https://easternpeak.com/blog/iot-in-agriculture-technology-use-cases-for-smart-farming-and-challenges-to-consider/.
43. Sarfraz, S., Ali, F., Hameed, A., Ahmad, Z. and Riaz, K. (2023). Sustainable agriculture through technological innovations. *In*: *Sustainable Agriculture in the Era of the OMICs Revolution*, 511–532. Springer International Publishing.

4 | Nourishing Nations: Harnessing Nutritional Food Security for Zero Hunger

Jyoti Rani[1*] (0000-0001-6334-4642), Sahil Singh Dhull[1], Deepak Beniwal[1], Deepali[1], Vibhuti Gulia[1] and Gagandeep Singh[2]

[1] Department of Botany, Chaudhary Devi Lal University, Sirsa, India
[2] Institute for Molecular Bioscience, University of Queensland, Australia

Global population growth and climate change make it harder to meet food demands, making hunger a serious issue. As we combat hunger, it is crucial to acknowledge and tackle environmental degradation, which negatively impacts agricultural productivity. Addressing both issues requires a dual focus on mitigating climate change and promoting ecological sustainability. This dual emphasis will guarantee that our efforts to mitigate hunger do not compromise the overall health of our planet. This chapter explores the complex interaction of agriculture, sustainability, and technology. 36 complex solution is developed through the integration of various approaches, such as adopting holistic methods, finding a balance between ecology and economy, identifying nutrient-rich foods, and adding variety to the diet. The variation of agricultural techniques, resulting from a transition from monoculture to polyculture and the implementation of agroecological models, highlights the significance of diversification. Local farmers are empowered and sustainable practices are fostered via community participation and education. Furthermore, international collaboration and inventive approaches are crucial in conquering global obstacles and guaranteeing food security for every individual. "Nourishing Nation" emphasizes the urgent requirement to revolutionize agricultural and food systems to ensure a hunger-free world.

1. Introduction

The task of ensuring global food security becomes more serious in a world where the population is projected to exceed 9 billion by 2050. Attaining the objective of

*Corresponding author: drjyotisangwan@cdlu.ac.in; jyotipanghal91286@gmail.com

"Zero Hunger" is not only desirable but also an ethical responsibility. Currently, the demand for innovative and environmentally friendly agricultural solutions is more urgent than ever before, due to the convergence of factors such as population growth, climate change, and the depletion of natural resources [1]. For a considerable duration, the United Nations and other organizations have focused on combating hunger by formulating the ambitious Sustainable Development Goals (SDGs) to deal with the issue of food scarcity; Goal 2, sometimes known as "Zero Hunger, is a particularly hopeful aim among the others. Despite the progress made throughout time, the battle to eradicate global hunger remains a significant challenge [2]. The challenges are numerous. Climate change has resulted in unpredictable weather patterns, prolonged periods of dryness known as droughts, and excessive rainfall leading to floods, all of which have significantly affected agricultural practices. Given the rise of the population, it is imperative to enhance food production on a smaller land area [3]. Traditional farming methods are facing challenges because of the depletion of natural resources, limited availability of water, and deterioration of soil quality. Furthermore, the issue is exacerbated by the uneven allocation of food, disparities in economic resources, and political instability. As a result of these challenges, traditional methods of agriculture are becoming more inadequate in providing food for the growing global population [4]. Smart agriculture, a transformative approach that integrates technology, data, and sustainable practices, is poised to revolutionize our food production, distribution, and consumption. This innovative concept combines a variety of modern techniques such as artificial intelligence (AI), data analytics, Internet of Things (IoT) sensors, and precision farming. The goal is to enhance crop productivity, bolster sustainability, and optimize the utilization of resources. In simple terms, smart agriculture is the key to unlocking future solutions in the agricultural Field [5].

At the heart of the "Nourishing Nations" initiative lies the principle of sustainability. Smart agriculture champions sustainable farming practices that minimize environmental impact, conserve natural resources, and reduce greenhouse gas emissions. Moreover, it fosters resilience against climate change by enabling adaptive strategies to address unexpected disruptions in the food supply chain [6].

Minimizing food waste is key to smart agriculture. Most food is wasted during harvesting, transporting, and storage. Smart agriculture reduces waste and improves distribution by monitoring and improving the food supply chain [7]. This attempt also seeks to empower smallholder farmers, who support global food production. We can help them increase revenue and productivity and minimize food insecurity and poverty by providing smart information, technology, and resources [8].

Small-scale farmers in developing countries may lack the necessary skills and technology. The "Nourishing Nations" strategy focuses on training and infrastructure to address this gap. One nation or institution cannot do this alone. "Nourishing Nations" promotes global collaboration and resource, policy, and knowledge coordination. It promotes global cooperation to sustain and complete smart agricultural efforts.

2. Understanding Nutritional Food Security

Nutritional food security is a critical aspect of the global goal of achieving Zero Hunger through various synchronized efforts, as depicted in Figure 1. It extends beyond the conventional notion of food security, which emphasizes the availability, access, and affordability of food, to encompass the nutritional excellence of the food ingested. Nutritional food security refers to the state in which individuals have consistent and open access to an adequate amount of food that is both safe and nutritious, meeting their dietary requirements and personal food preferences, thus enabling them to lead an active and healthy lifestyle. Availability refers to the actual existence of food in communities, requiring strong agricultural systems, effective distribution networks, and durable supply chains. Access examines the socioeconomic variables that influence an individual's capacity to acquire food, such as their income, job prospects, and the price of healthful choices. Utilization encompasses more than just access, highlighting the significance of knowledge and habits pertaining to nutrition, food preparation, and healthcare access [9]. This ensures that people are able to obtain the most nutritional benefits from the food they consume. The importance of resilience in the face of external shocks, such as natural disasters or economic crises, is highlighted by the necessity for stability in maintaining the constant availability of food across time. To achieve nutritional food security and eliminate hunger, a collaborative approach is required across several sectors such as agriculture, healthcare, education, and social welfare. This approach should aim to address the intricate relationship between factors that affect the availability, accessibility, use, and stability of food. To develop a world without hunger or malnutrition, we should prioritize policies and initiatives that provide fair access to healthy food, empower communities with information and resources, and enhance their ability to withstand external challenges [10].

The concept of nutritional food security recognizes that it's not enough to simply have enough food to eat. The concept of nutritional food security acknowledges that just food availability is insufficient. The meal must also be nutritionally balanced to fulfill the body's requirements for growth, development, and maintenance of optimal health. This entails consuming a well-proportioned combination of macronutrients (proteins, fats, and carbohydrates) and micronutrients (vitamins and minerals) [11]. Education and increasing awareness are essential elements of ensuring food security via nutrition. Individuals must comprehend the significance of maintaining a well-balanced diet and acquiring knowledge on selecting nourishing food options. This may be accomplished by implementing nutrition education programs, implementing food labeling systems, and conducting public health campaigns.

In the framework of achieving Zero Hunger, ensuring access to nutritious food is crucial for food security. Merely eradicating hunger is insufficient; it is vital

Figure 1: Synchronized efforts to attain global food and nutritional security.

to guarantee that every individual has the means to obtain nourishing sustenance in order to maintain a fit and vigorous lifestyle. This is a crucial milestone in the pursuit of the United Nations' objective of attaining Zero Hunger by 2030 as part of its Sustainable Development Goals. Achieving nutritional food security is therefore not just about feeding the world, but nourishing it [12].

3. Rediscovering Nutrient Powerhouses

The term "nutrient powerhouses" denotes food that possesses a high concentration of vital vitamins, minerals, and other nutrients that are fundamental for maintaining human health. Figure 2 highlights the implications of malnutrition and underscores the importance of addressing nutritional insecurity. To successfully combat malnutrition and food insecurity and eventually get closer to the goal of attaining zero hunger, we must reclaim and promote these nutrient-dense foods.

This method entails not only recognizing conventional foods that are high in nutrients but also investigating less familiar and underutilized food sources that might make a substantial contribution to fulfilling nutritional requirements, especially in groups that are more susceptible to health risks. In addition, efforts

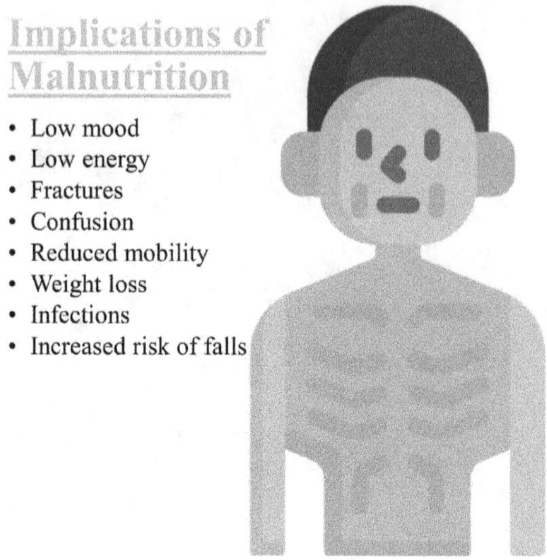

Figure 2: Addressing nutritional insecurity through implications of malnutrition.

to advance sustainable agriculture techniques can guarantee the presence and reachability of these nutrient-rich sources while protecting the environment and conserving biodiversity. In general, the rediscovery and promotion of nutrient-rich foods are seen as a viable approach in the worldwide effort to combat hunger [13]. This strategy offers a sustainable solution to enhance nutrition and provide food security for everyone.

The revival of millets and other neglected crops signifies a notable transition in agricultural paradigms towards practices that are more sustainable, nutrient-dense, and ecologically aware. This movement, sometimes referred to as the "Rediscovery of Nutrient Powerhouses", acknowledges the capacity of these crops to tackle current issues such as worldwide food poverty, climate change, and public health problems. Millets are a collection of small-seeded annual grasses that are grown as grain crops. They have the ability to flourish in poor-quality soils and dry areas found in temperate, subtropical, and tropical climates. Millets are well-known for their abundant fiber, vitamins, and minerals, making them a favorable choice for people with digestive problems and those looking to improve their general health [14]. The Food and Agricultural Organization and the United Nations have designated 2023 as the "International Year of Millets" to emphasize a focused endeavor in increasing knowledge about the nutritional excellence and health advantages of millets. Pearl millet, foxtail millet, proso millet, finger millet, kodo millet, tiny millet, and barnyard millet are some of the species belonging to the Poaceae family that play a role in this resurgence of nourishment. In addition to addressing hunger, millet has demonstrated pharmacological promise in the treatment of many diseases, as indicated by current studies [15]. The current focus on the nutritional content and environmental resilience of millet represents the result of extensive scientific research and policy efforts targeted at advancing sustainable agriculture and enhancing public health outcomes. Millets not only play a crucial role in addressing hunger but also exhibit pharmacological potential for treating various illnesses, as highlighted in Table 1.

However, the remarkable nutritional value and environmental resilience of millets have come to light due to current research and legislative initiatives.

Table 1: Nutritional properties and pharmacological potential of millets to treat various illnesses

S. No.	Nutritional Specifications	Healthcare
1.	Low glycemic index	Treat diabetes
2.	High fiber content	Reduce constipation Prevent gall-bladder stones Reduces the production of excessive bile juice
3.	High iron and zinc concentration	Increases hemoglobin
4.	High amounts of magnesium	Control blood pressure and heart stress Reduce respiratory problems Reduce migraine attack
5.	Presence of phosphorus	Bone growth, development, and repair
6.	Phytonutrients and lignin as strong antioxidants	Prevent heart disease
7.	Hypoallergenic property	Anti-allergic
8.	Inhibit tumor growth	Treat cancer
9.	Prevents formation of excess acidity	Convert the stomach alkaline and prevent the formation of ulcers

Source: [15]

The revival of millets and other neglected crops signifies a crucial transformation in agricultural methods, signaling a shift toward varied, robust, and environmentally friendly food systems. Millets, known for their high nutritional value including protein, fiber, antioxidants, and a wide range of vitamins and minerals, are a symbol of this movement. Their low glycemic index not only helps maintain stable blood sugar levels but also makes them acceptable for persons who are controlling diabetes. Furthermore, their exceptional ability to withstand unfavorable climatic circumstances, such as prospering in hot temperatures with little water and poor soil quality, makes them crucial partners in the fight against climate change and in strengthening food security in vulnerable countries. This renewal extends beyond millets, including a wider range of underused crops, including legumes, root vegetables, indigenous plants, and pseudocereals such as quinoa and amaranth. Adopting these many crops offers numerous advantages, including more variety in diets and improved human well-being, as well as more sustainable agricultural methods and increased environmental adaptability [16].

Continuing research and scholarly inquiries play a crucial role in the revival of agriculture, particularly in exploring the nutritional, economic, and agronomic aspects of millets and other crops that have been neglected. These efforts are crucial for properly incorporating these crops into modern agricultural practices and food systems. Millets and other neglected crops provide several advantages, including poverty reduction via the establishment of jobs and sustainable livelihoods, as well as improving the nutritional value of diets for low-income households. In addition, they have a crucial role in adjusting to unpredictable settings, sustaining agricultural ecosystems, and expanding the range of crop choices to satisfy changing market needs. By utilizing the capabilities of these hardy and nourishing crops, we are creating a path toward a more comprehensive, environmentally friendly, and resilient future in agriculture [17].

4. Role of Sustainable Agriculture

Sustainable agricultural practices are emerging as a key component of the worldwide effort to achieve Zero Hunger. In context with global issues such as climate change, resource depletion, and population expansion, it is crucial to promptly transition agricultural techniques to more sustainable and efficient ones. In smart agriculture for zero hunger, the spectrum of sustainability in agriculture requires a holistic and integrated approach that strikes a balance between social justice, environmental preservation, and economic feasibility in the food production process (see Figure 3) [18].

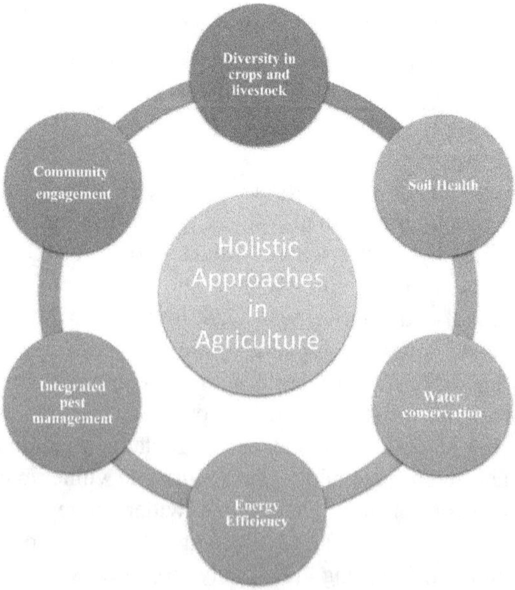

Figure 3: Holistic approaches in agriculture.

Through the promotion of diverse farming systems, crop rotations, and integrated pest management, holistic methods enhance the ability of agriculture to withstand the impacts of climate change and adverse weather conditions. This, in turn, protects food production and the livelihoods of vulnerable populations. The focus is on promoting economic sustainability by providing assistance to small-scale farmers, enhancing their access to markets, and developing inclusive value chains to ensure fair distribution of resources and opportunities. Moreover, social sustainability is of the highest priority, encompassing efforts to empower rural communities, enhance food security, and advocate for gender equality and social justice. Collaboration among governments, NGOs, farmers, researchers, and the commercial sector is essential for expanding comprehensive strategies to achieve sustainable farming [19]. This will provide the foundation for a future where hunger is eliminated and food systems flourish in harmony with the earth.

4.1. Agroecological Approaches and Agriculture Techniques for Nutritional Food Security

Agroecology provides a thorough framework for agricultural production that focuses on diversity and ecological principles. Agroecological models integrate biodiversity and natural processes into farming systems to enhance food security, strength, and sustainability. Crop diversification, agroforestry, integrated pest management, and soil conservation strategies are examples of agroecological activities. These methods improve yields, reduce environmental impact, and boost financial stability in addition to improving ecosystem health. Agroecological models can help achieve zero hunger by increasing food production, assisting small-scale farmers, and building resilient communities. They place a high value on diversity and ecological principles [20].

An interesting journey has been the evolution of agricultural techniques over time, shaped by a variety of circumstances. Understanding and reducing these socio-ecological effects—which include declines in ecosystem services, financial losses, and effects on human health—will be necessary to achieve agricultural sustainability. Evolution has shaped agricultural systems by including biological information from crop systems in the context of evolving agricultural practices. Evolution has shaped agricultural systems by including biological information from crop systems in the context of evolving agricultural practices. First, people manage agricultural fields where a limited number of species are grown, particularly in traditional agriculture where monocultures are common [21]. The realization of the drawbacks and threats of monoculture systems led to a major paradigm change in agriculture with the shift from monoculture to polyculture. The transition from monoculture to polyculture, which promotes a more linked and sustainable agricultural system, shows hope for attaining Zero Hunger.

Monoculture: Monoculture is the practice of growing a single primary crop throughout a large region with the goal of increasing yields and efficiency. But worries about degraded soil, heightened pest and disease pressures, and lost biodiversity are causing a review of agricultural methods.

Polyculture: A paradigm change in agriculture towards diversity is represented by polyculture. Contrarily, polyculture is cultivating several crops in one location, frequently using different rotations and combinations. This method minimizes dependency on artificial inputs like fertilizers and pesticides while promoting biodiversity, ecological resilience, and soil health [22].

4.2. Biofortification: Biofortified Crops – A Pathway to Nutritional Security

Biofortification is the process of improving the nutritional composition of crops using methods such as conventional breeding, genetic modification, or agronomic approaches as presented in Figure 4. Biofortification differs from traditional fortification by directly enhancing the nutritional composition of crops, rather than adding nutrients to food during processing. Biofortification, the natural enrichment of the nutritional value of crops, has significant potential to combat malnutrition in vulnerable populations, particularly in regions with limited access to diverse and nourishing foods. These internationally available fortified crops may be grown without any specific cultivation techniques, making them a realistic solution for nutritional deficiencies. The enrichment process does not compromise the productivity of the crops. Many of the targeted minerals are vital for the plants' own nutritional needs and can enhance their tolerance to environmental stressors. This can lead to improved growth and potentially higher yields, making biofortification a win-win strategy for both human health and agricultural productivity [23].

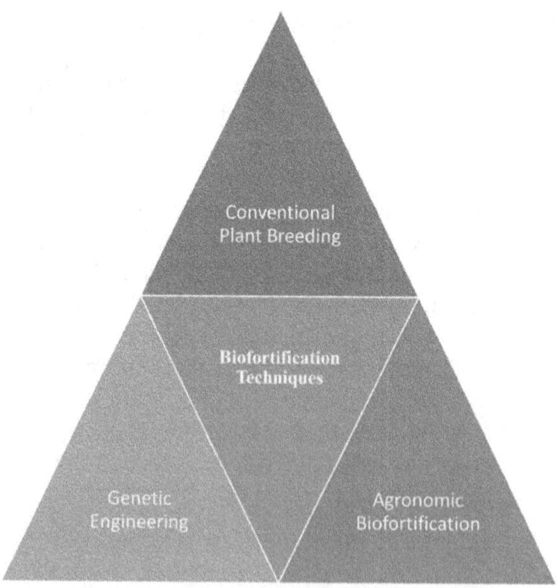

Figure 4: Types of biofortification techniques.

Various organizations have successfully introduced over 260 biofortified types of 11 essential crops in 30 countries. These organizations have provided biofortified seeds to more than 40 million individuals. Their objective is to facilitate the transformation of food systems to provide nutritious food to everyone. They plan to achieve this by promoting the growth and widespread adoption of staple crops that are high in essential nutrients. Their goal is to benefit 1 billion people by the year 2030. Ultimately, biofortification presents a hopeful, economically efficient, and enduring method to enhance the condition of malnutrition on a global scale. It serves as a means to enhance the ability of communities to withstand nutritional challenges and is a crucial stride toward attaining nutritional stability [24].

Biofortified crops encompass a wide range of staple crops, legumes, and vegetables that have been enhanced to contain higher levels of essential vitamins and minerals. These crops are developed through selective breeding or genetic engineering techniques aimed at increasing the bioavailability of key nutrients, such as vitamin A, iron, zinc, and folate. Even little increments in the concentrations of vital minerals in staple grains can have a significant impact on the health and productivity of millions of people worldwide. Biofortification of cereal crops can yield substantial health, social, and economic advantages by enhancing the mineral content in plants and subsequently benefiting consumers. Rice, wheat, maize, potato, and cassava are the five primary food crops in Africa, Asia, and Latin America, making them the ideal candidates for biofortification. Besides these, grain legumes, particularly pulses, form a significant part of daily diets in numerous Asian and African nations. Pulses are cultivated globally, with Asia leading in terms of area and production, closely followed by Africa. With the goal of making India self-sustaining in food production, the National Agricultural Research System (NARS), which consists of State Agricultural Universities (SAUs) and Indian Council of Agricultural Research (ICAR) institutes, has made significant contributions [25]. Table 2 displays various micronutrients, the crops in which they are found, and the origin country of each crop variety, along with the year.

Smarter farming employs IoT devices, big data analytics, automated machinery, Blockchain technology, and smart irrigation systems to enhance crop management techniques. IoT devices, including drones and soil sensors, offer up-to-the-minute information on weather patterns, crop health, and soil conditions. This data allows for accurate decision-making. Big data analytics enable the analysis of large volumes of data from many sources to foresee trends, maximize yields, and reduce risks. Automated technology, such as drones and self-driving tractors, decreases the need for human labor, enhances efficiency, and lessens the negative effects on the environment. Blockchain technology guarantees the safety and transparency of food in the supply chain, while smart irrigation systems optimize water consumption and reduce wastage [26].

The advancements in precision agriculture aim to optimize agricultural productivity while reducing adverse environmental impacts. Variable Rate Technology (VRT) enables the precise administration of inputs like water,

Table 2: Various micronutrients in crops; year and country of origin

Nutrient	Crop	Countries of First Release	Release Year
Iron	Cowpea 	India	2008
Iron and Zinc	Beans 	Rwanda, DRC	2012
	Lentils 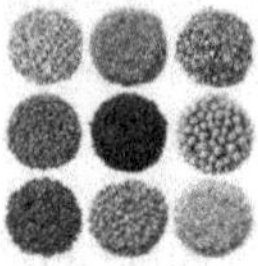	India, Nepal, Bangladesh	2012, 2013
	Mung Bean 	India	2013
	Pea 	USA	2012
	Pearl millet 	India	2013
	Rice 	Bangladesh, India	2013
	Wheat 	India, Pakistan	2013

Contd...

Provitamin A	Cassava 	Nigeria	2013
	Maize 	Zambia, Nigeria	2012
	Sweet potato 	Mozambique, Uganda	2007

herbicides, and fertilizers to particular zones within a field, minimizing resource inefficiency and environmental consequences. Geospatial technology, like GPS and GIS, empowers farmers to generate accurate field maps and make well-informed choices on planting, fertilization, and irrigation. Remote sensing technology, such as satellites and drones, offers crucial information on pest presence, plant vitality, and moisture levels, allowing for prompt actions to maximize agricultural productivity. Robotics, encompassing the use of robots for operations such as planting, harvesting, and weeding, streamlines repetitive activities, enhances precision, and liberates the human workforce to focus on more intricate decision-making duties [27].

5. Market Stability and Supply Chain Efficiency in the Agri-food Sector

Finding the Global Goal of zero hunger in the United Nations requires simplifying the agri-food sector. As summarized in Figure 5, there exist multiple methods that might be utilized to improve effectiveness and productivity within the agri-food industry, therefore reducing worldwide food insecurity and hunger [28].

5.1. Price Dynamics and Market Stability

To achieve the aim of zero hunger, price dynamics and market stability in the agri-food industry are essential. Price fluctuations can have a major effect on food accessibility, especially for disadvantaged groups. Rising prices can cause supply and demand mismatches by upsetting food supply systems. Stable prices promote growth in agriculture and guarantee that food items are affordable for both farmers and consumers. Achieving market stability and reducing the danger of food insecurity need several strategies such as price stabilization mechanisms, laws supporting fair trade practices, and market information systems [29].

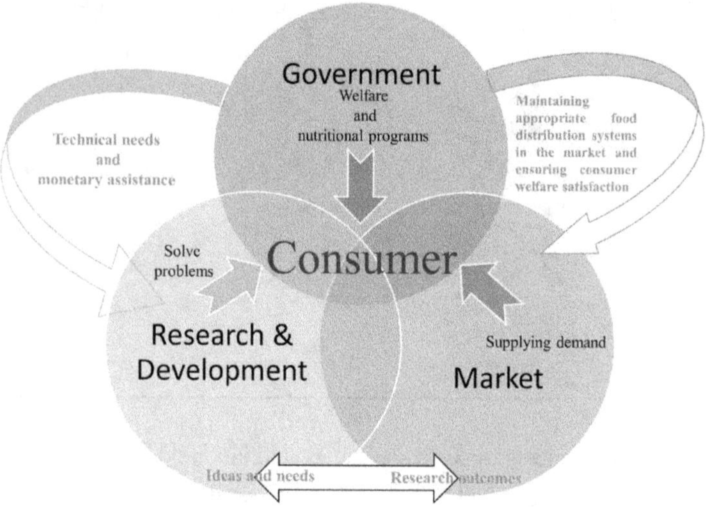

Figure 5: Synchronized efforts to attain global food and nutritional security.

5.2. Enhancing Supply Chain Efficiencies

Improving the agri-food market's supply chain efficiency is essential to solving the problem of zero hunger. To ensure that more food reaches those in need, food losses, and waste may be reduced by streamlining the supply chain from the farm to the table. Reducing post-harvest losses and increasing the supply of fresh produce in rural areas may be achieved by improving distribution networks, storage facilities, and transport networks. Supply chain operations may be improved by using technology-driven solutions like blockchain for tracking, connectivity for tracking perishable items, and data analytics for demand forecasting. These solutions encourage transparency and responsibility while allowing better use of resources. Working together, stakeholders—farmers, manufacturers, distributors, and legislators—will enable measures to be implemented that will improve supply chain efficiency and support the aim of ending hunger [30].

6. The Roadblocks to Zero Hunger

The objective of attaining zero hunger by 2030 is a challenging yet crucial aim, yet several important challenges hinder its achievement. In these regions, obtaining food becomes a daily challenge, worsened by violence that challenges the processes of food production and delivery. An endless cycle of hunger and instability is generally sustained by desperation, which frequently sparks new conflicts. The presence of economic inequality worsens the problem of hunger, as those living in poverty or experiencing income differences are unable to get adequate food. Approximately 2 billion individuals worldwide do not have consistent access to food that is safe and nutritious. This highlights the pressing necessity to tackle socioeconomic inequalities in the battle against hunger. Addressing hunger necessitates a collaborative endeavor to confront disparity and provide fair and impartial access to resources [31].

Climate change and environmental deterioration provide more obstacles to ensuring food security. Severe weather phenomena, such as extreme weather events, droughts, and floods, cause disruptions in food supply chains and reduce agricultural production. Environmental deterioration worsens these issues, resulting in soil erosion and reduced agricultural production. Ensuring food security in the long run requires us to prioritize efforts in addressing climate change and encouraging sustainable environmental practices [32].

Inadequate healthcare services worsen hunger and malnutrition since individuals have difficulties in obtaining enough medical care to address underlying health conditions. In the absence of sufficient medical treatment, individuals are unable to uphold nutritious dietary practices or maintain their general state of well-being, thus repeating the cycle of hunger and deteriorating health. Food scarcity and rising prices further compound the challenges of achieving zero hunger. Global hunger levels have experienced a significant increase, reaching levels that have not been observed since 2005. This rise can be attributed to several causes, including a decrease in food production, an increase in the cost of living, and civil instability. To mitigate hunger and poverty, it is crucial to make substantial investments in agriculture, while also implementing strategies to tackle the issues of food affordability and accessibility. The fight against hunger must prioritize the establishment of a sustainable food system. Reducing food waste, supporting small-scale farmers, and promoting diversified, nutrient-dense diets are crucial elements in constructing a food system that is more robust and fairer. By placing sustainability and justice as top priorities in the production and distribution of food, we can establish the groundwork for a future in which hunger is no longer a widespread danger to global welfare [33].

7. Community Engagement and Education

One of the SDGs of the United Nations is to end hunger globally. Multiple strategies are needed to achieve this aim, such as empowering local farmers and encouraging sustainable farming methods through community involvement and education. This study looks at how important it is to empower local farmers and increase awareness of sustainable farming methods to reduce hunger at the community level [34].

7.1. Empowering Local Farmers

To achieve food security and reduce hunger, local farmers must be empowered. To increase their output and revenue, this entails giving them access to markets, resources, and expertise. Local farmers are greatly helped by community-based programs including microfinance, agricultural extension services, and farmer cooperatives. Offering services for agricultural extension is an alternative strategy. Through these services, farmers may receive information, training, and technical support on better agricultural techniques, crop management, and post-harvest handling. Extension services contribute to increased production and climate change resistance by providing farmers with up-to-date information and methods. Through the provision of loans and financial services, microfinance programmers also empower local farmers. Farmers that have access to cash may use it to grow their companies and increase productivity by investing in technology, equipment, and supplies. Programs for microfinance designed specifically for rural populations have a big influence on food security and poverty reduction [35].

7.2. Building Knowledge for Sustainable Practices

Achieving zero hunger requires not just empowering local farmers but also expanding understanding of sustainable techniques. The goal of sustainable agriculture is to increase output while reducing the loss of resources and environmental damage. Programs for community education and training are essential for advancing sustainable farming methods. Farmer Field Schools (FFS), where farmers learn by practical experience and peer-to-peer teaching, are one successful strategy. Farmers may try out different methods, track results, and share information with other farmers in their community by using FFS. FFS contributes to long-term food security and environmental sustainability by supporting environmentally sound farming practices including organic and agroecological farming. Adding agroecology into official educational programs might also increase future generations' understanding and awareness of sustainable practices. Teachings on biodiversity protection, soil health, and water management can be implemented in schools and other educational establishments. Education may enable people to become environmental protectors and champions for sustainable food systems by developing an advanced understanding of ecological concepts and being environmentally friendly. Farming Online Resources make knowledge-sharing websites, applications, or social media platforms. Working together with non-governmental organizations for a larger consumer, collaborate with non-profit organizations [36].

8. The Future of Food: Innovations on the Horizon

In the pursuit of zero hunger, progress in food production and distribution are paramount. New developments have the potential to transform our approach to food scarcity, providing hopeful strategies to guarantee that no one goes hungry.

8.1. Next-generation Farming Technique

As technology continues to evolve, it is significantly transforming farming practices and emphasizing the increasing need for sustainable agriculture. Next-generation agricultural techniques, a collection of advancements, are designed to enhance the stability of food production systems, reduce their environmental impact, and boost productivity.

Smart Farming is one such technique that leverages sensors, GPS, and data analysis to improve crop management. This method allows farmers to apply pesticides, fertilizers, and water more efficiently, leading to increased yields and a reduced environmental footprint. The concept of Urban Agriculture, particularly vertical farming, is another significant theme in future farming. Vertical farms, using greenhouses and artificial lights, enable crops to grow in controlled conditions. These small-scale farms facilitate year-round production, reduce the land requirement, and conserve water [37].

Aquaponics is another innovative approach that uses fish waste or nutrient-rich water solutions as the primary nutrient source for soilless plant growth. These farming methods, which use less water and land than traditional agriculture, can be implemented in various environments, including deserts and cities. Lastly, Agroforestry, the practice of integrating trees or shrubs into agricultural landscapes, plays a crucial role in improving biodiversity, conservation, and production. By growing crops alongside trees or adding tree cover to farms, agroforestry enhances soil health, water absorption, and carbon storage. It also provides additional benefits such as shade and wind protection. These major themes are expected to shape the future of farming, making it more sustainable and productive [38].

8.2. The Role of Biotechnology in Future Food System

Biotechnology is positioned to profoundly transform upcoming food systems by tackling crucial obstacles such as improving nutrition, developing resistance against diseases and insects, and enhancing adaptability to climate change. Genetically Modified Organisms (GMOs) involve the manipulation of an organism's genetic material to incorporate certain characteristics such as enhanced nutritional content or the ability to withstand herbicides and pests. This process has the potential to enhance food security and conservation efforts. Gene editing technologies, such as CRISPR-Cas9, provide accurate adjustments to the genetics of plants and animals, improving traits like productivity, quality, and ability to withstand environmental pressures, without adding external DNA. Synthetic biology allows for the manipulation of biological systems to achieve many objectives, such as the creation of biofuels, enhancement of food flavors, and development of medications. This has the potential to result in the creation of novel ingredients and the implementation of sustainable agricultural practices that can be customized to meet evolving customer needs. Biopharming involves the use of genetically modified crops to manufacture commercial enzymes, vaccines, and pharmaceuticals. It addresses health and nutritional issues by creating biofortified crops and edible vaccinations. These developments in biotechnology provide potential for crafting a future in which food systems are more robust, nourishing, and environmentally friendly [39].

9. Conclusion

The intersection of agriculture, sustainability, and technology offers a promising path toward a hunger-free world. By integrating diverse approaches, we can enhance agricultural productivity while preserving our planet's health. Empowering local farmers, fostering sustainable practices, and promoting international collaboration is the key to overcoming global challenges and ensuring food security for all. The future of food lies in sustainable smart agriculture, a strategy that holds the promise of achieving zero hunger. By embracing holistic methods and recognizing the interdependence of ecology and economy, we can empower small-scale farmers, enhance food security, and foster resilient food systems. The resurgence of nutrient powerhouses such as millet and other forgotten crops presents a promising solution in the fight against global hunger and malnutrition. Recognizing their nutritional value, environmental resilience, and potential for sustainable agriculture, initiatives to promote and integrate these crops into farming practices are essential. By embracing these nutrient-rich alternatives, we can enhance food security, promote biodiversity, and build more resilient food systems, ultimately moving closer to the goal of achieving zero hunger. As the agricultural sector continues to evolve with technology, it holds immense potential to address global food insecurity and move closer to achieving the goal of zero hunger. Looking ahead, innovations in next-generation farming techniques and biofortification hold promise for transforming food production and distribution, offering sustainable solutions to address hunger and meet the nutritional needs of a growing population. Embracing these advancements will be crucial in shaping the future of food systems and achieving a world where everyone has access to an abundance of nutritious food. The journey highlights the significance of working together, using methods that can be maintained over time, and coming up with new ideas to tackle the complex problems related to ensuring that everyone has enough food worldwide.

References

1. Mc Carthy, U., Uysal, I., Badia-Melis, R., Mercier, S., O'Donnell, C. and Ktenioudaki, A. (2018). Global food security: Issues, challenges, and technological solutions. *Trends Food Sci. Technol.*, 77: 11–20. https://doi.org/10.1016/j.tifs.2018.05.002.
2. Okello, M., Lamo, J., Ochwo-Ssemakula., M. and Onyilo, F. (2021). Challenges and innovations in achieving zero hunger and environmental sustainability

through the lens of sub-Saharan Africa. *Outlook Agric.*, 50: 141–147. https://doi.org/10.1177/0030727020975778.

3. Hall, S. (2023). The climate disaster strikes: What the data say. *Nature*, 624: S26–S28. https://doi.org/10.1038/d41586-023-03910-w.
4. Ingrao, C., Strippoli, R., Lagioia, G. and Huisingh, D. (2023). Water scarcity in agriculture: An overview of causes, impacts, and approaches for reducing the risks. *Heliyon*, 9: e18507. https://doi.org/10.1016/j.heliyon.2023.e18507.
5. Dhanalakshmi, R., Jose Anand, Arun Kumar Sivaraman and Sita Rani. (2022, May). IoT-based water quality monitoring system using cloud for agriculture use. *In*: Pankaj Bhambri, Sita Rani, Gaurav Gupta, Alex Khang (Eds.), *Cloud and Fog Computing Platforms for Internet of Things*. Routledge Taylor & Francis Group.
6. Crona, B.I., Wassénius, E., Jonell, M., Koehn, J.Z., Short, R., Tigchelaar, M., Daw, T.M., Golden, C.D. et al. (2023). Four ways blue foods can help achieve food system ambitions across nations. *Nature*, 616: 104–112. https://doi.org/10.1038/s41586-023-05737-x.
7. Chauhan, C., Dhir, A., Akram, M.U. and Salo, J. (2021). Food loss and waste in food supply chains: A systematic literature review and framework development approach. *J. Clean. Prod.*, 295: 126438. https://doi.org/10.1016/j.jclepro.2021.126438.
8. Kapari, M., Hlophe-Ginindza, S., Nhamo, L. and Mpandeli, S. (2023). Contribution of smallholder farmers to food security and opportunities for resilient farming systems. *Front. Sustain. Food Syst.*, 7. https://doi.org/10.3389/fsufs.2023.1149854
9. Rosegrant, M.W., Sulser, T.B., Dunston, S., Mishra, A., Cenacchi, N., Gebretsadik, Y., Robertson, R., Thomas, T. and Wiebe, K. (2024). Food and nutrition security under changing climate and socioeconomic conditions. *Glob. Food Sec.*, 41: 100755. https://doi.org/10.1016/j.gfs.2024.100755.
10. Rukhsana, Alam, A. (2021). Agriculture, food, and nutritional security: An overview. *In*: *Agriculture, Food and Nutrition Security: A Study of Availability and Sustainability in India*. 3–13. Springer International Publishing.
11. Simelane, K.S. and Worth, S. (2020). Food and nutrition security theory. *Food Nutr. Bull.*, 41: 367–379. https://doi.org/10.1177/0379572120925341.
12. Hendriks, S., Soussana, J-F., Cole, M., Kambugu, A. and Zilberman, D. (2023). Ensuring access to safe and nutritious food for all through the transformation of food systems. *In*: *Science and Innovations for Food Systems Transformation*. 31–58. Cham: Springer International Publishing.
13. Alaimo, K., Chilton, M. and Jones, S.J. (2020). Food insecurity, hunger, and malnutrition. *In*: *Present Knowledge in Nutrition*. 311–326. Elsevier.
14. Sandhu, R., Chaudhary, N., Bindia, Shams R., Singh, K. and Pandey, V.K. (2023). A critical review on integrating biofortification in crops for sustainable agricultural development and nutritional security. *J. Agric. Food Res.*, 14. https://doi.org/10.1016/j.jafr.2023.100830.
15. Shree Charan, M., Vijay Prabha, V. and Jayanthi, M. (2023). Millets: A powerhouse for food security and combating malnutrition. *Journal of Experimental Agriculture International*, 45(10): 44–50. https://doi.org/10.9734/jeai/2023/v45i102199.
16. Paschapur, A.U., Joshi, D., Mishra, K.K., Kant, L., Kumar, V. and Kumar, A. (2021). Millets for life: A brief introduction. *In*: *Millets and Millet Technology*. 1–32. Singapore: Springer Singapore.
17. Anjula Pandey. (2017). *Underutilized Plant Species in Asia-Pacific Region*. 27–40. Asia-Pacific Association of Agricultural Research Institutions (APAARI).

18. Abdourahamane Illiassou, S. and Oeba V.O. (2020). Ecosystem-based approach for sustainable agricultural development in addressing food security and nutrition. 252–262. Part of Book Series *Encyclopedia of the UN Sustainable Development Goals* (ENUNSDG) .
19. Davila, F., Jacobs, B., Nadeem, F., Kelly, R. and Kurimoto, N. (2024). Finding climate smart agriculture in civil-society initiatives. *Mitig. Adapt. Strateg. Glob. Chang.*, 29. https://doi.org/10.1007/s11027-024-10108-6.
20. Gliessman, S.R., Méndez, V.E., Izzo, V.M. and Engles, E.W. (2022). *Agroecology*. Boca Raton: CRC Press.
21. Rodrigues, L.R., Montserrat, M. and Magalhães, S. (2022). Evolution in agricultural systems: Moving toward the understanding of complexity. *Evol. Appl.*, 15: 1483–1489. https://doi.org/10.1111/eva.13490.
22. Misra, S. and Ghosh, A. (2024). Agriculture paradigm shift: A journey from traditional to modern agriculture. *In*: *Biodiversity and Bioeconomy*. 113–141. Elsevier.
23. Dhaliwal, S.S., Sharma, V., Shukla, A.K., Verma, V., Kaur, M., Shivay, Y.S., Nisar, S., Gaber, A. et al. (2022). Biofortification: A frontier novel approach to enrich micronutrients in field crops to encounter the nutritional security. *Molecules*, 27.
24. Bouis, H.E. and Saltzman, A. (2017). Improving nutrition through biofortification: A review of evidence from HarvestPlus, 2003 through 2016. *Glob. Food Sec.*, 12: 49–58. https://doi.org/10.1016/j.gfs.2017.01.009.
25. Kiran, A., Wakeel, A., Mahmood, K., Mubaraka, R., Hafsa and Haefele, S.M. (2022). Biofortification of staple crops to alleviate human malnutrition: Contributions and potential in developing countries. *Agronomy*, 12. https://doi.org/10.3390/agronomy12020452.
26. Sundari, M.S., Jose Anand, A. and Nagarajan, T.S. (2022). Smart aerial imaging solution in precision agriculture. *In*: *2022 International Conference on Data Science, Agents & Artificial Intelligence* (*ICDsAAI*). 1–3. IEEE.
27. Durai Arasu, E., Mahalakshmi, S., Jose Anand, A., Manikandan, T. and El Rhalibi, A. (2023). Precision farming for crop prediction. *In*: K. Kalaiselvi, Jose Anand, A., Poonam Tanwar and Haider Raza, *Advanced Technologies for Smart Agriculture*. 195–217. Denmark: River Publishers.
28. Mechri, A., Lys, P. and Cachia, F. (2017). Productivity and efficiency measurement in agriculture: Literature review and gaps analysis. Technical report series GO-19-2017.
29. Olefir, V. (2021). Influence of external factors on the domestic price of agri-food products. *In*: *International Scientific Conference ITEMA on Recent Advances in Information Technology, Tourism, Economics, Management and Agriculture*. 161–167. https://doi.org/10.31410/itema.2021.161
30. Onwude, D.I., Chen, G., Eke-emezie, N., Kabutey, A., Khaled, A.Y. and Sturm, B. (2020). Recent advances in reducing food losses in the supply chain of fresh agricultural produce. *Processes*, 8: 1431. https://doi.org/10.3390/pr8111431.
31. Bongaarts, J. (2021). FAO, IFAD, UNICEF, WFP, and WHO: The State of Food Security and Nutrition in the World 2020. Transforming food systems for affordable healthy diets. FAO, 2020, 320 pp. *Popul. Dev. Rev.*, 47: 558–558. https://doi.org/10.1111/padr.12418.
32. Parfitt, J., Barthel, M. and Macnaughton, S. (2010). Food waste within food supply chains: Quantification and potential for change to 2050. *Philosophical Transactions of the Royal Society B: Biological Sciences*, 365: 3065–3081. https://doi.org/10.1098/rstb.2010.0126.

33. da Silva, J.G. (2017). *The Future of Food and Agriculture, and Challenges*. Report of the Food and Agriculture Organization of the United Nations.
34. Blesh, J., Mehrabi, Z., Wittman, H., Kerr, R.B., James, D., Madsen, S., Smith, O.M., Snapp, S. et al. (2023). Against the odds: Network and institutional pathways enabling agricultural diversification. *One Earth*, 6: 479–491. https://doi.org/10.1016/j.oneear.2023.03.004.
35. Tamako, N. (2020). Exploring agricultural knowledge systems and smallholder farmers' empowerment: Implication on household food security. Doctoral dissertation.
36. Francis, C.A., Nicolaysen, A.M., Lieblein, G. and Breland, T.A. (2020). Transformative education in agroecology: Student, teacher, and client involvement in co-learning. *International Journal of Agriculture and Natural Resources*, 47: 280–294. https://doi.org/10.7764/ijanr.v47i3.2262.
37. Basso, B. and Antle, J. (2020). Digital agriculture to design sustainable agricultural systems. *Nat. Sustain.*, 3: 254–256. https://doi.org/10.1038/s41893-020-0510-0.
38. Verma, A.K., Chandrakant, M.H., John, V.C., Peter, R.M. and John, I.E. (2023). Aquaponics as an integrated agri-aquaculture system (IAAS): Emerging trends and future prospects. *Technol. Forecast. Soc. Change*, 194: 122709. https://doi.org/10.1016/j.techfore.2023.122709.
39. Ranjha, M.M.A.N., Shafique, B., Khalid, W., Nadeem, H.R., Mueen-ud-Din, G. and Khalid, M.Z. (2022). Applications of biotechnology in food and agriculture: A mini-review. *Proceedings of the National Academy of Sciences, India Section B: Biological Sciences*, 92: 11–15. https://doi.org/10.1007/s40011-021-01320-4.

5 | Precision Farming in Urban Environments to Ensure Zero Hunger

Jose Anand A.[1], Mahalakshmi S.[2*] and Arjun S.[3]

[1] Department of ECE, KCG College of Technology, Chennai, Tamil Nadu, India
[2] Department of CSE, Chennai Institute of Technology, Chennai, Tamil Nadu, India
[3] Department of Cyber Security, Chennai Institute of Technology, Chennai, Tamil Nadu, India

This paper explores the implementation of precision farming techniques in urban environments as a means to address food insecurity and achieve zero hunger. Precision farming involves the use of advanced technologies to optimize crop production, resource utilization, and environmental sustainability. In urban settings, where space is limited and populations are dense, innovative farming methods are crucial for ensuring food security. The paper examines the concept of precision farming, its relevance in urban areas, various techniques applicable to urban farming, and presents case studies illustrating successful implementations. Additionally, it discusses the benefits of precision farming in combating hunger, the necessary policy and regulatory frameworks, potential future innovations, and concludes with recommendations for further action.

1. Introduction

1.1. Background of Urban Farming

1.1.1. Evolution and Significance of Urban Farming

Urban farming, a practice that dates back centuries, has seen a significant transformation over time. Originally, urban farming included small-scale, local food production to supplement urban diets and enhance food security during times of scarcity. In ancient civilizations, such as Mesopotamia and Rome, urban farming was integrated into city life to sustain growing populations. The

*Corresponding author: mahalakshmis@citchennai.net

modern evolution of urban farming can be traced back to the late 19th and early 20th centuries, with the emergence of victory gardens during World War I and II, which were cultivated to support war efforts by reducing pressure on public food supply. Post-war periods saw a decline in urban agriculture due to industrialization and the expansion of global food systems. However, the late 20th century and early 21st century marked a renaissance of urban farming, driven by factors such as increasing urbanization, climate change, and a growing awareness of sustainable living. Today, urban farming encompasses various forms, including community gardens, rooftop gardens, vertical farms, and hydroponic systems. It plays a crucial role in enhancing food security, promoting sustainable agricultural practices, and fostering community engagement. By localizing food production, urban farming reduces food miles, minimizes environmental impact, and provides fresh produce to urban dwellers [1-5].

1.1.2. Introduction to Precision Farming

Precision farming, also known as precision agriculture, refers to the application of technology and data-driven techniques to optimize agricultural practices. It involves the use of sensors, GPS, drones, and data analytics to monitor and manage crops with high accuracy. Precision farming aims to increase efficiency, reduce waste, and maximize yield by tailoring farming practices to specific conditions and needs. In the context of urban farming, precision farming technologies can significantly enhance productivity and sustainability. For example, sensors can monitor soil moisture levels, enabling precise irrigation that conserves water. Drones can survey crop health, identifying areas that need attention. Data analytics can optimize planting schedules and nutrient management, ensuring optimal growth conditions. By integrating precision farming into urban agriculture, cities can improve food production efficiency, reduce resource consumption, and contribute to sustainable urban development [6].

1.2. Importance of SDG2: Zero Hunger

1.2.1. Explanation of SDG2 and its Targets

Sustainable Development Goal 2 (SDG2), titled “Zero Hunger”, aims to end hunger, achieve food security and improved nutrition, and promote sustainable agriculture by 2030. SDG2 is a critical component of the United Nations’ 2030 Agenda for Sustainable Development, reflecting a global commitment to address food scarcity and malnutrition.

1.2.2. SDG2 Encompasses Several Specific Targets, Including:

- 2.1: End hunger and ensure access by all people, especially the poor and vulnerable, to safe, nutritious, and sufficient food all year round.
- 2.2: End all forms of malnutrition, including achieving targets on stunting and wasting in children under 5 years of age, and addressing nutritional needs of adolescent girls, pregnant and lactating women, and older persons.

- 2.3: Double the agricultural productivity and incomes of small-scale food producers, particularly women, indigenous peoples, family farmers, and fishers.
- 2.4: Ensure sustainable food production systems and implement resilient agricultural practices that increase productivity, maintain ecosystems, and strengthen capacity for adaptation to climate change.
- 2.5: Maintain genetic diversity in seeds, cultivated plants, and farmed and domesticated animals, and promote access to fair and equitable sharing of benefits arising from their use.

1.2.3. Relevance to Global Food Security

Achieving SDG2 is crucial for global food security, as hunger and malnutrition remain pervasive challenges affecting millions of people worldwide. Urban areas, with their dense populations and limited space for traditional agriculture, face unique challenges in ensuring food security [7-10]. Urban farming, supported by precision farming technologies, can play a vital role in addressing these challenges by:

- **Increasing Local Food Production:** Urban farming can supplement food supplies, reducing reliance on distant agricultural areas and enhancing local resilience.
- **Improving Nutrition:** Access to fresh, locally grown produce can improve dietary diversity and nutritional outcomes for urban residents.
- **Supporting Sustainable Practices:** Precision farming can optimize resource use, reduce waste, and promote environmentally friendly agricultural methods.
- **Fostering Community Involvement:** Urban farming initiatives often engage local communities, fostering social cohesion and education about sustainable food practices.

1.3. Objective of the Chapter

The primary objective of this chapter is to establish a clear link between precision farming and the achievement of SDG2: Zero Hunger, particularly within urban contexts. By exploring the evolution and significance of urban farming, the introduction of precision farming technologies, and the importance of SDG2, this chapter aims to:

- **Highlight the potential of precision farming** to revolutionize urban agriculture, making it more efficient, productive, and sustainable.
- **Demonstrate how urban farming** can contribute to the targets of SDG2, particularly in terms of increasing food security, improving nutrition, and promoting sustainable agricultural practices.
- **Set the stage for subsequent chapters** that will delve deeper into specific case studies, technologies, and strategies for integrating precision farming into urban agricultural systems to achieve Zero Hunger.

By establishing this foundation, the chapter aims to inspire and inform policymakers, urban planners, and stakeholders about the transformative potential of precision farming in urban environments and its critical role in achieving global food security goals. Figures 1 and 2 show examples of precision farming and terrace farming.

Figure 1: Example of precision farming.

Figure 2: Example of terrace farming.

2. The Concept of Precision Farming

2.1. Definition and Principles

2.1.1. Core Principles and Technologies in Precision Farming

Precision farming, also known as precision agriculture, is an advanced farming management concept that utilizes digital technology to monitor and optimize agricultural production processes [11]. The core principles of precision farming focus on maximizing efficiency, productivity, and sustainability by leveraging detailed, real-time data, and tailored farming practices. The main principles include:

- **Site-Specific Management:** Precision farming treats different areas within a field as unique entities, recognizing that soil properties, moisture levels, and crop performance can vary significantly. This principle involves tailoring inputs and management practices to the specific conditions of each area.
- **Data-driven Decision-making:** Utilizing data collected from various sources (e.g., sensors, satellites, drones), farmers can make informed decisions about planting, irrigation, fertilization, and pest control. Data analytics helps identify patterns and predict outcomes, leading to optimized resource use and increased yields.
- **Integrated Information Systems:** Precision farming relies on a range of technologies and systems working together. These include Geographic Information Systems (GIS) for mapping and analyzing spatial data, Global Positioning System (GPS) for accurate field positioning, and Internet of Things (IoT) devices for real-time monitoring of environmental conditions.
- **Sustainability and Environmental Stewardship:** By applying inputs (water, fertilizers, pesticides) only where and when they are needed, precision farming minimizes waste and environmental impact. This principle promotes sustainable agricultural practices that protect natural resources and biodiversity.

2.2. Historical Development

2.2.1. Timeline and Milestones in the Evolution of Precision Farming

The concept of precision farming has evolved significantly over the past few decades [12-14]. Key milestones in its development include:

- **1980s:** The early use of computers and satellite imagery in agriculture marked the beginning of precision farming. Farmers began experimenting with variable rate technology (VRT) for applying fertilizers and other inputs.
- **1990s:** The introduction of GPS technology revolutionized precision farming by providing accurate location data. This decade saw the development of GIS software, allowing farmers to create detailed field maps and analyze spatial variability.
- **2000s:** The advent of affordable sensors and data collection tools enabled more widespread adoption of precision farming practices. This period also saw the emergence of automated machinery and the integration of GPS with farm equipment, facilitating precise field operations.
- **2010s:** The proliferation of IoT devices, drones, and cloud computing further advanced precision farming. Farmers could now monitor crops and environmental conditions in real time, while big data analytics offered deeper insights into farm management.
- **2020s:** Precision farming has become more sophisticated with the use of artificial intelligence (AI) and machine learning. These technologies help predict crop performance, optimize resource use, and enhance decision-making processes. Additionally, advancements in biotechnology and remote sensing continue to expand the capabilities of precision agriculture.

2.3. Key Technologies

GPS (Global Positioning System): GPS technology provides precise location data, essential for mapping fields, monitoring machinery, and managing farm operations [15]. It enables site-specific management by guiding equipment for accurate planting, fertilizing, and harvesting.

GIS (Geographic Information System): GIS integrates spatial data from various sources, allowing farmers to create detailed maps of their fields. These maps help analyze soil properties, crop health, and other factors, facilitating informed decision-making and efficient resource allocation.

IoT (Internet of Things): IoT devices, such as soil moisture sensors, weather stations, and remote cameras, collect real-time data on environmental conditions. This data is transmitted to central systems for analysis, enabling precise control over irrigation, fertilization, and pest management.

Sensors: Various types of sensors (e.g., soil, crop, climate) are used in precision farming to monitor conditions and gather data. Soil sensors measure moisture, nutrient levels, and pH, while crop sensors assess health and growth. Climate sensors track temperature, humidity, and other atmospheric variables.

Drones: Drones equipped with cameras and sensors provide aerial imagery and data on crop health, growth patterns, and pest infestations. They offer a bird's-eye view of fields, helping identify issues early and enabling targeted interventions.

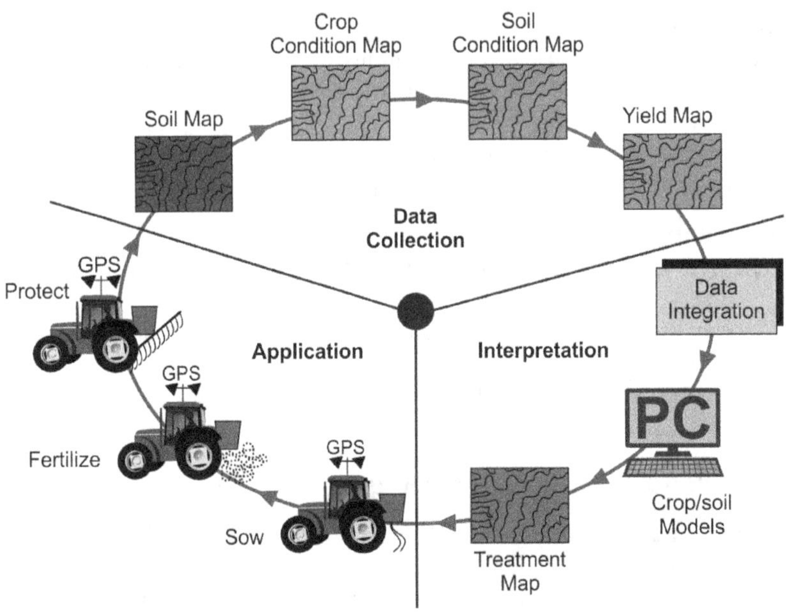

Figure 3: Cycle of precision agriculture.

Data Analytics: Advanced data analytics tools process and analyze vast amounts of data collected from different sources. These tools use algorithms and machine learning to uncover patterns, predict outcomes, and provide actionable insights. Farmers can use this information to optimize planting schedules, manage resources efficiently, and improve overall farm productivity.

The cycle of precision agriculture is shown in Figure 3. In conclusion, precision farming represents a transformative approach to agriculture, integrating advanced technologies and data-driven practices to enhance productivity, sustainability, and profitability. The evolution of precision farming, from its early beginnings to the sophisticated systems of today, highlights the ongoing innovation in agricultural practices and the critical role of technology in addressing the challenges of modern farming [16-17].

3. Urban Environments and Farming

3.1. Characteristics of Urban Agriculture

Types of Urban Farming

- **Rooftop Gardens**
  - *Description:* Gardens established on the rooftops of buildings.
  - *Benefits:* Efficient use of unused space, insulation for buildings, reduction of urban heat island effect.
  - *Examples:* Rooftop farms in cities like New York (e.g., Brooklyn Grange).
- **Vertical Farming**
  - *Description:* Farming practices that involve growing crops in vertically stacked layers, often in controlled indoor environments.
  - *Benefits:* Maximizes space usage, allows year-round production, reduces water usage through hydroponics or aeroponics.
  - *Examples:* Vertical farms like AeroFarms in New Jersey.
- **Community Gardens**
  - *Description:* Shared plots of land where local residents can grow food together.
  - *Benefits:* Promotes community engagement, provides access to fresh produce, educational opportunities.
  - *Examples:* Community gardens in urban neighborhoods, such as P-Patch gardens in Seattle.

3.2. Challenges of Urban Farming

- **Space Limitations**
  - *Issue:* Urban areas have limited open spaces for traditional farming.
  - *Solutions*: Innovative use of rooftops, abandoned lots, and vertical spaces; promotion of micro-farming techniques.

- **Pollution**
  - *Issue:* Urban environments often have higher levels of air and soil pollution.
  - *Solutions:* Use of raised beds with clean soil, implementation of air-purifying plants, regular monitoring of pollutant levels.
- **Water Scarcity**
  - *Issue:* Cities may face water shortages, making irrigation a challenge.
  - *Solutions:* Adoption of water-efficient irrigation systems (e.g., drip irrigation), rainwater harvesting, recycling graywater for agricultural use.

3.3. Opportunities in Urban Farming

- **Accessibility**
  - *Advantage:* Urban farming brings food production closer to consumers, reducing transportation costs and emissions.
  - *Implementation:* Development of urban agriculture zones, integration of urban farms into city planning.
- **Community Engagement**
  - *Advantage:* Urban farming can foster community spirit and provide educational and recreational opportunities.
  - *Implementation:* Programs for community gardening, workshops on sustainable farming practices, school gardens to educate children about agriculture.
- **Sustainability**
  - *Advantage:* Urban farming contributes to environmental sustainability by reducing food miles, promoting biodiversity, and enhancing green spaces.
  - *Implementation:* Encouraging organic farming practices, utilizing renewable energy sources for farming operations, integrating permaculture principles.

4. Precision Farming Techniques in Urban Settings

4.1. Soil and Crop Monitoring

Use of Sensors for Soil Moisture and Nutrient Levels

In urban farming, space constraints and the variability of soil conditions necessitate precise monitoring and management of soil and crop health [18]. Sensors play a crucial role in this context, providing real-time data that enables urban farmers to make informed decisions.

- **Soil Moisture Sensors:** These sensors measure the volumetric water content in the soil, helping to maintain optimal moisture levels for different crops. By providing real-time data, they prevent over- or under-watering, thus conserving water and enhancing crop health. For example, soil moisture sensors can be placed at various depths to monitor the moisture profile of the soil, ensuring that water penetrates to the root zone of plants.

- **Nutrient Sensors:** These sensors detect the levels of essential nutrients such as nitrogen, phosphorus, and potassium in the soil. Accurate nutrient management is crucial in urban farming due to limited soil volume and the need to avoid nutrient runoff, which can contaminate urban water supplies. Nutrient sensors help in applying fertilizers precisely where needed, reducing waste and enhancing plant growth. For instance, in a hydroponic system, nutrient sensors continuously monitor the nutrient solution, ensuring plants receive the optimal nutrient mix.
- **Multi-Parameter Sensors:** These advanced sensors can simultaneously measure various parameters, including pH, electrical conductivity (EC), and temperature. By integrating these sensors with data analytics platforms, urban farmers can receive comprehensive insights into soil health and make data-driven adjustments to their farming practices.

4.2. Water Management

Drip Irrigation and Automated Watering Systems

Water management is a critical component of urban farming, especially in regions with limited water resources [19]. Precision farming techniques, such as drip irrigation and automated watering systems, enhance water use efficiency and ensure that crops receive the right amount of water at the right time.

- **Drip Irrigation:** Drip irrigation systems deliver water directly to the root zone of plants through a network of valves, pipes, tubing, and emitters. This method minimizes water loss due to evaporation and runoff, making it highly efficient. In urban settings, drip irrigation can be tailored to fit small plots, rooftop gardens, and vertical farms. For example, a rooftop garden can be equipped with a drip irrigation system that adjusts the flow rate based on real-time soil moisture data, ensuring optimal water use.
- **Automated Watering Systems:** These systems use sensors and timers to automate the irrigation process, reducing manual labor and ensuring consistent watering schedules. Automated systems can be integrated with weather data to adjust watering based on rainfall forecasts, further enhancing water conservation. For instance, a community garden might use an automated watering system that schedules irrigation cycles early in the morning or late in the evening to minimize evaporation.
- **Smart Irrigation Controllers:** These controllers use real-time data from soil moisture sensors, weather stations, and other sources to dynamically adjust irrigation schedules. They can be programmed to apply water only when needed, preventing over-irrigation and promoting sustainable water use. In an urban farm, smart controllers can help maintain the health of diverse crops by providing tailored irrigation based on specific crop needs.

4.3. Pest and Disease Control

Integrated Pest Management and Drone Surveillance

Urban farming faces unique pest and disease challenges due to the proximity to urban infrastructure and varying microclimates [20]. Precision farming techniques offer effective solutions through integrated pest management (IPM) and drone surveillance.

- **Integrated Pest Management (IPM):** IPM is an ecological approach to pest control that combines biological, cultural, physical, and chemical methods to manage pest populations at acceptable levels. In urban farming, IPM involves regular monitoring of pest activity, using natural predators (like ladybugs for aphid control), and applying organic pesticides when necessary. For example, urban farmers can use pheromone traps to monitor and control insect populations, reducing the need for chemical interventions.
- **Drone Surveillance:** Drones equipped with high-resolution cameras and multispectral sensors can monitor crop health and detect early signs of pest infestations and diseases. Drones provide a bird's-eye view of the farm, identifying affected areas quickly and accurately. This allows for targeted treatments, reducing the use of pesticides and minimizing crop damage. In an urban vertical farm, drones can navigate through narrow spaces to inspect crops at different levels, ensuring comprehensive monitoring.
- **Remote Sensing Technologies:** In addition to drones, other remote sensing technologies, such as satellite imagery and ground-based sensors, can be used to monitor large urban farms. These technologies help in detecting changes in plant health, identifying stress factors, and assessing the effectiveness of pest and disease control measures. For instance, thermal imaging cameras can detect temperature variations in crops, indicating potential disease outbreaks.

4.4. Crop Management and Yield Optimization

Data-driven Decision-making and Predictive Analytics

Precision farming in urban settings leverages data-driven decision-making and predictive analytics to optimize crop management and maximize yields [21]. By analyzing vast amounts of data, urban farmers can implement strategies that enhance productivity and sustainability.

- **Data-driven Decision-making:** Urban farmers can use data collected from sensors, drones, and other sources to make informed decisions about planting schedules, irrigation, fertilization, and pest control. For example, analyzing soil moisture and weather data can help determine the best times for watering, while nutrient data can guide fertilization practices.
- **Predictive Analytics:** Predictive analytics involves using historical and real-time data to forecast future conditions and outcomes. In urban farming,

predictive models can anticipate crop growth patterns, pest outbreaks, and weather impacts, allowing farmers to take proactive measures. For instance, a predictive model might indicate a high likelihood of pest infestation during a particular season, prompting the farmer to implement preventive IPM strategies.

- **Precision Planting and Variable Rate Technology (VRT):** Precision planting involves using GPS-guided equipment to plant seeds at optimal depths and spacing, ensuring uniform growth and reducing competition for resources. VRT allows for the application of inputs (e.g., seeds, fertilizers) at variable rates based on soil and crop conditions, enhancing efficiency and yield. In an urban greenhouse, precision planting can optimize space utilization, while VRT ensures that each plant receives the appropriate amount of nutrients.
- **Crop Modeling and Simulation:** Crop modeling tools simulate various growth scenarios based on different inputs and environmental conditions. These tools help urban farmers experiment with different management practices and predict their impact on yield and resource use. For example, a crop model might simulate the effects of different irrigation schedules on tomato yield, helping the farmer choose the most effective strategy.

Figure 4 shows the demand of precision farming in future. In conclusion, precision farming techniques tailored for urban settings offer significant advantages in terms of resource efficiency, crop health, and yield optimization. By integrating advanced technologies such as sensors, automated systems, drones, and data analytics, urban farmers can overcome the challenges of limited space and diverse environmental conditions, contributing to sustainable urban agriculture and enhanced food security [22].

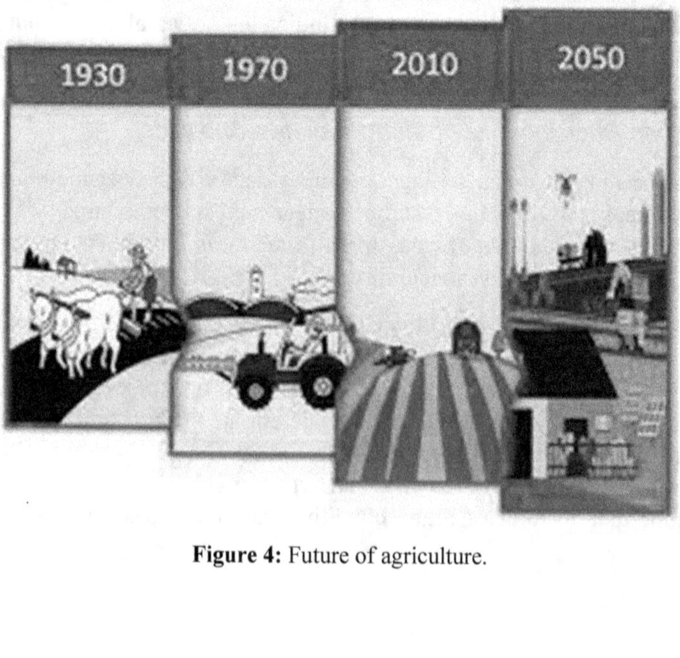

Figure 4: Future of agriculture.

5. Case Studies of Urban Precision Farming

5.1. Successful Projects Worldwide

5.1.1. Case Study 1: Rooftop Farming in New York City

Background: New York City, with its dense urban environment and limited space for traditional agriculture, has embraced rooftop farming as a viable solution for local food production. A prime example of this is the Brooklyn Grange, the world's largest soil-based rooftop farm, which spans over 5.6 acres across multiple rooftops in the city.

Implementation: Brooklyn Grange uses a combination of precision farming techniques to maximize productivity and sustainability. These techniques include:

- **Soil and Crop Monitoring:** Soil sensors are used to continuously monitor soil moisture and nutrient levels, ensuring that crops receive optimal conditions for growth.
- **Water Management:** The farm employs drip irrigation systems controlled by automated watering schedules based on real-time soil moisture data. This conserves water and reduces manual labor.
- **Pest and Disease Control:** IPM strategies are used, combining natural predators, organic pesticides, and regular monitoring to control pest populations without harmful chemicals.
- **Crop Management:** Data analytics tools are utilized to plan crop rotations, planting schedules, and harvest times, optimizing the yield and quality of produce.

Outcomes: Brooklyn Grange produces over 50,000 pounds of organically-grown vegetables each year, supplies local restaurants and markets, and engages the community through educational programs and events. The project demonstrates the potential of rooftop farming to contribute to urban food security and sustainability.

5.1.2. Case Study 2: Vertical Farming in Singapore

Background: Singapore, with its limited land area and heavy reliance on food imports, has turned to vertical farming to enhance food security. Sky Greens, one of the pioneers in this field, operates a large-scale vertical farm that uses advanced precision farming techniques.

Implementation: Sky Greens employs a range of precision farming technologies to optimize vertical farming operations:

- **Vertical Growing Systems:** The farm uses multitiered towers to grow leafy greens in a controlled environment. This maximizes space utilization and allows for year-round production.

- **Automated Systems:** Automated irrigation and nutrient delivery systems ensure that plants receive the precise amounts of water and nutrients needed for optimal growth. These systems are controlled by real-time data from sensors monitoring environmental conditions.
- **Energy Efficiency:** The vertical farm incorporates energy-efficient LED lighting and renewable energy sources to minimize the carbon footprint.
- **Pest and Disease Management:** The controlled environment of the vertical farm reduces the risk of pests and diseases. Any outbreaks are managed with targeted biological controls and minimal use of chemicals.

Outcomes: Sky Greens produces high-quality, pesticide-free vegetables with significantly higher yields per square foot compared to traditional farming. The success of Sky Greens has inspired other vertical farming initiatives in Singapore and other densely populated cities around the world.

5.2. Lessons Learned

Key Takeaways from Successful Implementations

- **Adaptability:** Both rooftop and vertical farming projects demonstrate the importance of adapting precision farming techniques to the specific constraints and opportunities of urban environments. Tailoring technology to fit limited spaces and diverse conditions is crucial for success.
- **Resource Efficiency:** Effective use of resources, such as water, nutrients, and energy, is a common theme in successful urban farming projects. Precision farming technologies that optimize resource use not only enhance productivity but also contribute to sustainability.
- **Community Engagement:** Engaging the local community through education, participation, and transparency helps build support for urban farming initiatives. Projects like Brooklyn Grange have successfully integrated community involvement, which has been key to their success.
- **Scalability:** Successful urban farms often start with pilot projects that demonstrate feasibility and build a case for scaling up. Gradual expansion, backed by data and proven results, facilitates larger-scale implementation and attracts investment.

5.3. Scaling and Replication

Potential for Replication in Other Urban Areas

The successful implementation of precision farming in urban settings offers a blueprint for replication in other cities facing similar challenges. Key factors to consider for scaling and replication include:

- **Feasibility Studies:** Conducting thorough feasibility studies to assess the local environment, available infrastructure, and potential market demand is crucial. Understanding local conditions helps tailor precision farming solutions to fit specific urban contexts.

- **Public-Private Partnerships:** Collaboration between government agencies, private companies, and community organizations can provide the necessary resources, expertise, and support for urban farming projects. Public policies and incentives can also play a significant role in promoting urban agriculture.
- **Investment in Technology:** Investing in the latest precision farming technologies, such as sensors, automated systems, and data analytics, is essential for replicating success. Access to affordable and scalable technology solutions can lower the entry barriers for new projects.
- **Knowledge Sharing:** Sharing best practices, lessons learned, and success stories from existing projects can inspire and guide new initiatives. Establishing networks and platforms for knowledge exchange can facilitate the spread of innovative urban farming practices.
- **Customization and Flexibility:** Urban farming solutions must be customized to meet the unique needs and constraints of each city. Flexibility in design and implementation allows projects to adapt to changing conditions and evolving challenges.

In conclusion, the case studies of Brooklyn Grange and Sky Greens illustrate the potential of precision farming to transform urban agriculture. By leveraging advanced technologies and data-driven practices, urban farms can enhance food security, promote sustainability, and engage communities. The lessons learned from these successful projects provide valuable insights for scaling and replicating precision farming in urban settings worldwide. Figure 5 shows the use cases for precision farming using IoT.

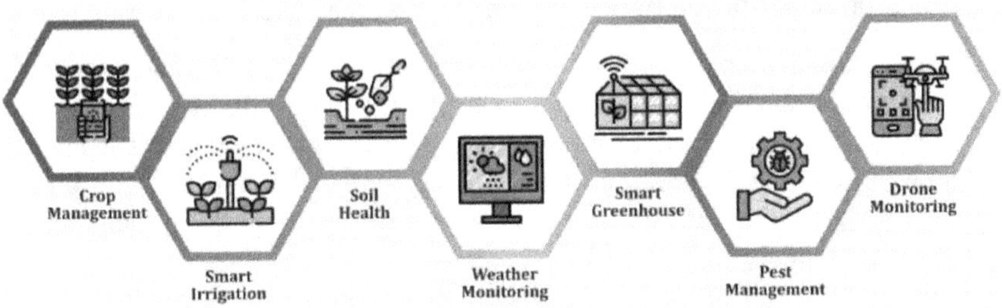

Figure 5: Use cases of precision farming using IoT.

6. Benefits of Precision Farming in Achieving Zero Hunger

6.1. Increased Productivity and Yield

Efficiency Gains through Technology

Precision farming employs advanced technologies to significantly enhance agricultural productivity and yield [23]. By utilizing tools such as GPS-guided equipment, automated irrigation systems, and real-time monitoring sensors, farmers can achieve remarkable efficiency gains:

- **Optimized Planting and Harvesting:** GPS-guided machinery ensures precise planting and harvesting, maximizing field utilization and reducing seed wastage. For instance, GPS-enabled tractors can plant seeds at optimal depths and spacing, ensuring uniform crop growth and higher yields.
- **Tailored Inputs:** VRT allows for the precise application of fertilizers, pesticides, and water based on the specific needs of different field zones. This precision reduces input waste and enhances crop health, leading to increased yields.
- **Real-Time Monitoring:** Sensors and drones provide continuous data on soil conditions, crop health, and environmental factors. This information enables timely interventions, such as adjusting irrigation schedules or applying targeted treatments, to address issues before they affect productivity.

Overall, the integration of these technologies in precision farming ensures that every aspect of crop production is optimized, resulting in higher outputs and more efficient use of resources.

6.2. Resource Efficiency

Sustainable Use of Water and Nutrients

One of the core principles of precision farming is the sustainable and efficient use of resources [24]. By leveraging technology, precision farming minimizes waste and promotes the conservation of vital resources like water and nutrients:

- **Water Management:** Precision irrigation systems, such as drip and automated irrigation, ensure that water is delivered directly to the plant roots in the right amounts. Soil moisture sensors provide data that helps avoid over-irrigation and water logging, conserving water, and maintaining soil health. For example, a drip irrigation system can reduce water usage by up to 60% compared to traditional methods.
- **Nutrient Management:** Precision farming technologies enable the precise application of fertilizers based on real-time soil nutrient data. This reduces the risk of over-fertilization and nutrient runoff, which can lead to environmental degradation. Nutrient sensors help farmers apply the exact amounts needed, improving crop uptake and reducing wastage.
- **Energy Efficiency:** Automated systems and smart controllers optimize the use of energy in farming operations. Energy-efficient practices, such as using renewable energy sources and LED lighting in vertical farms, further reduce the environmental footprint of urban agriculture.

By improving resource efficiency, precision farming contributes to sustainable agricultural practices that support long-term food production without depleting natural resources.

6.3. Food Security and Accessibility

Local Food Production Reducing Dependency on External Supply

Precision farming plays a vital role in enhancing food security and accessibility by promoting local food production in urban areas [25]. This localized approach reduces dependency on external food supplies and ensures a more resilient food system:

- **Urban Agriculture:** Implementing precision farming in urban settings, such as rooftop gardens and vertical farms, increases local food production. This not only provides fresh produce to urban residents but also shortens the supply chain, reducing transportation costs and carbon emissions.
- **Year-round Production:** Controlled environment agriculture (CEA) techniques, including hydroponics and aquaponics, allow for year-round crop production. This consistent supply of fresh produce enhances food security, especially in regions with seasonal limitations or adverse climatic conditions.
- **Reduced Food Waste:** Precision farming minimizes food waste through efficient resource use and targeted interventions. By addressing issues such as pest infestations or nutrient deficiencies early, farmers can prevent crop losses and ensure more produce reaches the market.

Localizing food production through precision farming ensures that urban populations have access to nutritious, fresh, and affordable food, contributing to the broader goal of achieving zero hunger.

6.4. Economic and Social Impacts

Job Creation, Community Empowerment

Precision farming not only improves agricultural productivity and sustainability but also has significant economic and social impacts that contribute to community empowerment and economic growth:

- **Job Creation:** The adoption of precision farming technologies creates new job opportunities in various sectors, including technology development, equipment maintenance, data analysis, and farm management. For instance, urban farms employing advanced hydroponic systems need skilled technicians to operate and maintain the equipment.
- **Community Empowerment:** Urban farming initiatives often involve local communities, providing education and training in sustainable farming practices. This empowerment fosters community engagement, enhances local knowledge, and promotes a sense of ownership and pride.
- **Economic Growth:** By increasing local food production and reducing reliance on imported food, precision farming can boost local economies. The revenue generated from urban farming can be reinvested in community development projects, further stimulating economic growth.

- **Health and Well-being:** Access to fresh, locally grown produce improves the nutritional status of urban residents, contributing to better health outcomes. Community-based urban farms also offer social benefits, such as providing spaces for recreation and community gatherings, which enhance overall well-being.

7. Policy and Regulatory Framework

7.1. Government Initiatives and Support

Policies Promoting Urban Agriculture and Precision Farming

Governments around the world are increasingly recognizing the critical role that urban agriculture and precision farming can play in ensuring food security, promoting sustainability, and fostering economic development [26-29]. Consequently, a range of policies and initiatives have been implemented to support these practices:

- **Financial Incentives:** Many governments offer subsidies, grants, and low-interest loans to encourage the adoption of urban farming and precision agriculture technologies. These financial incentives help offset the high initial costs associated with setting up urban farms and purchasing advanced equipment. For example, the U.S. Department of Agriculture (USDA) provides grants for urban agriculture initiatives through programs like the Urban Agriculture and Innovative Production Grants.
- **Tax Breaks and Incentives:** Some jurisdictions provide tax incentives for urban farming projects, such as property tax deductions or exemptions for land used for urban agriculture. These incentives can make it more financially viable for landowners to convert unused or underutilized urban spaces into productive farms.
- **Research and Development Funding:** Governments invest in research and development to advance precision farming technologies and sustainable agricultural practices. For instance, the European Union's Horizon 2020 program has funded numerous projects focused on smart farming and urban agriculture innovations.
- **Educational and Extension Programs:** Extension services and educational programs are crucial for disseminating knowledge and skills related to urban farming and precision agriculture. Governments often collaborate with universities, agricultural institutions, and nonprofits to offer workshops, technical assistance, and online courses to urban farmers. For example, the Cooperative Extension System in the United States provides educational resources and support to urban farmers.

7.2. International Standards and Guidelines

FAO Guidelines and International Best Practices

International organizations like the Food and Agriculture Organization (FAO) of the United Nations play a pivotal role in establishing standards and guidelines that promote sustainable and effective urban agriculture and precision farming practices:

- **FAO Guidelines:** The FAO has developed comprehensive guidelines on urban agriculture and precision farming that emphasize sustainability, resource efficiency, and integration with urban planning. These guidelines provide a framework for implementing best practices and ensuring that urban agriculture contributes to food security and environmental sustainability.
- **Codex Alimentarius:** This international food standards body, jointly run by the FAO and the World Health Organization (WHO), provides guidelines on food safety and quality. Compliance with Codex standards ensures that urban farms produce safe, high-quality food that can be confidently consumed by local populations.
- **Global Best Practices:** The FAO and other international organizations compile and disseminate best practices from successful urban farming projects worldwide. These best practices cover a range of topics, including soil health management, water conservation, pest control, and the use of technology in urban farming. By sharing these practices, the FAO helps urban farmers and policymakers implement effective strategies tailored to their specific contexts.
- **ISO Standards:** The International Organization for Standardization (ISO) has developed standards relevant to precision farming, such as ISO 22000 for food safety management and ISO 14001 for environmental management systems. Adherence to these standards helps ensure that urban agriculture practices are safe, sustainable, and environmentally responsible.

7.3. Challenges and Barriers

Regulatory Hurdles and Funding Issues

Despite the numerous benefits and growing support for urban agriculture and precision farming, several challenges and barriers need to be addressed to fully realize their potential:

- **Regulatory Hurdles:** Urban farmers often face complex and restrictive regulations related to zoning, building codes, and health standards. Navigating these regulations can be time-consuming and costly. For instance, some cities have stringent regulations that limit the types of structures that can be used for urban farming or impose strict requirements on food production and sales [30].
- **Funding Issues:** Securing adequate funding for urban farming projects is a significant challenge. The initial setup costs for precision farming

technologies, such as sensors, automated systems, and CEA infrastructure, are high. Additionally, ongoing operational costs, including maintenance and energy, can be substantial. Accessing loans or grants may be difficult due to bureaucratic processes and the perceived risks associated with urban farming ventures.

- **Infrastructure Limitations:** Urban areas may lack the necessary infrastructure to support large-scale precision farming. Issues such as inadequate water supply, limited space for expansion, and insufficient waste management systems can hinder the growth of urban agriculture. Retrofitting existing buildings for urban farming or constructing new facilities requires significant investment and planning.
- **Technical Expertise:** The successful implementation of precision farming technologies requires a high level of technical expertise. Urban farmers need to be proficient in using advanced equipment, analyzing data, and managing automated systems. Training and retaining skilled personnel can be challenging, especially in regions where access to agricultural education and technical support is limited.
- **Market Access:** Urban farmers may struggle to access markets for their produce. Competing with established suppliers, navigating distribution channels, and meeting consumer demand can be difficult. Developing robust local markets and distribution networks is essential for the economic viability of urban farming.

8. Future Prospects and Innovations

8.1. Emerging Technologies

AI, Machine Learning, Blockchain in Precision Farming

The integration of emerging technologies into precision farming is set to revolutionize the agricultural landscape, enhancing efficiency, productivity, and sustainability [31-34].

- **Artificial Intelligence (AI):** AI algorithms can analyze large datasets to optimize farming practices. For instance, AI-powered systems can predict crop yields based on weather patterns, soil conditions, and historical data, allowing farmers to make data-driven decisions. AI can also assist in identifying pest infestations and diseases early through image recognition technologies, enabling timely interventions.
- **Machine Learning:** Machine learning models improve over time by learning from new data, offering increasingly accurate predictions and recommendations. These models can be used to optimize irrigation schedules, nutrient applications, and harvest times. For example, machine learning algorithms can analyze soil moisture data to determine the precise amount of water needed, reducing wastage and improving crop health.

- **Blockchain:** Blockchain technology ensures transparency and traceability in the food supply chain. Every transaction and process can be recorded on a decentralized ledger, enhancing food safety, reducing fraud, and improving supply chain efficiency. For urban farms, blockchain can verify the origin and quality of produce, building consumer trust and potentially opening new markets.

8.2. Integration with Smart Cities

Urban Farming as Part of Smart City Initiatives

Smart cities leverage advanced technologies to improve urban living, and integrating urban farming into these initiatives can enhance sustainability and food security [35-38].

- **IoT Integration:** IoT connects various devices and systems, facilitating real-time monitoring and control. IoT sensors in urban farms can monitor environmental conditions like temperature, humidity, and soil moisture, automating irrigation and climate control systems. This integration ensures optimal growing conditions and reduces resource wastage.
- **Energy Efficiency:** Smart cities prioritize energy efficiency and the use of renewable energy sources. Urban farms can incorporate solar panels, wind turbines, and energy-efficient LED lighting to power their operations sustainably. These measures reduce the carbon footprint of urban farming and contribute to the overall energy efficiency goals of smart cities.
- **Data-driven Decision-making:** Smart cities rely on data analytics to make informed decisions. Urban farms can contribute valuable data on food production, resource use, and environmental impact. This data can be integrated into city-wide analytics platforms to optimize food systems, enhance urban planning, and improve resilience to climate change.
- **Community Engagement:** Smart city platforms can facilitate community engagement in urban farming projects. Online portals and mobile apps can provide information on local urban farms, volunteer opportunities, and marketplaces for purchasing fresh produce. Engaging the community fosters a sense of ownership and support for urban agriculture initiatives.

8.3. Global Trends and Future Directions

Predictive Trends in Urban Agriculture and Food Security

Several global trends and innovations are shaping the future of urban agriculture and precision farming:

- **Increased Urbanization:** With more people moving to cities, the demand for local food production is rising. Urban agriculture will become increasingly important in meeting this demand sustainably. Vertical farming, rooftop gardens, and other space-efficient methods will proliferate in densely populated urban areas.

- **Climate Change Adaptation:** Precision farming techniques will be crucial in adapting to climate change. CEA technologies, such as hydroponics and aquaponics, allow for year-round production regardless of external weather conditions. Urban farms can serve as models for climate-resilient agriculture.
- **Technological Advancements:** Continuous advancements in AI, robotics, and biotechnology will drive innovation in urban farming. For example, autonomous robots can perform tasks like planting, weeding, and harvesting, reducing labor costs and increasing efficiency. Biotechnology advancements, such as CRISPR gene editing, can develop crops that are more resilient to urban growing conditions.
- **Policy Support:** Governments and international organizations are increasingly recognizing the benefits of urban agriculture and precision farming. Supportive policies, funding opportunities, and research initiatives will continue to grow, fostering the development and expansion of urban farming projects.
- **Sustainability Focus:** There will be a stronger emphasis on sustainability, with urban farms adopting practices that conserve resources, reduce waste, and promote biodiversity. Circular economy principles, such as recycling organic waste into compost or bioenergy, will become integral to urban farming operations.
- **Food Security Initiatives:** Urban agriculture will play a key role in global food security initiatives. By providing local, fresh produce, urban farms reduce dependency on long supply chains and mitigate the risks associated with food import disruptions. This localized approach enhances food security, especially in times of crisis.

9. Conclusion

Precision farming in urban areas offers a promising solution to combat hunger. By employing advanced technologies and innovative farming methods, cities can boost local food production, improving food security and nutrition. Precision farming provides various benefits such as higher crop yields, efficient resource use, and resilience to climate change. Case studies highlight its success in urban contexts, but realizing these benefits requires supportive policies, resource access, and stakeholder collaboration. Future advancements and integration into urban planning are crucial for scaling up initiatives. In conclusion, precision farming is a vital step towards zero hunger in cities, promoting sustainable food systems and ensuring access to nutritious food for all.

References

1. Uthayakumar, C., Sarukesi, K. and Jose A. (2009). A review on knowledge management and knowledge audit. *International Journal on The Technology World*, Malaysia, V(I): 228–233, Oct–Nov.

2. Jose Anand and J. Raja Paul Perinbam. (2014). Automatic irrigation system using fuzzy logic. *AE International Journal of Multidisciplinary Research*, 2(8): 1–9.
3. Ramya, E., Jose, A., Renugha Devi, R., Roshni Prasenth, K. and Neethu Anna Issac. (2021, Oct.). Solar grass cutter with water spraying vehicle. *Proceedings of IEEE International Conference on Advancements in Electrical, Electronics, Communication, Computing and Automation* (*ICAECA*), pp. 1–6.
4. Dhanalakshmi, R., Jose, A., Arun Kumar Sivaraman and Sita Rani. (2022, May). IoT-based water quality monitoring system using cloud for agriculture use. *In*: Pankaj Bhambri, Sita Rani, Gaurav Gupta, Alex Khang (Eds.), *Cloud and Fog Computing Platforms for Internet of Things*. Routledge Taylor & Francis Group.
5. Jose, A., Shobika, M. and Sneha, T. (2022, Sep.). Programmable stickers to monitor perishable goods using blockchain. *Proceedings of IEEE 2022 Second International Conference on Computer Science, Engineering and Applications* (*ICCSEA*), 1–5.
6. Sundari, M.S., Anand, A.J. and Nagarajan, T.S. (2022). Smart aerial imaging solution in precision agriculture. *In*: *2022 International Conference on Data Science, Agents, and Artificial Intelligence* (*ICDSAAI*), 1–3. Chennai, India.
7. Althaf Shaik, Nafisa Badruzaman, Devadharshini Gajendran, Geethalakshmi, M. and Anand, J. (2022). Organic farming in drainage system with advanced automation through robotics and IoT. *In*: *2022 International Conference on Data Science, Agents, and Artificial Intelligence* (*ICDSAAI*), 592–597. Chennai, India.
8. Jose, A.A., Pawan, K., Narendran, M., Johnson, M. and Rakesh, S. (2022). Real time livestock rearing. *In*: *2022 International Conference on Computer, Power, and Communications* (*ICCPC*), 603–606. Chennai, India.
9. Jose, A.A., Sandhiya, S., Sweety Sharon, A. and Vinisha, R. (2023). Disease detection of various plant leaf using image processing technique. *In*: *2023 Eighth International Conference on Science Technology Engineering and Mathematics* (*ICONSTEM*), 1–5. Chennai, India.
10. Resmi R. Nair, Senthamizh Selvi, R., Harshaa Vartini, G., Jayashri, B., Suruthi, S. and Jose, A.A. (2023). Maturity level classification of moringa leaf using DNN algorithm. *In*: *2023 4th International Conference on Smart Electronics and Communication* (*ICOSEC*), 1667–1672. Trichy, India.
11. Anuradha, T., Suganya, A., Muthalagu, R., Narender, M. and Jose, A.A. (2023). Evaluation of a wireless sensor network for the detection of forest fires. *In*: *2023 2nd International Conference on Edge Computing and Applications* (*ICECAA*), 1021–1027. Namakkal, India.
12. Kavitha Balamurugan, Jose, A.A., Rammya, B., Dinesh Kumar, R., Kalanithy, K. and Mathan, J.N. (2023). Implementation of effective rainfall forecast model using machine learning. *In*: *2023 4th International Conference on Smart Electronics and Communication* (*ICOSEC*), 1655–1660. Trichy, India.
13. Duraiarasu, E., Mahalakshmi, S., Jose, A.A., Manikandan, T. and El Rhalibi, A. (2023). Precision farming for crop prediction. *In*: K. Kalaiselvi, Jose Anand, A., Poonam Tanwar and Haider Raza, *Advanced Technologies for Smart Agriculture*. 195–217. Denmark: River Publishers.
14. Nisha, K., Augustince, C., Jose, A.A. and Kamesh, M. (2023). Deep learning in smart agriculture. *In*: K. Kalaiselvi, Jose Anand, A., Poonam Tanwar and Haider Raza, *Advanced Technologies for Smart Agriculture*. 153–175. Denmark: River Publishers.
15. Jose, A.A., Nirmala Devi, K., Vijayakumar, S. and Shiju, C. Chacko. (2023). Decision-making support in smart farming. *In*: K. Kalaiselvi, Jose Anand, A., Poonam Tanwar

and Haider Raza, *Advanced Technologies for Smart Agriculture*. 219–239. Denmark: River Publishers.

16. Augustine, C., Balaji, K., Dharani Kumar, S.V. and Jose, A.A. (2023). Urban farming: Case study. *In*: K. Kalaiselvi, Jose Anand A., Poonam Tanwar and Haider Raza, *Advanced Technologies for Smart Agriculture*. 321–338. Denmark: River Publishers.
17. Vittorio, M., Comba, L., Khaliq, Aleem, Chiaberge, M. and Gay, P. (2020). UAV and machine learning-based refinement of a satellite-driven vegetation index for precision agriculture. *Sensors* (Basel, Switzerland), 20.
18. Zhao, Yan, Potgieter, A., Miao, Z., Bingfang, W. and Hammer, G. (2020). Predicting wheat yield at the field scale by combining high-resolution Sentinel-2 Satellite imagery and crop modeling. *Remote Sens*., 12: 1024.
19. Ouhami, Maryam, Hafiane, A., Es-saady, Youssef, Hajji, M. and Canals, R. (2021). Computer vision, IoT and data fusion for crop disease detection using machine learning: A survey and ongoing research. *Remote Sens*., 13: 2486 .
20. Ndikumana, Emile, Dinh, H.T.M., Baghdadi, N., Courault, D. and Hossard, L. (2018). Deep recurrent neural network for agricultural classification using multitemporal SAR Sentinel-1 for Camargue, France. *Remote Sens*., 10: 1217 .
21. Filho, Hugo Crisóstomo de Castro, Júnior, O.C., Carvalho, O., Bem, Pablo Pozzobon de, Moura, Rebeca dos Santos de, Albuquerque, Anesmar Olino de, Silva, Cristiano Rosa E., Ferreira, Pedro Henrique Guimarães, Guimarães, R. and Gomes, R. (2020). Rice crop detection using LSTM, Bi-LSTM, and machine learning models from Sentinel-1 time series. *Remote Sens*., 12: 2655.
22. Leroux, L., Castets, Mathieu, Baron, C., Escorihuela, M., Bégué, A. and Seen, D. Lo. (2019). Maize yield estimation in West Africa from crop process-induced combinations of multi-domain remote sensing indices. *European Journal of Agronomy*, 108: 11–26.
23. Kim, Nari and Lee, Yangwon. (2016). Machine learning approaches to corn yield estimation using satellite images and climate data: A case of Iowa State. *Journal of the Korean Society of Surveying, Geodesy, Photogrammetry, and Cartography*, 34: 383–390.
24. Sazib, N., Mladenova, I. and Bolten, J. (2018). Leveraging Google Earth engine for drought assessment using global soil moisture data. *Remote Sensing*: 10.
25. Maimaitijiang, M., Sagan, V., Sidike, P., Daloye, Ahmad M., Erkbol, H. and Fritschi, F. (2020). Crop monitoring using Satellite/UAV data fusion and machine learning. *Remote Sens*., 12: 1357.
26. Fieuzal, R., Baup, F. and Marais-Sicre, C. (2013). Monitoring wheat and rapeseed by using synchronous optical and radar satellite data: From temporal signatures to crop parameters estimation. *ARS*, 2: 162–180.
27. Aghighi, H., Azadbakht, M., Ashourloo, Davoud, Shahrabi, H. and Radiom, Soheil. (2018). Machine learning regression techniques for the silage maize yield prediction using time-series images of Landsat 8 OLI. *IEEE Journal of Selected Topics in Applied Earth Observations and Remote Sensing*, 11: 4563–4577 .
28. Sagawa, T., Yamashita, Y., Okumura, T. and Yamanokuchi, T. (2019). Satellite derived Bathymetry using machine learning and multi-temporal satellite images. *Remote. Sens*., 11: 1155.
29. Senthil Kumar Swami Durai and Mary Divya Shamili. (2019, Mar.). Smart farming using machine learning and deep learning techniques. *In*: *Proceedings of Decision Analytics Journal*, 3(2022): 100041. ISSN: 2772- 6622.

30. Shafiulla Shariff, Shwetha R.B., Ramya, O.G., Pushpa, H. and Pooja, K.R. (2022). Crop recommendation using machine learning techniques. *In*: *Proceedings of International Journal of Engineering Research and Technology*. ISSN: 2278-0181.
31. Vishal Kumar, Raushan Kumar, Shubham Kumar and Ajinkya. (2021, Apr.). Soil analysis for suitable crop prediction using machine learning and image processing. *In*: *Proceedings of International Journal of Advanced Research in Science, Communication and Technology*, 4(2). ISSN (Online): 2581-9429.
32. Myat Mon Kyaw, San San Nwe and Myint Myint Yee. (2019, Aug.). Pest classification and pesticide recommendation system. *In*: *Proceedings of International Journal of Trend in Scientific Research and Development*, 3(5). e-ISSN: 2456–6470.
33. Devdatta, A. Bondre and Santosh Mahagaonkar. (2019, Sep.). Prediction of crop yield and fertilizer recommendation using machine learning algorithms. *In*: *Proceedings of International Journal of Engineering Applied Sciences and Technology, 2019*, 4(5): 371–376. ISSN: 2455-2143.
34. Guan Wang, YuSun and Jianxi Wang. (2017, 5 Jul.). Automatic image-based plant disease severity estimation using deep learning. *In*: *Proceedings of Hindawi Journal, 2017*, Article ID 2917536.
35. Dhruvi Gosai, Chintal Raval, Rikin Nayak, Hardik Jayswal and Axat Patel. (2021, May-June). Crop recommendation system using machine learning. *In*: *Proceedings of International Journal of Scientific Research in Computer Science, Engineering and Information Technology*, 7(3), 554–557. ISSN: 2456-3307. https://doi.org/10.32628/CSEIT2173129 554.
36. Archana, K. and Saranya, K.G. (2020, May). Crop yield prediction, forecasting and fertilizer recommendation using voting based ensemble classifier. *In*: *Proceedings of International Journal of Computer Science and Engineering*, 7(5). ISSN: 2348–8387.
37. Akhil Venkataraju, Dharanidharan Arumugam, Calvin Stepan, Ravi Kiran and Thomas Peters. (2023). A review of machine learning techniques for identifying weeds in corn. *In*: *Proceedings of Smart Agricultural Technology*, 3: 100102. Elsevier B.V. Available online 1 August 2022 2772-3755/© 2022.
38. Anis Ahmad, Dharmendra Saraswat and Aly El Gamal. (2023). A survey on using deep learning techniques for plant disease diagnosis and recommendations for development of appropriate tools. *In*: *Proceedings of Smart Agricultural Technology*, 3: 100083. Elsevier B.V. Available online 16 June 2022 2772-3755/© 2022. Accepted 14 June 2022.

6 | Blockchain for Transparent Food Supply Chains

Jothy C.R.[1*], Judith J.E.[1] and Nandita Sengupta[2]

[1] Department of Computer Science and Engineering, Noorul Islam Centre for Higher Education, Kumaracoil, Tamil Nadu

[2] Department of Information Technology, University College of Bahrain, Manama, Kingdom of Bahrain

"Food for All" is the main objective of SDG2. Understanding the complexities of hunger is crucial, as it is not merely about empty stomachs. It is a multifaceted problem with various root causes and widespread impacts. The chapter unlocks the possibilities of Blockchain Technology (BCT) and explores the benefits it gained while used along with Food Supply Chain (FSC) for enhancing transparency and efficiency of it, thereby contributing to the achievement of Sustainable Development Goal 2 (SDG2): Zero Hunger. Transparency, decentralization, and immutability, the major key features of blockchain, offer solutions to the inefficiencies and lack of traceability in the traditional food supply chain. Monitoring the flow of food products, recording each and every transaction securely and also transparently helps the food supply chain to be traceable by all its consumers, which aids in the outbreak management. Smart contracts automate tasks such as quality control and payment settlement, it would also help in reducing costs and eliminating middlemen interface. Moreover, blockchain-based authentication systems help combat food fraud by verifying the authenticity of products through QR codes or RFID (radio frequency identification technology) tags connected to immutable blockchain records. In addition to this, blockchain facilitates monitoring of environmental conditions in real time during transportation and storage, enabling quick responses to deviations that could lead to food spoilage. Transparent nature of BCT with the integration of FSC is capable of eradicating hunger and by this achieve the SDG2 which is the main objective of this chapter as stated in the 2030 agenda. This chapter enumerates the role of BCT in FSC which has the ability to

*Corresponding author: crjothiesh@gmail.com

reduce food waste, combat fraud, promote sustainability, and ensure safe, secure, and quality food products for all its users so as to achieve the second goal of Sustainable Development Goal (SDG2). The chapter concludes that blockchain holds promise for transforming food supply systems and advancing SDG2.

1. Introduction

"Food is not a word; it's a life for every living organism." Food forms an essential component for all living organisms in orbit. For humans it not only provides sustenance but also serves as a cultural, social, and economic cornerstone in societies worldwide. However, as global food systems become increasingly complex, concerns surrounding food safety, authenticity, and sustainability have risen to the forefront. Before the food reaches our plates it undergoes several stages, i.e., from farm to fork, the journey of food products involves numerous processes. Farmers, stakeholders, retailers, distributors, and customers significantly contribute to the progress along the path of the FSC from the farm to plate transformation. It is crucial to ensure transparency and accountability across the entire supply chain especially in the food supply chain as food forms the essential one in guaranteeing one's lives [7]. Hence everyone has his/her own right to know how the food products are cultivated, processed, transported, distributed, and marketed. Food supply chain may face several challenges. In the present scenario, in response to these challenges, innovative technologies come into existence in bringing a solution to make the food supply chain transparent, ensuring that all customers have access to information regarding food products they consume from different retailers [2]. The source of the food, the production process involved, the manufacturing details, etc. can be easily available for their convenience. In recent years, concerns about food safety, authenticity, and sustainability have prompted a growing demand for greater transparency, traceability, and accountability within food supply chains. Lack of transparency and accountability within a traditional food supply chain system paves way to overall performance inefficiencies. Blockchain (BC), being a smart cutting edge technology recently being used in various fields, has broadened its role and is offering the capability to transform how we monitor, verify, and manage food supply chains to achieve the SDG2 [1]. Effective management and collaboration among stakeholders are essential to address challenges such as food security, sustainability, and resilience in the face of emerging threats and disruptions [3, 11]. Blockchain, a decentralized, immutable, and tamper-resistant digital ledger, provides a transparent record of transactions and information exchanges. By integrating BC in FSC, stakeholders can enhance traceability, reduce food fraud, minimize waste, and empower consumers with real-time access to reliable data about the consumed products [28]. In the three major sectors, i.e., production, distribution, and consumption, BT can make remarkable enhancements towards FSC traceability. By leveraging blockchain in food supply chains, stakeholders can achieve greater visibility and trust throughout the entire supply chain ecosystem [13].

Section 2 describes the basics of the FSC and discusses the basic processes involved with its architecture. Section 3 gives a detailed study of the importance of transparency which should be maintained within an FSC so as to maintain the sustainability in an FSC in order to attain the objective of SDG2. This section also describes the role of smart technologies which are being a part of the FSC to make it transparent. Section 4 gives a detailed overview of the integration of the BCT with FSC transparency enhancement with the intention for achieving SDG2 which aims at "Zero Hunger". Section 5 states the overall inference of the whole chapter. Finally, the chapter concludes by stating the BCT–FSC integration which brings customer satisfaction and provides food safety, thereby minimizing the food loss and wastage so as to achieve SDG2 [9].

2. All About Food Supply Chain

Food Supply Chain (FSC) encompasses the entire process in bringing food products from the initial production stage to the end, i.e., consumption phase. It aims at efficiently delivering safe and nutritious food to every consumer. FSC is a dynamic and interconnected system which has a great role in ensuring safety, nutritious, and affordable food to consumers worldwide. The components of a food supply chain typically include [25]:

- **Producers/Farmers:** These are the individuals or organizations responsible for growing crops, raising livestock, or cultivating other food products.
- **Suppliers:** Suppliers provide necessary inputs to farmers such as seeds, fertilizers, pesticides, and equipment.
- **Processors/Packers:** Once harvested, raw food materials are processed, cleaned, sorted, and packaged by processors or packing facilities.
- **Distributors/Wholesalers:** Distributors purchase food products from processors in bulk and distribute them to retailers, restaurants, and other establishments.
- **Retailers:** These are the outlets where consumers purchase food products directly. Retailers can include grocery stores, supermarkets, online platforms, etc.
- **Transportation and Logistics:** This component involves the movement of food products from farmers to end users, including transportation, storage, and handling.
- **Consumers:** The unavoidable as well as the final component is the consumers, who acquire and utilize the food products.

FSC is a multifaceted network comprising interconnected activities like production, processing, distribution and consumption of food products. Figure 1 illustrates the flow of a food supply chain, delineating the various stages through which a produced food progresses from the farm to the plate.

FSC encompasses a range of stages that map the journey of food products from farm to fork, as depicted in Figure 1. These stages include the production

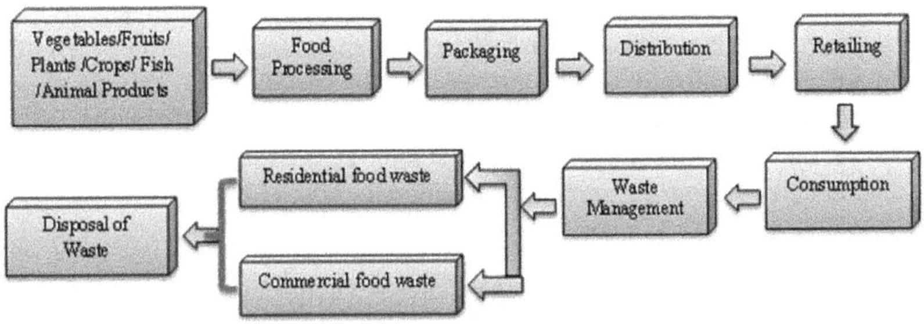

Figure 1: Processes in the food supply chain.

of food products, their processing, packaging, distribution, retailing, consumption, waste management, and disposal of waste [4]. These various stages are discussed as follows.

i. **Production:** Initial stage in an FSC is the production phase, which begins from the agricultural field. In this stage the activities such as planting, cultivating, harvesting, and raising livestock are involved. Farmers play a crucial role, employing various techniques to ensure the quality and quantity of food produced.

ii. **Food processing:** Once harvested or slaughtered the next stage is the food processing where the raw agricultural products undergo pre- and post-processing methods to transform them into value-added food products. Cleaning, sorting, packaging, preserving, and manufacturing are the activities involved in this phase.

iii. **Packaging:** Packaging is essential for protecting food products from physical damage, contamination, and spoilage during storage, transportation, and display. It plays a vital role in safeguarding food items. It also serves as a means of branding, marketing, and providing information to consumers about product attributes, ingredients, and usage instructions. Sustainable packaging initiatives aim to reduce environmental impact by minimizing waste, promoting recyclability, and exploring innovative solutions.

iv. **Distribution:** After pre-processing and packaging of the produced food products are over the next stage is the distribution where the food products are transported from production facilities to distribution centres, warehouses, and ultimately to retailers or food service providers. Distribution networks may involve multiple intermediaries such as wholesalers, distributors, and logistics companies, each responsible for verifying timely and efficient delivery of finished food products to reach the destinations on time [8].

v. **Retail:** Before the food products reach the hands of the consumers they pass through various retail channels, including grocery stores, supermarkets, restaurants, online platforms etc. Retailers have a significant role in merchandising, marketing, and selling food products to consumers, catering to diverse preferences and demands.

vi. **Consumption:** Consumption, the final phase in an FSC, means the food products were consumed by the consumers. Consumer choices and behaviours influence the entire FSC, driving demand for specific products and shaping trends in food production, distribution, and marketing. Even though this stage forms the final stage of FSC, the journey never ends here as the waste management forms a significant role in ending the FSC more efficiently.

vii. **Waste management:** Throughout the supply chain, there are challenges related to food waste and loss. Efficient waste management schemes are essential to minimize the environmental impact on production and distribution of food products, as well as to ensure economic sustainability so that to achieve the SDG2 with an objective of "Zero Hunger" which forms the second goal of the sustainable development goal which is targeted by 2030 [29].

  a. *Residential waste management:* Residential and commercial food waste refers to the discarding the unused or expired food products or organic material by households, including kitchen scraps, spoiled produce, and expired leftovers [11]. This includes both edible and inedible food waste. Effective waste management strategies aim at minimizing waste generation, divert organic waste from landfills, and promote recycling, composting, and energy recovery. Tackling residential food waste not only mitigates environmental issues but also preserves resources and decreases the emission of greenhouse gases linked to waste disposal [6].

  b. *Commercial waste management:* Commercial food waste refers to surplus or unsold food items generated by businesses, including restaurants, supermarkets, and food service providers. It arises from various factors such as overproduction, spoilage, expiration, and consumer preferences, posing economic and environmental challenges [24]. Businesses have the opportunity to reduce commercial food waste through various strategies like managing inventory, controlling portions, and establishing donation programmes to redirect excess food to those who require it. Effective collaboration among stakeholders and non-profit organizations is crucial for developing comprehensive solutions to tackle the multifaceted issue of commercial food waste [10].

viii. **Disposal of waste:** Waste disposal marks the concluding stage of the FSC. If due to any situation, there happens to be excess in the food production then one should adopt proper disposal methods, such as composting or recycling, and ensure that waste is handled in an environmentally responsible manner.

3. Significance of Transparency in Food Supply Chain for Achieving SDG2

To achieve SDG2, we need to make our nation hunger-free, and hence we need to concentrate on food supply chains. One of the key aspects to make an FSC more efficient is making it more transparent. In simple terms 'Transparency' provides 'Trust', that's what every customer expects [9]. Transparency being a fundamental feature of FSC plays a significant role in ensuring accountability, trust, and safety throughout the entire process from farm to fork. At its core, transparency in an FSC provides consumer openness, visibility, and accessibility of information regarding the food production, further processing, product distribution, and final consumption of food products [12]. It's the right of every consumer to know the source of the food, how it is being prepared, what are all the ingredients used, how long the food can be consumed etc. This transparency helps to build trust among consumers and food producers, manufacturers, and retailers, ultimately bringing trust on the quality and safety of the food products they choose. Figure 2 shows the visibility of processes within an FSC to be available to the consumers from its production to consumption. Moreover, transparency is essential for ensuring food safety and security. By providing visibility to its customers in each step of the FSC, from the sourcing of raw materials to the finished products, transparency allows for better traceability and accountability in the event of foodborne illnesses or contamination incidents. Furthermore, transparency fosters innovation and improvement within the food industry. When information is openly shared among stakeholders, including suppliers, producers, distributors, and regulators, it creates opportunities for collaboration and knowledge exchange. This collaborative environment will result in new technological development, best practices, and standards that enhance efficiency, sustainability, and resilience across the FSC. By using this cutting edge technological development, sustainability can be maintained in an FSC so as to achieve the SDG2 by 2030 [14, 26].

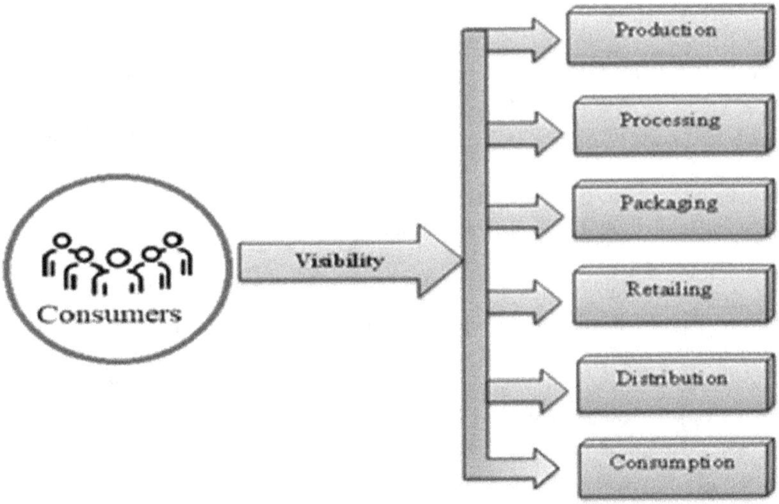

Figure 2: Transparent food supply chain.

3.1. Enhancing Transparency in FSC by Making Use of SMART Technologies

As the growth of technological developments increases day by day, it's time to integrate these new smart technologies with FSC so that it can be more efficient. In today's interconnected world, FSC faces numerous challenges such as lack of transparency, traceability, and accountability. The introduction of smart technologies into FSC offers promising solutions to address these issues and enhance transparency throughout the entire FSC in order to achieve SDG2 which aims at bringing a nation with "Zero Hunger" [26]. Table 1 shows the benefits gained by integrating the different smart technologies with the FSC and also it describes the transparency attributes of using such technologies. By increasing the transparency in an FSC the benefits of the traditional FSC gets multiplied. The smart technologies provide enhanced performance when integrated with FSC. The main objective of this chapter sticks with transparency while using the blockchain technology. Use of smart technologies enables consumers to access detailed information about the origin of their food, including where it is grown or raised, how it is processed, and when it is distributed and how long it can be preserved. Internet of Things (IoT), a smart technology, has the capacity to enhance transparency in the FSC [9]. RFID tags and various IoT sensors collect real-time data of the food products such as temperature, humidity, and location. By integrating IoT devices into the FSC, stakeholders can monitor and trace the food product movement from its production to consumption, so that they ensure whether they are handled and stored properly to maintain quality and safety.

Table 1: Smart technologies role in Food Supply Chain (FSC)

Smart Technology	Applications towards FSC	Advantages	Transparency Attributes
Internet of Things (IoT)	Product traceability, monitoring, and control	Real-time monitoring of food materials	Provides real-time visibility of movement of products from farm to the fork
Blockchain (BC)	Secure digital ledger transaction, provides trust among the customers	Enhanced transparency and traceability	Provides immutable records of transactions and product information
Big Data	Data-driven pricing decision-making	Predictive analytics for market demand	Huge data analysis based on market demand and trends.
Digital Twins	Modelling and creating virtual replicas of physical processes	End-to-end visibility and risk-free virtual visualization	Provides a virtual representation of the supply chain

Furthermore, Artificial Intelligence (AI) and data analytics marks an unavoidable role in enhancing transparency by analyzing large quantities of data to identify the patterns, trends, and other anomalies within the food supply chain. AI algorithms can detect potential issues such as food fraud, contamination, or spoilage, allowing stakeholders to take proactive measures to mitigate risks and ensure compliance with regulatory standards. Additionally, QR codes on food packets and Near Field Communication (NFC) tags in smart packaging enable consumers to access detailed information about food products directly from packaging [32]. Scanning QR code or tapping NFC tags, one can get the information such as ingredient lists, nutritional content, allergen warnings, and sustainability certifications, empowering them to make informed choices regarding the products they buy. Overall, the integration of smart technologies offers unprecedented opportunities to enhance transparency in the food supply FSC. By leveraging blockchain, IoT, AI, and smart packaging, stakeholders can create a more transparent, traceable, and trustworthy food system that prioritizes consumer safety, quality, and sustainability [16]. As the adoption of these technologies continues to grow, we can expect to see significant improvements in transparency and accountability across the entire FSC. Also these smart technologies pave the way towards the sustainable development goals [7, 9].

3.2. Blockchain's Contribution to Food Security and Access

Food security programmes often face significant disruptions from both human-made and natural disasters, leading to losses in food production, accessibility, and utilization. In such scenarios, crop insurance emerges as a viable solution to mitigate the above said challenges. Figure 3 shows the number of potential food crises considered and the number of food crises identified from 2016 to 2023.

Blockchain technology's role in reducing food wastage is significant, as it combats food fraud and identifies the source of dangerous products, thus minimizing wastage. Given the global food crisis and environmental concerns related to food wastage, blockchain's benefits in FSC are undeniable. Consumers can now verify product authenticity by scanning items and accessing digital records through blockchain technology. Figure 4 shows the number of severely food-secure people in different parts of Asia [12].

Blockchain technology enables the secure recording and accessibility of data concerning the quality and safety of food products. Smart contracts can automate quality assurance procedures, guaranteeing compliance with regulatory standards and certifications. This automation serves to prevent the distribution of inferior or hazardous food items, thereby protecting public health and bolstering food security.

Blockchain technology stands out as a promising solution for achieving sustainable food security using its decentralized immutable databases, peer-to-peer (P2P) networking, distribution across networks, and the implementation of smart contract mechanisms [11]. Multiple Criteria Decision Making (MCDM)

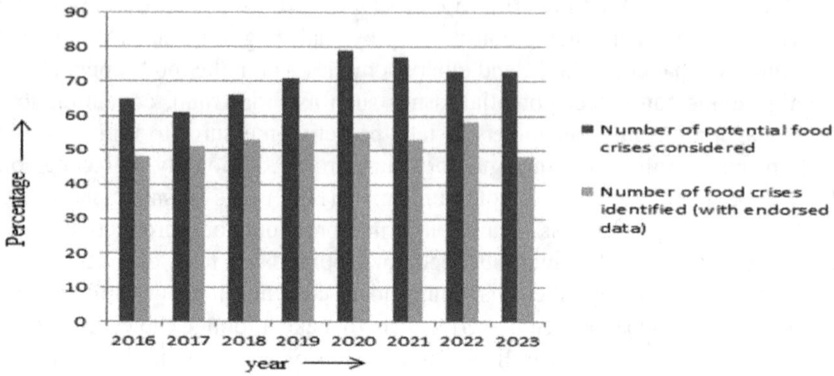

Figure 3: Charting the evolution of food crises: 2016–2023.

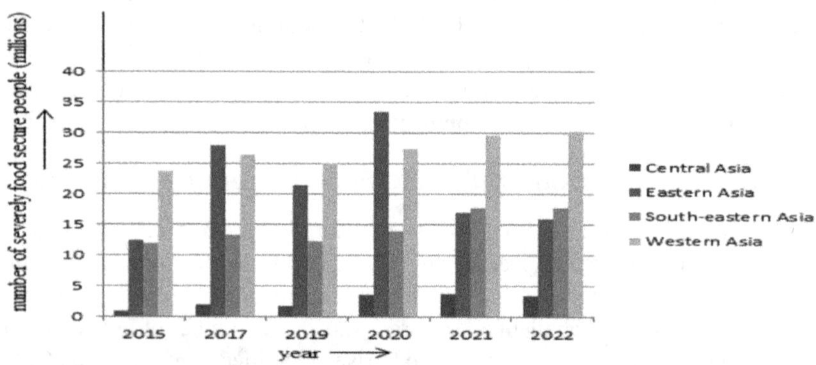

Figure 4: Changes in severely food-secure population percentage over time.

approach along with the blockchain technology to develop a secure Agriculture Food Supply Chain. Blockchain is revolutionizing food security compliance by effectively managing food quality and security by addressing food wastage, traceability, and real-time monitoring [1, 2].

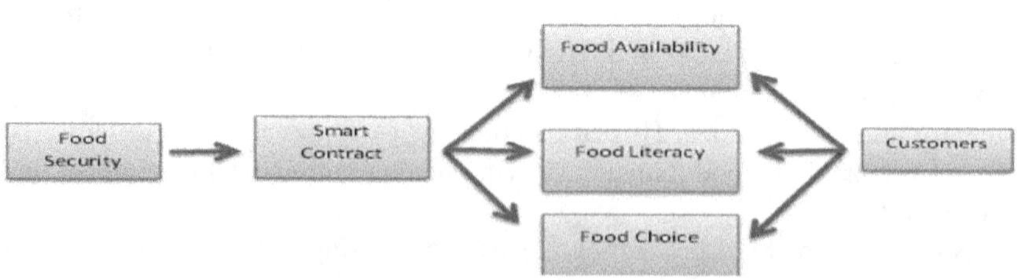

Figure 5: Secured food supply chain.

Traceability, particularly within the FSC ensures safety and quality of food. It establishes a seamless internal trust between farmers and consumers. Figure 5 shows a secured food supply chain with the smart contract concept which aims at securing food products based on their availability, literacy, and choice. Food security, a crucial component of SDG2, encompasses not only the availability, accessibility, and affordability of nutritious food but also the need to eradicate hunger and ensure food sovereignty. In an FSC, food availability becomes paramount, addressing the adequacy of food supplies to meet dietary requirements and support sustainable agricultural practices. Food literacy has a significant role in achieving SDG2 by empowering individuals with the knowledge and skills required to make healthy and sustainable food choices, thereby promoting nutrition security and reducing food-related health disparities. Together, these concepts underscore the multifaceted approach required to address the complex challenges of hunger, malnutrition, and food insecurity while advancing progress towards SDG2 targets. Blockchain and IoT are being utilized to enhance traceability within the FSC, allowing consumers and third parties to monitor the food products flow instantly [10, 19]. Consumers can now verify product authenticity by scanning items and accessing digital records through blockchain technology. Moreover, blockchain implementation proves to be more cost-effective compared to existing food tracking systems. It enables the tracking of ownership records and provides tamper-resistant data, which is crucial in addressing urgent food supply chain concerns such as safety recalls and inefficiencies. For FSC operators, accessing business-critical data is essential for making informed decisions. With many FSC organizations operating globally or nationally, digitizing these data sources becomes imperative to support the broader food industry's decision-making processes. Exact condition of data sources is fragmented and isolated, presenting hurdles for food supply chain operators to analyse and utilize data efficiently. This data is pivotal for determining pesticide and herbicide levels and ensuring the integrity of purchased food items. Blockchain-based traceability systems offer a solution to optimize large-scale food supply chains. Employing a consensus process with blockchain to store transaction data can enhance food integrity, safety, and facilitate smart farming practices. Initial costs may deter industry players from participation in some cases. As of now the major problem in the food industry is food safety, plagued by persistent inefficiencies. Implementing blockchain holds promise for major improvements in food safety by its immutable transaction history and bringing stakeholders together for accountability. This transparency fosters customer trust and brand loyalty by enabling easy tracing of food items' origins, enhancing overall quality and reliability. Blockchain technology also promotes accountability among food manufacturers and suppliers, ultimately benefiting consumers through improved safety and profitability. By enabling transparency throughout the food supply chain, blockchain provides real-time updates and facilitates supply chain management processes. It maintains a digital record of all transactions, enabling ease of traceability efforts and enhancing food safety in an increasingly critical landscape.

4. Blockchain in Food Supply Chain for Transparency: A Key to Achieve SDG2: Zero Hunger

As we all know blockchain is one of the smart technologies which is being used in most of the fields with the intention to enhance security and transparency in the system, it can also be integrated with the food supply chain to make the process transparent which is the main objective of this chapter. By utilizing blockchain technology, stakeholders in the different stages of an FSC can create transparent and traceable records of each and every step of manufacturing, processing, distributing, and marketing of food products. In an FSC, by enhancing the transparency, it is possible to achieve SDG2 which is aimed to eradicate hunger [22, 27]. Figure 6 shows how a blockchain can be integrated with FSC to form a transparent customer friendly FSC.

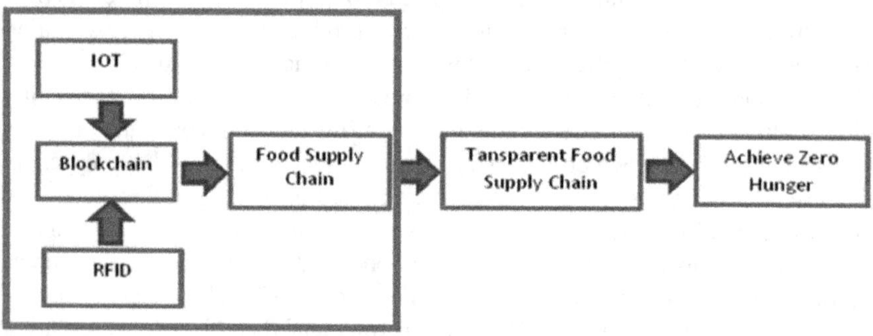

Figure 6: Integration of BC in FSC for SDG2.

Blockchain, a decentralized digital record system, offers a secure and highly transparent way to record transactions and trace information of products from its production to the end user consumption [1]. RFID technology integrated with blockchain improves transparency and traceability. By tagging food items with RFID chips and recording their movements on a blockchain ledger, stakeholders can trace product journeys from the farm to the plate [2, 3]. It ensures authenticity, reduces food fraud, and enables swift identification of contaminated or expired goods. Moreover, blockchain's immutable nature prevents tampering, fostering a more resilient and efficient food supply chain [4, 11]. Table 2 shows the benefit gained by the blockchain technology towards the FSC so the food loss can be minimized by tracking the food flow.

IoT sensors, trackers, and smart appliances collect on-time data throughout the supply chain process. Those collected information will be securely recorded on a blockchain ledger, providing immutable records of every step, from production to distribution. By monitoring factors like temperature, humidity, and location,

Table 2: Consumer benefits on integrating BC with FSC

Technology	Aspect	Consumer Benefits
Blockchain	Traceability	Ease of trace of movement of food products
	Product information	Visibility of product origin, transportation history etc.
	Authenticity verification	Verify the authenticity of food products
	Supply chain visibility	Real-time visibility
	Compliance monitoring	Smart contracts automate compliance checks

stakeholders can ensure optimal conditions for food safety and quality [19]. Additionally, consumers get provision to view detailed information regarding the products they buy, fostering trust and safety in the supply chain. This seamless integration enhances efficiency, reduces waste, and mitigates risks in the food industry. Each time a product changes hands or moves through a different stage of the FSC, a new block gets added to the blockchain, containing information about the origin of the product, processing methods, and distribution ways. This transparent record allows stakeholders, including consumers, to trace the journey of a food product back to its source, providing unprecedented visibility into its entire lifecycle. Furthermore, blockchain technology enhances transparency by facilitating real-time data sharing and collaboration among supply chain participants [13, 14]. Figure 7 depicts how FSC can be used along with BC to improve its transparency.

Through a shared blockchain platform, stakeholders have the ability to both access and provide input to a shared database of information. This transparent and decentralized approach reduces the risk of data manipulation or fraud, and provides the right to access up-to-date accurate information about the products they consume. Moreover, BCT helps to improve food safety and quality by enabling faster and more efficient traceability in the event of a foodborne illness outbreak or product recall. By quickly identifying the source of contamination or adulteration, stakeholders can take targeted action to remove affected products from the market, preventing further harm to consumers and protecting public health. In addition to transparency and traceability improvement, BCT can also drive efficiency and cost savings within the FSC. By streamlining processes such as inventory management, logistics, and payments, blockchain can reduce inefficiencies and delays, ultimately lowering costs for businesses and consumers alike. By facilitating real-time data sharing and collaboration, blockchain can help create a more transparent, trustworthy, and resilient food system for all stakeholders involved. As the adoption of BCT continues to grow, we can expect to see significant advancements in supply chain transparency and accountability across the food industry which in turn will help to achieve the SDG2 [15, 18].

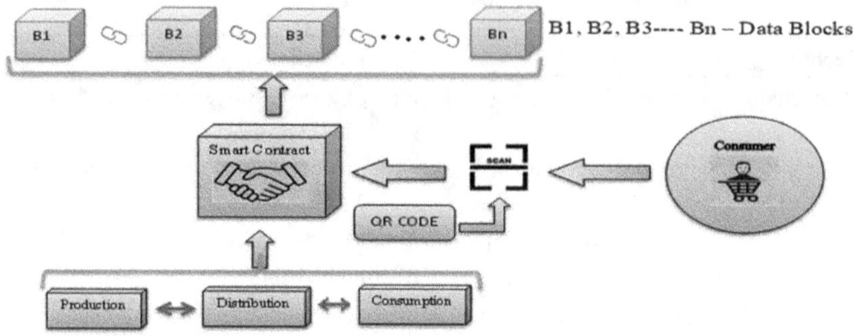

Figure 7: Blockchain technology based transparent food supply chain.

4.1. Blockchain in Reducing Food Waste and Loss: Step Towards Zero Hunger

Achieving zero hunger, the goal of SDG3 can be made possible only by minimizing food waste [10]. In most of the situations we face significant waste of food as the production goes surplus due to many factors. So to avoid such food waste we need to find an alternate solution to minimize the wastage of food [11]. SmartNoshWaste extends the capabilities of the FoodSQRBlock framework by addressing two crucial aspects: one is accessibility of data across all phases in an FSC and the other is reduction of food waste based on those collected data. Unlike its predecessor, FoodSQRBlock, which focused solely on traceability, SmartNoshWaste ensures that consumers can easily access information from production to consumption and implements strategies to minimize food wastage. The framework comprises two main layers: one is the Data System Architecture and the other one is the Machine Learning Module. With the help of the smart technologies like blockchain, QR codes, and cloud computing, SmartNoshWaste digitizes and stores food data efficiently, enabling effective utilization in the Machine Learning Module for waste reduction purposes [9].

BCT provides a fruitful solution for tackling food loss and waste within the agri-food supply chain [9]. The SmartNoshWaste implements blockchain-based multisensor monitoring systems to track storage conditions, guaranteeing both quality preservation and data integrity. By leveraging smart contracts enabled by blockchain, traceability to the product's origin is enhanced, facilitating the identification of contamination sources and mitigating additional waste. Methodologies employed in the study include case studies, interviews with industry experts, and semi-structured interviews aimed at exploring the connection between food loss management and digital innovations [13]. Figure 8 illustrates how blockchain can be integrated with an FSC so as to reduce food loss and waste considerably by making use of QR codes and also with smart contacts. The SmartNoshWaste framework is designed to tackle food waste by leveraging BCT. Its initial step involves converting food production data into a digital format to improve traceability and easy accessibility by analysing the various phases and processes within a typical FSC. The aim of SmartNoshWaste is to digitize data from all the five different phases such as Production, Processing, Distribution, Retailing, and Consumption allowing stakeholders such as farmers and consumers to access it easily. Once the supply chain data is digitized, it is stored on the blockchain, ensuring traceability and accessibility.

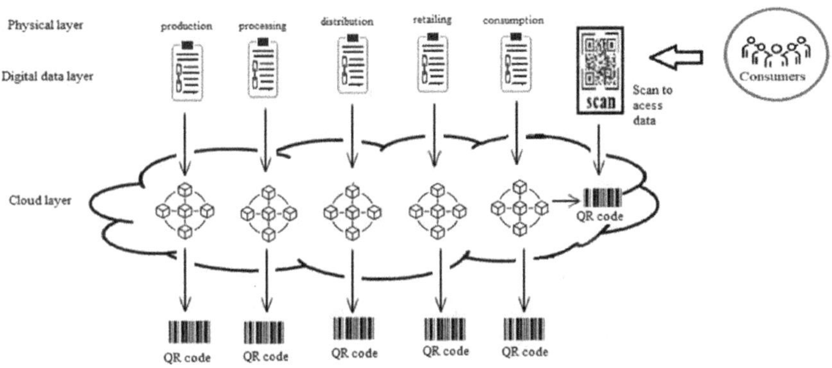

Figure 8: Food waste reduction using SmartNoshWaste framework.

All digital data related to these phases is stored and organized on the data blocks of blockchain in the cloud. Services of cloud computing were preferred for the two major reasons: firstly, to improve processing speed and enhance transaction efficiency on the blockchain, and secondly, to facilitate easy deployment of software updates across different regions by developers [23].

Data System Architecture: The blockchain-based framework features a multilayered system, outlined in Figure 8. This system comprises of three layers:

i. **Physical layer:** This layer includes a variety of products sourced from various farmers or producers from different parts within the supply chain.
ii. **Digital data layer:** Digital data associated with the products will be stored here in this layer, facilitating traceability and accessibility. These digital data include the food product's expiry date, the identity of the farm from where it has been manufactured, and the batch of production, identity number, and so on.
iii. **Cloud layer:** Cloud layer processes the digital data on the cloud using BCT and made available to the consumers via QR codes at different phases, enabling traceability and accessibility.

How it Works: As a food item being produced by the farmer it undergoes through various stages in the FSC, at each and every stage of the FSC the relevant information will be added or stored in data blocks as unchangeable records with the BCT. These stored data contain the basic information such as item type, farm ID and production date and these data are placed in the initial block called the genesis block (or block 0). As the product moves through different stages, additional relevant data will be added, and with this additional information a new data block will be created for each phase, containing the exact data along with the hash value of the previous block. The data collected during consumption phases include farm ID, item type, weight, manufacturing date, expiry date, batch number, distribution centre ID, retailer shop ID, and the location of sale. These details may vary depending on the specific requirements of farms, processing centres, distributors, or countries of origin. To protect user privacy, only geographical location data of consumers is stored alongside the food item data in the consumption phase, ensuring that the true identity of the user remains concealed. Consumption and wastage information related to a particular food item can be easily tracked by this framework [12, 13]. With the help of this framework the exact data stored can be accessed by everyone and so that the excess food can be distributed to those who are in need in order to achieve SDG2.

4.2. Ensuring Food Safety and Quality through Blockchain Traceability

The main objective of a transparent FSC is to retain the food safety and quality. For ensuring food safety and quality a framework has been proposed in [15] which comprises a main layer, employing a consortium blockchain for information sharing among different organizations, and a sub-layer, which utilizes a private blockchain within individual enterprises for storing product information. To ensure data integrity and secure communication between the main layer and the sub-layer, mechanisms such as hash pointers and application programming interfaces (APIs) are utilized. Additionally, privacy protection measures are implemented to safeguard sensitive enterprise data, including access control measures and the storage of sensitive data within private blockchains. Scalability concerns are addressed through features enabling the easy integration of third-party quality inspection agencies and new enterprises into the product traceability system. Performance optimization strategies focus on enhancing operational efficiency by segregating data entry and storage operations in the sub-layer from data querying in the main layer. Consumers can utilize product scanning to authenticate items, while blockchain ensures access to digital records for inquiries. Moreover, compared to existing tracking systems, blockchain is a more cost-effective solution. Its ability to track ownership records and provide tamper-resistant data is pivotal in mitigating food safety concerns and streamlining operations. However, blockchain adoption in an FSC faces challenges, primarily due to the initial implementation costs [16]. Nevertheless, the benefits it offers, particularly in enhancing food safety and reducing wastage, are substantial. By providing transparent and immutable transaction histories, blockchain fosters collaboration among stakeholders and boosts consumer trust. This transparency also encourages accountability among manufacturers and suppliers, leading to better compliance with food safety regulations. Additionally, blockchain aids in minimizing food wastage by reducing the likelihood of mislabelling and enabling swift identification of unsafe products. Given the alarming rate of food wastage globally, particularly in light of population growth and industry mismanagement, blockchain's role in optimizing supply chain efficiency is crucial [18]. Despite initial hurdles, the benefits it provides in terms of transparency, accountability, and efficiency cannot be overstated [20].

Double layer framework [15], a novel approach for enhancing product traceability systems by utilizing a permissioned blockchain was shown in Figure 9. This framework aims to address key challenges in traceability by separating data entry and read functions, thus improving scalability and system performance [31]. By segregating these functions, the framework optimizes blockchain resources and enhances operational efficiency. One primary feature of the double-layer framework is the division of responsibilities between the two layers [8, 17]. The sub-layer handles data entry and storage operations, which often consume significant system resources and impact efficiency. In contrast, the main layer focuses on providing query services for traceability information, ensuring quick access to accurate data for consumers and regulatory agencies [31].

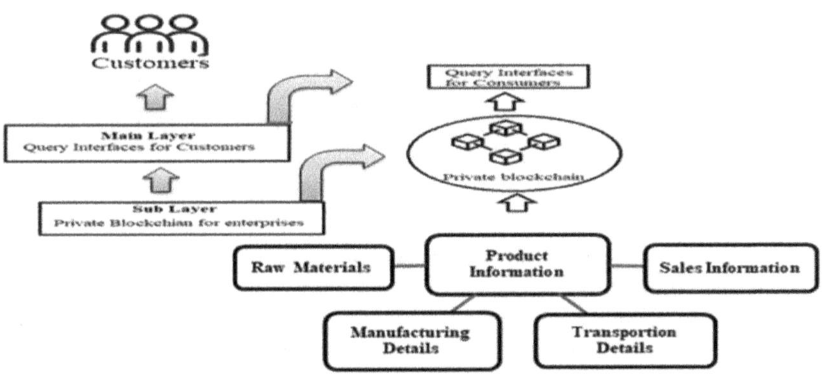

Figure 9: Food traceability using double layer framework.

Furthermore, the framework facilitates regulatory agency participation through smart contracts and establishes a consortium blockchain for data sharing among stakeholders. This collaborative approach enhances transparency, accountability, and regulatory oversight within the supply chain [5, 6]. Additionally, mechanisms are incorporated to prevent malicious activities such as tampering with traceability information, ensuring data integrity and authenticity within the private blockchain. Overall, the double-layer framework provides a comprehensive solution for improving product traceability systems by enhancing security, scalability, and performance. By leveraging BCT and smart contracts, this framework offers a robust foundation for building efficient and transparent traceability systems that meet the evolving needs of various industries [21].

5. Transparent Food Supply Chain with Blockchain: A Sustainable Development to Achieve Zero Hunger

Food serves as the primary source of energy for all living organisms. BCT, known for its security in transactions, has gained traction in various industries, including food supply chains. Its decentralized and distributed database eliminates intermediaries, enhancing security and preventing cyber-attacks.

Introducing BCT and integrating it with the FSCs fosters transparency, supporting sustainable development endeavours aimed at eradicating hunger [29]. By securely documenting each stage of the supply chain, blockchain guarantees traceability, thereby diminishing food waste and fraudulent activities while upholding food safety standards. This heightened transparency cultivates trust among the stakeholders, empowering enhanced decision-making processes and resource management [25]. Ultimately, a transparent food supply chain driven by blockchain facilitates the effective distribution of food resources, making substantial strides towards global hunger eradication and the attainment of sustainable development objectives [2]. This technology's use in food traceability has surged due to its numerous benefits for the food sector:

- **Improved quality:** Blockchain enables comprehensive tracking of food items from production to sale, ensuring optimal quality throughout the supply chain. Consumers receive products in pristine condition, reducing health risks associated with compromised food items.
- **Fraud prevention:** Blockchain's immutable records deter fraudulent activities throughout the supply chain. Each transaction is associated with the entity responsible, making tampering evident. Integration of authenticity verification measures further strengthens fraud prevention efforts.
- **Optimized production:** Enhanced understanding of the production process facilitates cost and time savings. With the proliferation of the IoT and upcoming 5G technology, companies can leverage vast amounts of data to optimize processes and enhance sustainability.
- **Environmental sustainability:** Consumers increasingly prioritize environmentally sustainable practices. Blockchain's transparency and security empower companies to adopt eco-friendly practices in agriculture, livestock, and fisheries. However, the widespread use of blockchain may contribute to environmental concerns [30]. The decentralized nature of blockchain requires consensus among numerous agents, potentially increasing environmental impact. Nonetheless, advancements in information management systems are expected to mitigate these concerns over time.

- **Government support:** Government agencies, such as the United States Food and Drug Administration (FDA), recognize the potential of blockchain in enhancing food safety. Blockchain's ability to trace food origins rapidly and deploy advanced data analysis techniques strengthens disease prevention efforts, safeguarding consumers from contaminated or mislabelled foods in real-time. Hence the achievement of the second goal of sustainable development can be possible by these cutting edge technological developments.

6. Conclusion

Food serves as the primary source of energy for all living organisms. As 2030 is not so far away, it is the responsibility of every individual to commit themselves in achieving the 17 goals as stated in the agenda 2030. Smart technologies mark its presence in making a zero hunger world through smart technologies to achieve those sustainable development goals. Implementation of blockchain on an FSC provides a transformative opportunity to advance Sustainable Development Goal 2: Zero Hunger. By fostering transparency, traceability, and accountability throughout the entire FSC process, blockchain offers a powerful tool to enhance food security and safety, reduce food waste and loss, and improve customer accessibility to vital information about their food. Firstly, blockchain guarantees transparency by creating an unchangeable and auditable ledger of transactions which enables stakeholders to trace the journey of food items from farm to plate. Transparency gives trust among consumers and also enables them to make informed choices about the products they consume, thereby promoting food security. Secondly, blockchain enhances traceability by recording every step in the FSC. This enables swift identification of issues such as contamination or spoilage, facilitating targeted interventions to mitigate risks and ensure food safety. Furthermore, blockchain improves customer accessibility to critical information about their food, such as origin, production methods, and handling practices. This enables consumers to make ethical and sustainable buying decisions while promoting accountability among producers and suppliers. Moreover, by reducing inefficiencies and redundancies in the supply chain, blockchain helps to minimize food loss and food waste, thereby maximizing the utilization of resources and reducing the environmental impact of food production and consumption. In essence, the integration of BCT in FSC represents a significant step towards achieving Zero Hunger by addressing key challenges such as food security, traceability, food safety, customer accessibility, and reducing food waste and loss. As we continue to harness the potential of blockchain and collaborate across sectors, enabling a sustainable FSC for the purpose of providing information related to bulk production, excess wastage, and where the excess food needs to be provided, and thereby each and every individual can have the benefit of having the access to nutritious and safe food, thus advancing the goal of Zero Hunger for the present and future generations.

References

1. Gupta, Shankar R. (2024). Managing food security using blockchain-enabled traceability system. *Benchmarking: An International Journal*, 31(1): 53–74.
2. Chandan, A., John, M. and Potdar, V. (2023). Achieving UN SDGs in food supply chain using blockchain technology. *Sustainability*, 15(3): 2109.
3. Omar, I., Hassan, H., Jayaraman, R., Salah, K. and Omar, M. (2023). *Using Blockchain Technology to Achieve Sustainability in the Hospitality Industry by Reducing Food Waste*. SSRN: 4530352.
4. Balezentis, T., Zickiene, A. Volkov, A., Streimikiene, D., Morkunas, M., Dabkiene, V. and Ribasauskiene, E. (2023). Measures for the viable agri-food supply chains: A multi-criteria approach. *Journal of Business Research*, 155: 113417.
5. Sridhar, A., Ponnuchamy, M., Kumar, P.S., Kapoor, A., Vo, D.V. and Rangasamy, G. (2023). Digitalization of the agro-food sector for achieving sustainable development goals: A review. *Sustainable Food Technology*, 1: 783–802.
6. Trevisan, C. and Formentini, M. (2023). Digital technologies for food loss and waste prevention and reduction in agri-food supply chains: A systematic literature review and research agenda. *IEEE Transactions on Engineering Management*, 71(2024).
7. Jose Anand, A., Pawan, K., Narendran, M., Johnson, M., Duraiarasu, E.S., Mahalakshmi, S., Manikandan, T. and El Rhalibi, A. (2023). Precision farming for crop prediction. *In*: *Advanced Technologies for Smart Agriculture*, 195–217. Denmark: River Publishers.
8. Nisha, K., Augustine, C., Jose Anand, A. and Kamesh, M. (2023). Deep learning in smart agriculture. *In*: *Advanced Technologies for Smart Agriculture*, 153–175. Denmark: River Publishers.
9. Kumar, M., Choubey, V.K., Raut, R.D. and Jagtap, S. (2023). Enablers to achieve zero hunger through IoT and blockchain technology and transform the green food supply chain systems. *Journal of Cleaner Production*, 405: 136894.
10. Kör, B., Krawczyk, A. and Wakkee, I. (2022). Addressing food loss and waste prevention. *British Food Journal*, 124(8): 2434–2460.
11. Dey, S., Saha, S., Singh, A.K. and McDonald-Maier, K. (2022). SmartNoshWaste: Using blockchain, machine learning, cloud computing, and QR code to reduce food waste in decentralized web 3.0 enabled smart cities. *Smart Cities*, 5(1): 162–176.
12. De Boni, A., Ottomano Palmisano, G., De Angelis, M. and Minervini, F. (2022). Challenges for a sustainable food supply chain: A review on food losses and waste. *Sustainability*, 14(24): 16764.
13. Centobelli, P., Cerchione, R., Del Vecchio, P., Oropallo, E. and Secundo, G. (2022). Blockchain technology for bridging trust, traceability, and transparency in circular supply chain. *Information & Management*, 59(7): 103508.
14. Jiang, S., Jakobsen, K., Bueie, J., Li, J. and Haro, P.H. (2022). A tertiary review on blockchain and sustainability with focus on sustainable development goals. *IEEE Access*, 10: 114975–115006.
15. Amentae, T.K. and Gebresenbet, G. 2021. Digitalization and future agro-food supply chain management: A literature-based implication. *Sustainability*, 13(21): 12181.
16. Nirmala Devi, K., Jose Anand, A., Kothai, R., Ajai Krishna, J.M. and Muthurampandian, R. (2022). Sensor-based posture detection system. *Materials Today Proceedings*, 55(2): 359–364.
17. Jose Anand, A., Nirmala Devi, K., Edward Kennedy, Z. and Dhanalakshmi, R. (2022).

Processing techniques for sensor materials: A review. *Materials Today Proceedings*, 55(2): 430–433.

18. Joshi, P., Singh, A., Kumar, S., Joshi, G., Aggarwal, A. and Gupta, S.K. (2022). When sustainable development embraces blockchain: A systematic intelligent systems and applications. *Proceedings of ICISA 2022*, 959: 467.
19. Dhanalakshmi, R., Jose Anand, A., Arun Kumar Sivaraman and Sita Rani. (2022. Apr.). IoT-based water quality monitoring system using cloud for agriculture use. *In*: *Cloud and Fog Computing Platforms for Internet of Things*, 183–196. Routledge. Taylor & Francis Group.
20. Jose Anand, A., Shobika, M. and Sneha, T. (2022). Programmable stickers to monitor perishable goods using blockchain. *In*: *Proceedings of IEEE 2022 Second International Conference on Computer Science, Engineering, and Applications* (*ICCSEA 2022*), 1–5.
21. Saha, A.S., Raut, R.D., Yadav, V.S. and Majumdar, A. (2022). Blockchain changing the outlook of the sustainable food supply chain to achieve net zero. *Sustainability*, 4(24): 16916.
22. Yadav, V.S., Singh, A.R., Raut, R.D. and Cheikhrouhou, N. (2021). Blockchain drivers to achieve sustainable food security in the Indian context. *Annals of Operations Research*, 1–39.
23. Amentae, T.K. and Gebresenbet, G. (2021). Digitalization and future agro-food supply chain management: A literature-based implications. *Sustainability*, 13: 12181.
24. Stroumpoulis, A., Kopanaki, E. and Oikonomou, M. (2021). The impact of blockchain technology on food waste management in the hospitality industry. *ENTRENOVA-Enterprise Research Innovation*, 7(1): 419–428.
25. Abideen, A.Z., Sundram, V.P., Pyeman, J., Othman, A.K. and Sorooshian, S. (2021). Food supply chain transformation through technology and future research directions: A systematic review. *Logistics*, 5(4): 83.
26. He, C.C. (2021). Promoting food system transformation for SDGs. *J. of Global Change Data and Discovery*, 5(3): 230–236.
27. Behnke, K. and Janssen, M.F. (2020). Boundary conditions for traceability in food supply chains using blockchain technology. *International Journal of Information Management*, 52: 101969.
28. Barbosa, L. (2020). Social entrepreneurship and digital technology: Case study of the blockchain for zero hunger initiative. *Economic and Social Development: Book of Proceedings*, 193–202.
29. Medaglia, R. and Damsgaard, J. 2020. Blockchain and the United Nations Sustainable Development Goals: Towards an Agenda for IS Research. *In*: *Pacific Asia Conference Information Systems* (*PACIS*), 36.
30. Zhao, G., Liu, S., Lopez, C., Lu, H., Elgueta, S., Chen, H. and Boshkoska, B.M. (2019). Blockchain technology in agri-food value chain management: A synthesis of applications, challenges, and future research directions. *Computers in Industry*, 109: 83–99.
31. Ding, Q., Gao, S., Zhu, J. and Yuan, C. (2019). Permissioned blockchain-based double-layer framework for product traceability system. *IEEE Access*, 8: 6209–6225.
32. Hasan, I., Habib, M.M., Mohamed, Z. and Tewari, V. (2023). Integrated agri-food supply chain model: An application of IoT and blockchain. *American J. of Industrial and Business Management*, 13(2): 29–45.

7 | Mobile Solutions for Food Access and Distribution

Aasma Iqbal[1], Tehseen Mazhar[2*], Tariq Shahzad[3], Humera Amin[4] and Jose Anand A.[5]

[1] Department of Computer Science, Wuhan Textile University, China
[2] Department of Computer Science, Virtual University of Pakistan, Lahore, Pakistan
[3] COMSATS University Islamabad, Sahiwal Campus, Pakistan
[4] Department of Agricultural Extension and Rural Studies, College of Agriculture, University of Sargodha Pakistan
[5] Department of ECE, KCG College of Technology Chennai, Tamil Nadu, India

Information and communication technology (ICT) plays an essential part in improving food accessibility and distribution while dealing with serious issues in the worldwide food supply chain. ICT enables effective agricultural methods, real-time information on markets, simplified logistics through digital platforms, statistical analysis of big data, and mobile technology. This advancement in technology improves the accessibility and traceability of food items, thereby ensuring their safety and quality. Furthermore, ICT solutions provide small-scale farmers with access to critical information such as predictions of the weather, prices for crops, and most effective farming practices, boosting productivity and revenue. Online marketplaces and payment methods broaden market reach and increase customers' access to various food products. Overall, the use of ICT in the agricultural industry promotes an improved, resilient, inclusive, and sustainable delivery network, thereby reducing the shortage of food while contributing to the worldwide struggle against hunger. This paper will look at technological advances in agriculture, access to markets via e-commerce, supply chain and logistics enhancements, food security and transparency, visible case studies that demonstrate effective ICT applications in food chains, and make suggestions for future research.

*Corresponding author: tehseenmazhar719@gmail.com

1. Introduction

Food access is related to an individual's capacity to obtain sufficient, nutritious, and secure food [1], whereas food distribution equates to the processes that transport food from producers to consumers. Food access inequality is frequently caused by geographical, economic, and societal variables, leaving numerous communities, particularly those in neglected rural and urban regions, without dependable sources of nutritious food. Effective distribution of food systems is critical for closing these gaps by ensuring that nutritious food is attained to every part of the general population in an effective and timely way [2]. What if food and agricultural systems were influenced by the goal of improving human and environmental well-being? Concerns regarding the sustainability of the global food system are expressed in the United Nations General Assembly's (UN) "Zero Hunger" Sustainable Development Goal, adopted in 2015. Recognizing the link between health, the environment, and agriculture represents a developing change in educational and general awareness. Specifically, by incorporating sustainable agriculture focuses into the general effort to eliminate hunger, the Zero Hunger goal (SDG2) demonstrates a long-overdue understanding that commercial farming endangers critical ecosystem processes that support food production [3]. These extensively documented effects involve decreases in biodiversity, elevated pest threat, erosion of soil, decreases of organic matter in the soil, emissions of greenhouse gasses, and eutrophication and contaminants of water bodies.

While environmental compromises were long regarded as a Faustian bargain in initiatives to address malnutrition and starvation, agricultural production has failed to deliver on its promise of eradicating hunger. While the world is awash in 'calories', the Food and Agricultural Organization of the United Nations has barely transformed its description of food security to "food and nutrition security", indicating an increase in deficiencies in micronutrients. Currently, 815 million people are malnourished, and up to 2 billion individuals are affected by deficiency in micronutrients [4]. In 2003, the World Health Organization (WHO) declared obesity an 'epidemic' in both poor and rich countries. Obesity and diet-related chronic illnesses are the leading causes of failed years of good health, and rather than affecting society, sick people are an enormous strain on the economy [5]. These environmental and healthcare consequences are exacerbated by political and economic imbalances as corporations consolidate control of worldwide markets and agri-food administration. SDG2's call to "end hunger, accomplish food security, strengthen nutrition, and encourage sustainable agriculture" crosses disciplinary boundaries. The overall framework of the Zero Hunger SDG indicates that dealing with global sustainability and nutritional security problems at the same time will necessitate transforming changes in politics and the economy.

SDG2's particular goals include issues related to the environment (biodiversity, productivity in agriculture, resilient and sustainable systems of production, and climate change adaptation), health issues (which include malnutrition and starvation, focuses on stunting and wasted development, nutritional requirements for women and girls), and economic variables (farmer earnings, opportunities and markets for adding value, agricultural extension and research, trade distortions). SDG2, which addresses the worldwide food system, represents an intrinsically multifaceted goal. A recent paper applied the Zero Hunger goal to demonstrate how individual SDGs are parts of an overall framework that must be regarded as a whole [6]. Food security relates to the processes of transformation required to achieve sustainable food systems. Digital technology constitutes one of the most significant revolutionary processes in global agriculture, food production, and retail systems [7].

2. The Role of Mobile (Information and Communication Technologies [ICTs]) Technology in Addressing Food Security

There are numerous reasons for optimism that ICTs could have an important effect on the outcome of agricultural markets. ICTs can help various industry agents convey more effectively, thereby improving information flow. This is particularly significant for rural areas in countries that are developing, in which markets are less integrated because of a lack of infrastructure. This addresses some of the most important prospective gains from utilizing data in agricultural markets. In particular, information can boost market efficiency. Prices in basic terms, indicate beneficial possibilities for manufacturers, consumers, and investors: opportunities in which excessive demand develops more profitable selling opportunities or when excessive supply leads to lower buying prices. In the absence of information and consequently limited arbitrage, prices tend to fluctuate based on present local supply. Nevertheless, as communication methods become better and more opportunities for arbitrage develop, effectively reducing the impact of local variations by more closely coordinating market prices with (less unpredictable) aggregate supply, we looked into the adoption of mobile phones by fishers in Kerala, India. Fish prices in coastal markets were volatile before the implementation of cellular coverage. The author discovers that whenever fishers utilized cell phones to find out about different prices in markets, regional prices converged (i.e., following the law of one price), and inefficiency was eliminated. Information enabled an improved distribution of fishermen's catch between markets, resulting in higher amounts of surpluses for consumers as well as producers. Similarly, [8] investigates the development of cell phones between grain traders in Niger. She discovers that phones enable traders to conduct searches for price information over greater distances and sell grains in more markets. The enhanced ability of five discovers comparable findings for fishermen in Kerala. Traders' ability to arbitrage among markets resulted in a 10–16% reduction in grain price variation. Aker's (2010) [8] research was conducted during a time of food shortages and rising prices for grains in some Nigerian communities. With this mentality, more effective grain distribution across markets is expected to improve food availability in the most affected regions. According to [9], the widespread use of cell phones among Niger farmers reduced the dispersal of cowpea farm gate sales prices. Furthermore, the authors discover that the availability of cellular phones resulted in a decline in the inter-annual value of variation for sales prices. It further indicates that having access to more accurate information is additionally correlated with reduced profit-risk among farmers. Although market prices indicate greater earning opportunities, having access to cellular phones may additionally assist farmers in determining how much to plant every growing season and what types of investments are likely to be successful according to demand and supply dynamics [10] evaluated an initiative in India that provides soybean farmers with current market price information via internet kiosks known as e-couples. Farmers' land dedicated to soybean cultivation increased by 19% as a result of the intervention, which allowed them to obtain higher prices. Figure 1 illustrates the use of ICT technology in food security. ICT plays a crucial role in enhancing food security by improving agricultural productivity, market access, and supply chain efficiency. Here are key areas where ICT contributes to food security:

Agricultural Productivity

- **Precision Farming:** ICT tools like GPS, sensors, and drones enable precision farming, which optimizes field-level management regarding crop farming.
- **Data Analytics:** Farmers can use data analytics to understand crop patterns, soil health, weather forecasts, and pest outbreaks, leading to informed decision-making.
- **Mobile Apps:** Agricultural apps provide farmers with real-time information on best farming practices, crop management, and weather conditions.

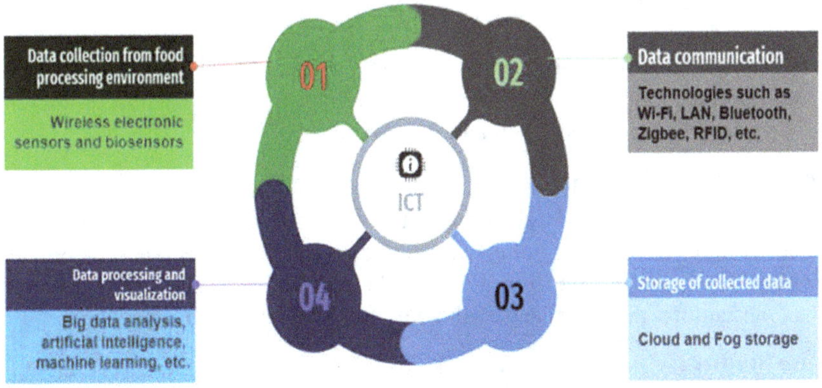

Figure 1: ICT technology in food security.

Market Access

- **E-Marketing Platforms:** Platforms such as eNAM (National Agriculture Market in India) connect farmers directly with buyers, eliminating intermediaries and ensuring better prices.
- **Mobile Payments:** Mobile banking and payment systems enable farmers to conduct transactions efficiently and securely, increasing financial inclusion.
- **Market Information Systems:** These systems provide farmers with current market prices, demand forecasts, and trading opportunities, helping them make better selling decisions.

Supply Chain Efficiency

- **Traceability Systems:** ICT enables the tracking of food products from farm to table, ensuring transparency, reducing food fraud, and improving food safety.
- **Inventory Management:** Technologies like RFID (radio-frequency identification) and IoT (internet of things) help in real-time tracking of inventory, reducing waste, and optimizing supply chains.
- **Logistics Optimization:** ICT solutions help in optimizing transportation routes, reducing transit times, and improving the overall efficiency of food distribution networks.

Knowledge Sharing and Capacity Building

- **Online Training and Resources:** Farmers can access online courses, webinars, and digital libraries to enhance their knowledge and skills.
- **Extension Services:** ICT facilitates the dissemination of agricultural research and innovations to farmers through SMS, apps, and social media platforms.
- **Community Platforms:** Online forums and social networks allow farmers to share experiences, advice, and collaborate on common issues.

Climate Resilience

- **Weather Forecasting:** ICT tools provide accurate and timely weather forecasts, enabling farmers to plan their activities and mitigate the impact of adverse weather conditions.
- **Disaster Management:** Early warning systems and disaster response platforms help in preparing for and responding to natural calamities, thereby safeguarding food production and storage.

Policy and Decision Making

- **Data Collection and Analysis:** ICT facilitates the collection and analysis of agricultural data, aiding policymakers in developing evidence-based policies.
- **Monitoring and Evaluation:** ICT tools help in monitoring the implementation of agricultural programs and evaluating their impact on food security.

Case Studies

- **Digital Green:** An initiative that uses videos to train farmers on best practices in agriculture, leading to significant improvements in crop yields and income.

- **M-Farm:** A mobile platform in Kenya that provides market prices, connects farmers with buyers, and offers weather updates, enhancing market access and productivity.
- **AgriON:** A platform that offers comprehensive farm management solutions, including crop monitoring, financial management, and market linkage, empowering farmers with ICT tools.

3. Advancements in Mobile Technology for Food Distribution

Innovations in mobile devices are transforming the distribution of food by increasing effectiveness, accountability, and accessibility throughout the supply chain. Mobile marketplaces allow farmers to establish relationships directly with buyers, offering actual-time price information while decreasing reliance on agents, resulting in more equitable pricing and fresher products. Mobile payment solutions enable secure, cashless transactions, thereby increasing financial access and lowering fraud risks. Management of supply chain apps enhances monitoring and managing stock, maintaining timely and open food deliveries [11]. Furthermore, climate and consultative apps provide farmers with critical information for more effective crop management, whereas delivery optimization apps improve transportation, minimizing waste as well as expenses. Food rescue and redistribution platforms help minimize food waste by connecting surplus food from businesses to consumers or charities. Blockchain integration in mobile platforms enhances traceability, allowing consumers to verify the origins and journey of their food.

3.1. M-Farm in Kenya

M-Farm is an application for mobile devices that offers Kenyan producers real-time market prices, farming recommendations, and an outlet for selling their crops. Farmers can obtain important crop cost information and communicate directly with consumers by sending an easy SMS. This availability has given farmers the ability to negotiate higher prices and reduce middleman exploitation. As a consequence, numerous small-scale farmers have witnessed their incomes and living conditions improve [12].

3.2. E-Choupal in India

ITC, a private company, launched e-Choupal, a programme that provides agricultural knowledge and resources to farmers in rural areas through internet-connected machines. While not entirely mobile, the project does use mobile devices to share predictions for the weather, standards of excellence, and prices on the market. The program has increased the productive capacity of the agriculture supply chain, minimized transaction expenses, and strengthened farmers' power of negotiation. E-Choupal has benefited more than 4 million farmers in 40,000 villages [13].

3.3. Hello Tractor in Nigeria

Hello Tractor, formerly known as the “Uber for Tractors”, uses an application for smartphones to connect tractor owners via small-scale farmers who need cost-effective tractor services. The invention discusses the problem of low automation in African agriculture, thereby increasing efficiency and productivity. Farmers can inquire about tractor amenities by SMS or the app, resulting in on-time and effective preparation for cultivation and planting. Many Nigerian farmers have seen an increase in agriculture profitability and productivity thanks to the platform [14].

3.4. TaniHub in Indonesia

TaniHub is a mobile-based platform that brings together Indonesian farmers directly with customers, merchants, and food manufacturers. By eliminating conventional channels of distribution, TaniHub minimizes the number of middlemen, which leads to lower costs for producers and more nutritious goods for consumers. The app additionally offers information on the funding and inputs for agriculture. Since its inception, TaniHub has substantially enhanced access to markets for thousands of farmers and increased the distribution of food productivity in Indonesia.

3.5. Too Good To Go in Europe

Too Good To Go is an application for mobile devices that reduces wasted food by connecting customers with eateries and retail establishments that have excess food. Users can buy leftovers at a discount, minimizing waste and making meals more affordable. The app is now widely used throughout Europe, preventing hundreds of millions of food from being thrown away and increasing awareness about wasted food problems.

4. IoT in Smart Agriculture

The IoT is a knowledgeable and declaring technology that provides novel and useful alternatives in a variety of fields, as illustrated in Figure 2, including smart cities, intelligent homes, traffic management, medical care, agricultural automation, and so on. In the field of agriculture, IoT technology has led to major advances in the management of crops. This technology connects all agriculture machinery and devices to make informed decisions about the fertilizer and irrigation supply. Smart systems improve the precision as well as the effectiveness of devices used to track the development of plants and raise livestock. Wireless sensor networks (WSNs) collect data from various gadgets. Furthermore, cloud services must be incorporated with IoT to analyze and deal with remote data, allowing for better decision-making and execution. Smart management of farms necessitates the use of ICT, momentum sensors, and

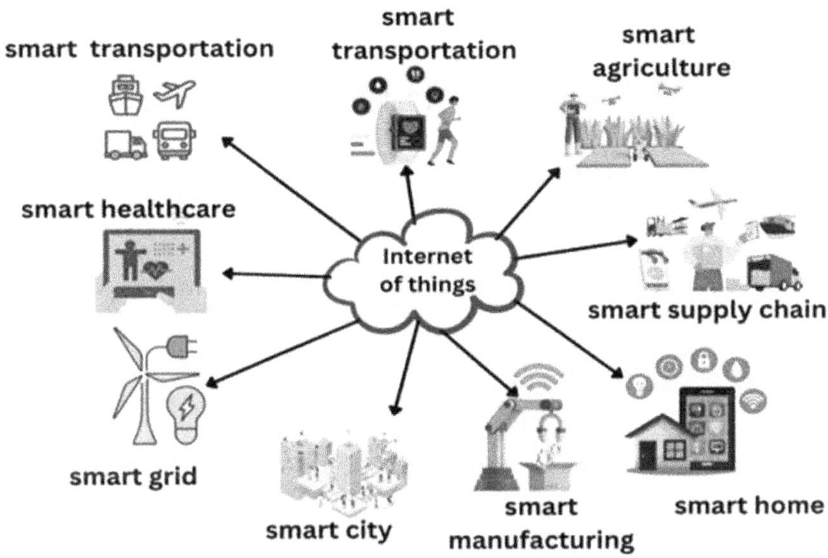

Figure 2: Application of IoT in different areas.

management systems set up on autonomous vehicles, robots, as well as other automated machinery.

The achievement of smart systems is dependent on fast internet connections, developed mobile devices, as well as satellites to provide images and positioning. James et al. (2019) [15] successfully utilized IoT immediate time monitoring and diagnosis of diseases of leaves that hinder crop development utilizing many satellite imagery and sensors placed on farms (rice and bananas crops); this technology assisted in analyzing the data and making the decision, which was subsequently transmitted back to the farmers, via the web server. According to the Food and Agricultural Organization's report, approximately 20–40% of agricultural products are lost each year because of diseases and pests, as well as a lack of proper crop monitoring. As a result, the application of sensors and automated systems enables the monitoring of weather conditions, and fertility status, and determining the precise amount of fertilizer required for crop growth. The excessive application of fertilizers hurts soil fertility. Sixty-seven research papers were published between 2006 and 2019 on the use of IoT in various agricultural inquiries; instead, they discovered that approximately 16% of the research papers were on precision agriculture, 16% on irrigation monitoring, 13% on soil tracking, 12% on temperature, livestock monitoring, and moisture monitoring were 11% each, air and disease tracking were 5% each, and monitoring of water was 7.7%. Lastly, fertilization monitoring constituted only 4% of the research papers. Figure 3 illustrates the application of IoT on Smart Agricultural practices.

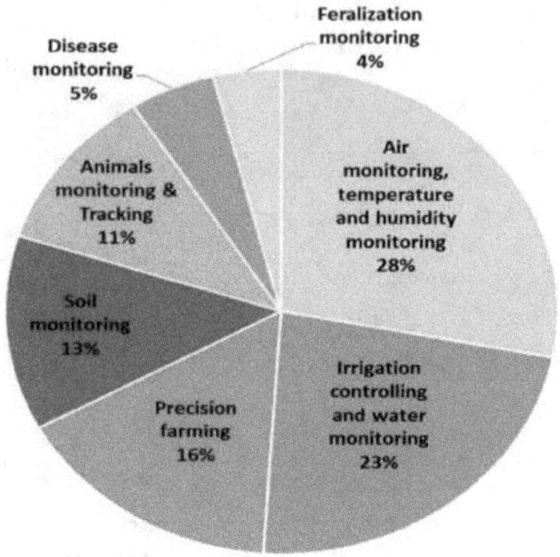

Figure 3: Application of IoT on smart agricultural practices.

5. Policy and Stakeholder Engagement

Government initiatives are important in promoting mobile solutions and encouraging collaboration among technological advances programmers, farmers, and suppliers to improve food availability and distribution [16]. Policies like these can help to foster creativity, ensure equal treatment, and tackle the agricultural industry's particular issues. Here is an in-depth look into how government policies can support these efforts.

5.1. Regulatory Frameworks

5.1.1. Data Privacy and Security

Governments can enact regulations to protect the safety and confidentiality of data collected using mobile solutions. This promotes trust between producers and suppliers, encouraging them to take up novel technologies. Policies can be developed that guarantee that mobile technologies encourage ethical trade practices, thus avoiding farmer exploitation by ensuring fair prices for what they produce [17].

5.1.2. Grants and Subsidies

Financial assistance to tech programmers and stakeholders in agriculture can help reduce the costs associated with implementing novel mobile technologies [18]. Providing tax reductions or incentives for agricultural equipment investments can encourage the development and implementation of creative approaches.

5.2. Infrastructure Development

Providing widespread and dependable access to the Internet in isolated regions is critical for maximizing the potential of mobile technology in agriculture [19]. Mobile phones and tablets alongside supporting the infrastructure require secure and inexpensive power.

5.3. Training and Capacity Building

Governments may promote educational initiatives that improve farmers' and suppliers' technical skills, allowing them to utilize mobile technology more effectively [18]. Raising awareness among stakeholders regarding mobile solutions' benefits and applications can increase the rate of adoption.

5.4. Research and Development Support

Offering funding for development and research can result in more efficient and adapted mobile solutions for agriculture. Working together between government agencies and business entities can result in creative approaches and resources that are shared [19].

5.5. Market Access and Information

Applying systems which offer real-time market data may assist farmers make sensible choices about where and when to market what they grow. Promoting the development of online platforms which connect producers directly to customers and distributors can help improve the chain of supply while minimizing ineffectiveness [20].

5.6. Regulatory Support for Startups

Streamlining the procedure for startups to get into the farming technology market can boost creativity and competitiveness [21]. Creating innovation centers and incubation facilities can help startups gain the resources and guidance they need to create and commercialize their ideas [22].

6. The Challenge of Food Distribution

The global distribution of food experiences serious difficulties, resulting in frequent malnutrition and a shortage of food. These issues are caused by an array of economic, administrative, and political variables. Inequalities in resources and wealth among nations indicate that certain areas can purchase extra food while others suffer from scarcity [23]. Logistically, insufficient infrastructure in nations that are developing hinders the effective movement and storage of nutrition, causing considerable losses because of waste and spoilage. Conflicts over politics and trade practices may interfere with the supply of food, making it difficult to deliver food to underserved regions. Furthermore, climate change contributes to these issues by causing unpredictable weather patterns, which reduce crop yields. Combating these issues necessitates international collaboration to enhance facilities encourage ethical trade practices, and establish policies which make sure fair distribution of food [24], and inadequate food access contributes to starvation by preventing healthy food from accomplishing individuals who need it the most. This inefficiency can be attributed to poor infrastructure, such as insufficient roads and storage spaces, which cause significant food loss during transport and storage. In developing countries, this leads to a limited availability of affordable, fresh produce. Furthermore, economic barriers, such as prohibitive costs and disparate wealth distribution, restrict susceptible populations' access to food. Political conflict and instability additionally disrupt the supply chain, making it difficult to ensure a consistent supply of food. As a consequence, millions of people suffer from chronic malnutrition and starvation, which affects their condition, efficiency, and general standard of life. Addressing such inefficiencies is critical to reducing starvation and maintaining that food supplies are used effectively to feed global populations.

7. Mobile Technology as a Solution

Mobile technology can provide creative approaches for improving food availability and distribution, especially in areas afflicted by inefficiency and logistical difficulties. Farmers can improve the productivity of crops and minimize post-harvest losses by utilizing mobile platforms that provide immediate access to marketplace prices, predictions for the weather, and optimal farming techniques [25]. Mobile payment systems facilitate seamless banking transactions, allowing small-scale producers to market their agricultural products directly to customers or local markets, avoiding intermediaries who frequently inflate costs. Furthermore, mobile apps can help coordinate the distribution of food networks by maximizing delivery vehicle routes to minimize inefficiency and ensure effective access to nutritious foods in remote areas. These innovations also improve supply chain monitoring and transparency, allowing for the rapid identification and resolution of bottlenecks. Furthermore, mobile-based alerting systems may alert communities regarding the availability of food as well as distribution plans, guaranteeing that nutritious food reaches those who are in need quickly [26]. Overall, the use of mobile technology in food distribution and accessibility shows great promise for increasing productivity, decreasing hunger, and encouraging global food security. Figure 4 illustrates Mobile Technology in food distribution.

Mobile technology has significantly transformed food distribution by enhancing efficiency, transparency, and connectivity throughout the supply chain. Through mobile applications, farmers, distributors, retailers, and consumers can access real-time information, facilitating better decision-making and streamlined operations. For instance, mobile apps allow farmers to track the movement of their produce from farm to market, ensuring transparency and reducing losses due to

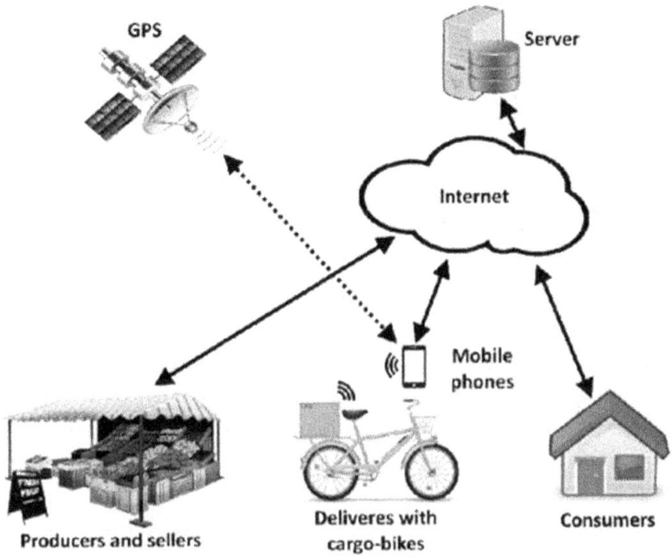

Figure 4: Mobile technology in food distribution.

spoilage or mismanagement. Distributors can use GPS-enabled mobile devices to optimize delivery routes, monitor vehicle locations, and manage inventory more effectively, resulting in faster and more reliable deliveries. Retailers benefit from mobile technology through improved inventory management systems that update stock levels in real-time, preventing overstocking or stockouts and ensuring that fresh produce is always available to consumers. Mobile platforms also enable direct communication between suppliers and retailers, facilitating quick adjustments to orders based on demand fluctuations. Consumers, on the other hand, can use mobile apps to track the origin of their food, check its quality, and make informed purchasing decisions, fostering trust and satisfaction.

Moreover, mobile payment systems have revolutionized transactions within the food distribution network, providing secure, fast, and cashless payment options that enhance convenience and reduce the risk of financial discrepancies. These systems are particularly beneficial in rural areas where traditional banking infrastructure may be lacking. Additionally, mobile technology supports the integration of blockchain in food distribution, ensuring that every transaction is recorded immutably, further enhancing traceability and accountability. In emergency food distribution scenarios, such as during natural disasters or pandemics, mobile technology plays a crucial role in coordinating relief efforts. Mobile platforms can quickly disseminate information about food availability and distribution points, ensuring that aid reaches those in need promptly. Overall, mobile technology has become an indispensable tool in modern food distribution, driving improvements in efficiency, transparency, and accessibility across the entire supply chain.

7.1. Mobile Markets

Mobile markets are smartphone grocery stores that deliver fresh produce along with additional groceries to communities where services are lacking. These markets are typically run by vehicles or trucks and may be planned to visit various locations on particular days [27].

7.2. Food Delivery Apps

Food delivery services like Door-Dash, Uber Eats, and others and their local alternatives can collaborate with supermarkets and dining establishments to provide food to regions that have restricted access. These apps can also collaborate with charities that distribute food and other non-profit organizations to provide food to individuals in need [28]. Mobile payment methods make it easier to make food purchases, particularly in areas with limited banking services. Services such as M-Pesa in Kenya enable mobile money transactions, which may be used in connection with distributing food initiatives. Farm-to-table applications connect customers with their local farmers, facilitating them to buy fresh produce right from where it comes from. These apps frequently offer delivery or planned transportation at the hubs.

7.3. Food Rescue Apps

Food rescue apps such as Too Excellent to Take and Olio assist in minimizing the waste of food by enabling users to purchase surplus food from eateries and grocery stores at discounted prices. These mobile applications also help with the distribution of food by connecting surplus food to charity organizations and shelters [29]. Mobile inventory control systems enable food pantries and distribution centers to remain in track of their supply in real time. This makes certain effective distribution and reduces waste by determining which items must be dispersed quickly. These mobile applications offer consumers methods and plan meals according to what food products they have available. They can also provide healthy details and recommendations for healthy eating based on readily accessible ingredients [30].

7.4. Crowdsourcing Platforms

Crowdsourcing platforms may assist in recognizing food insecurity hotspots and rapidly mobilize resources. Consumers can identify locations with limited food access, and groups may react by controlling supplies where they are most needed. Blockchain technology might be used to generate accurate and impermeable documentation of food manufacturing and distribution. This improves food safety and ensures that food attains its intended location. Geographic Information Systems (GIS) and satellite mapping can aid in the identification of food-insecure areas and the monitoring of agricultural production. This information can be used to improve routes for distribution and ensure food is available to those who need it [31].

7.5. Challenges in Food Access and Distribution

7.5.1. Infrastructure Deficiencies

Poor transportation infrastructure in rural areas can impede the effective delivery of goods, resulting in putting off and higher prices. Lack of proper warehouse and refrigeration facilities causes significant post-harvest losses, particularly for perishable products [32]. The farmers in rural areas have restricted availability of internet access and mobile devices, making it difficult to gain advantages from mobile technology. Many farmers and suppliers may lack the skills required for successfully utilizing modern technology and smartphone apps. Farmers on a small scale frequently struggle to get credit, which is required for purchasing novel equipment and expanding production. Without adequate financial support, small-scale farmers may find it prohibitively expensive for them to utilize new technologies [33].

7.5.2. Regulatory and Policy Challenges

Incompatible or inadequately executed agricultural policies can cause uncertainty and impede future planning and expenditures [34]. Taxes, limitations on exports, and various other barriers to trade may interfere with the transportation of agricultural products, affecting domestic as well as global markets. The market fluctuations in prices may render it challenging for farmers to properly organize and handle their finances [35]. The existence of several middlemen in the food supply chain may decrease farmer earnings while increasing consumer costs.

7.5.3. Environmental and Climate Challenges

Uncertain weather patterns as well as severe storms can harm crops, interfere with supply chains, and threaten the availability of food. Excessive use of renewable resources, like water and soil, can pose long-term sustainability challenges [36].

8. Future Directions for Improving Food Access and Distribution

8.1. Enhancing Infrastructure

The government and entrepreneurs ought to make investments in rural transportation systems to ensure goods are delivered efficiently and on time. Establishing improved storage facilities, such as cold rooms and warehouse infrastructure, will minimize post-harvest losses. Implementing modern storage solutions, such as cold rooms and warehouse infrastructure, may substantially decrease post-harvest losses while also improving food quality.

8.2. Leveraging Technology

Increasing internet and cell phone service in rural areas to help farmers connect the digital gap and gain access to mobile technology and market data [18]. Creating easier-to-use and simple-to-use mobile programs that address the distinctive requirements and educational attainment of farmers and suppliers [37].

8.3. Financial Innovations

Presenting microfinance programs and agriculture insurance options to assist small-scale agricultural producers in managing financial risks and investing in emerging innovations. Subsidies and monetary incentives for the adoption of environmentally friendly agriculture methods and equipment [38, 39].

8.4. Policy and Regulatory Reforms

Creating stable and consistent policies regarding agriculture that promote investment while offering long-term support to farmers and technology developers [38]. Minimizing restrictions on trade and facilitating export-import processes would enhance farmers' market access and the distribution of agricultural products [40].

8.5. Market Efficiency Improvements

Encouragement of online marketplaces that interact farmers directly with customers and merchants, thereby minimizing the function of middlemen and maintaining fairer prices. Applying complete market information platforms which offer current information on prices, consumer demand, and supply to assist farmers in making informed choices [41, 42].

8.5.1. Climate-smart Agriculture

Encouraging the use of environmentally friendly agricultural practices like conservation agriculture, rotation of crops, and organic agriculture. Climate-resilient plant varieties and farming techniques are being developed and disseminated to help farmers change to altering weather patterns Shrestha [43, 44].

8.5.2. Capacity Building and Education

Setting up educational events and seminars to help farmers and distributors improve their technical and commercial abilities, allowing them to successfully employ novel methods and technologies. The development extension services in agriculture aim to offer farmers ongoing assistance, advice, as well as access to the latest studies and developments [45–47]. Promoting collaboration among government departments, private businesses, academic institutions, and nonprofit organizations in order will encourage creativity and magnitude of effective interventions [48]. Developing agricultural creativity hubs and incubation facilities to help startups and develop novel approaches for food distribution and accessibility [49–53].

9. Conclusion

ICT (information and communication technology) has come to serve as a driving force in transforming food access and distribution, providing unprecedented possibilities for tackling global sustainability and food security issues. As stakeholders committed to community and environmental well-being, we must acknowledge and accept ICT's potential for transformation in the agricultural sector. We must commit to creating a climate beneficial to the widespread use and application of ICT tools as well as platforms, maintaining they can reach and gain all actors together in the food distribution chain, particularly small-scale farmers and communities that are marginalized. This necessitates substantial investments in ICT infrastructure, specific strategies that encourage digital access and literacy, and the formation of collaborative organizations that use technology to achieve proper and adaptable food systems. By being involved collectively to maximize the potential of ICT, we can create a future in which everybody has a choice of nutritious, secure food and sustainable natural resource management is prioritized. Let us take advantage of this opportunity to effect beneficial change and open the way for future generations to enjoy greater prosperity and food security.

References

1. Leroy, J.L., Ruel, M., Frongillo, E.A., Harris, J. and Ballard, T.J. (2015). Measuring the food access dimension of food security: A critical review and mapping of indicators. *Food and Nutrition Bulletin*, 36(2): 167–195.
2. Li, C., Mirosa, M. and Bremer, P. (2020). Review of online food delivery platforms and their impacts on sustainability. *Sustainability*, 12(14): 5528.
3. IPCC. (2013). Climate Change 2013. *The Physical Science Basis. Summary for Policymakers*. United Nations, http://www.ipcc.ch/.
4. Initiative, M. and UNICEF. (2009). World Bank, USAID, Flour Fortification Initiative, Global Alliance for Improved Nutrition: Investing in the Future. A United Call to Action on Vitamin and Mineral deficiencies. *Global Report 2009*.
5. Murray, C.J., Vos, T., Lozano, R., Naghavi, M., Flaxman, A.D., Michaud, C., ... Haring, D. (2012). Disability-adjusted life years (DALYs) for 291 diseases and injuries in 21 regions, 1990–2010: A systematic analysis for the Global Burden of Disease Study 2010. *The Lancet*, 380(9859): 2197–2223.
6. Nilsson, M., Griggs, D. and Visbeck, M. (2016). Policy: Map the interactions between Sustainable Development Goals. *Nature*, 534(7607): 320–322.

7. Serbulova, N., Kanuri, S., Gorodyansky, A. and Persiyanova, A. (2019, Dec.). Sustainable food systems and agriculture: The role of information and communication technologies. *In*: *IOP Conference Series: Earth and Environmental Science*, 403(1): 012127. IOP Publishing.
8. Aker, J.C. (2010). Information from markets near and far: Mobile phones and agricultural markets in Niger. *American Economic Journal: Applied Economics*, 2(3): 46–59.
9. Aker, J.C. and Fafchamps, M. (2010). *How Does Mobile Phone Coverage Affect Farm-gate Prices? Evidence from West Africa*, 1–50. Department of Economics and the Fletcher School, Tufts University.
10. Goyal, A. (2010). Information, direct access to farmers, and rural market performance in central India. *American Economic Journal: Applied Economics*, 2(3): 22–45.
11. Uña, G., Verma, A., Bazarbash, M. and Griffin, M.N.N. (2023). *Fintech Payments in Public Financial Management: Benefits and Risks*. International Monetary Fund.
12. Bradford, F. (2022). *Innovating Kenya's Trading System Through Mobile Technology: A Case Study of M-Farm in Nairobi, Kenya*. Cornell University.
13. Mukerji, M. (2020). Re-examining strategic and developmental implications of e-Choupal, India. *The Electronic Journal of Information Systems in Developing Countries*, 86(4): e12132.
14. Daum, T., Villalba, R., Anidi, O., Mayienga, S.M., Gupta, S. and Birner, R. (2021). Uber for tractors? Opportunities and challenges of digital tools for tractor hire in India and Nigeria. *World Development*, 144: 105480.
15. James, A., Saji, A., Nair, A. and Joseph, D. (2019). Cropsense: A smart agricultural system using IoT. *Journal of Electronic Design Engineering*, 5(3).
16. Das Nair, R. and Landani, N. (2020). *Making Agricultural Value Chains More Inclusive through Technology and Innovation* (No. 2020/38). UNU WIDER working paper.
17. Emeana, E.M., Trenchard, L. and Dehnen-Schmutz, K. (2020). The revolution of mobile phone-enabled services for agricultural development (m-Agri services) in Africa: The challenges for sustainability. *Sustainability*, 12(2): 485.
18. Zhang, Y., Love, D.J., Krogmeier, J.V., Anderson, C.R., Heath, R.W. and Buckmaster, D.R. (2021). Challenges and opportunities of future rural wireless communications. *IEEE Communications Magazine*, 59(12): 16–22.
19. Mapiye, O., Makombe, G., Molotsi, A., Dzama, K. and Mapiye, C. (2023). Information and communication technologies (ICTs): The potential for enhancing the dissemination of agricultural information and services to smallholder farmers in sub-Saharan Africa. *Information Development*, 39(3): 638–658.
20. Leckel, A., Veilleux, S. and Dana, L.P. (2020). Local open innovation: A means for public policy to increase collaboration for innovation in SMEs. *Technological Forecasting and Social Change*, 153: 119891.
21. Nedumaran, S., Selvaraj, A., Nandi, R., Suchiradipta, B., Jyosthnaa, P. and Bose, D. (2020). Digital integration to enhance market efficiency and inclusion of smallholder farmers: A proposed model for fresh fruit and vegetable supply chain. *International Food and Agribusiness Management Review*, 23(3): 319–337.
22. Gorowara, N., Yadav, S. and Kumar, V. (2024). Sustainable future: Government initiatives in the adoption of emerging sustainable technologies by Startups in India. *In*: *Fostering Innovation in Venture Capital and Startup Ecosystems*, 286–305. IGI Global.

23. Mwantimwa, K., Ndege, N., Atela, J. and Hall, A. (2021). Scaling innovation Hubs: Impact on knowledge, innovation and entrepreneurial ecosystems in Tanzania. *Journal of Innovation Management*, 9(2): 39–63.
24. Lipton, M. (2023). Why poor people stay poor: A study of urban bias in world development. *In*: *Rural Development*, 66–81. Routledge.
25. Raza, A., Razzaq, A., Mehmood, S.S., Zou, X., Zhang, X., Lv, Y. and Xu, J. (2019). Impact of climate change on crops adaptation and strategies to tackle its outcome: A review. *Plants*, 8(2): 34.
26. Kamal, M. and Bablu, T.A. (2023). Mobile applications empowering smallholder farmers: An analysis of the impact on agricultural development. *International Journal of Social Analytics*, 8(6): 36–52.
27. Lin, K., Chavalarias, D., Panahi, M., Yeh, T., Takimoto, K. and Mizoguchi, M. (2020). Mobile-based traceability system for sustainable food supply networks. *Nature Food*, 1(11): 673–679.
28. Weissman, E., Robinson, J. and Cecio, W. (2020). The promise and pitfalls of mobile markets: An exploratory survey of mobile food retailers in the United States and Canada. *Agriculture and Human Values*, 37: 895–906.
29. Ahuja, K., Chandra, V., Lord, V. and Peens, C. (2021). Ordering in: The rapid evolution of food delivery. *McKinsey & Company*, 22: 1–13.
30. Pielow, Z. (2022). *The Contribution of Platform-based Food Sharing Start-ups to Food Waste Reduction in Developed Countries* (Doctoral dissertation). Universidade Católica Portuguesa.
31. Mauch, C.E., Wycherley, T.P., Laws, R.A., Johnson, B.J., Bell, L.K. and Golley, R.K. (2018). Mobile apps to support healthy family food provision: Systematic assessment of popular, commercially available apps. *JMIR mHealth and uHealth*, 6(12): e11867.
32. Quamar, M.M., Al-Ramadan, B., Khan, K., Shafiullah, M. and El Ferik, S. (2023). Advancements and applications of drone-integrated geographic information system technology. A review. *Remote Sensing*, 15(20): 5039.
33. Rajapaksha, L., Gunathilake, D.M.C.C., Pathirana, S.M. and Fernando, T. (2021). Reducing post-harvest losses in fruits and vegetables for ensuring food security: Case of Sri Lanka. *MOJ Food Process Technols.*, 9(1): 7–16.
34. Kuhl, L. (2020). Technology transfer and adoption for smallholder climate change adaptation: Opportunities and challenges. *Climate and Development*, 12(4): 353–368.
35. Matthan, T. (2023). Beyond bad weather: Climates of uncertainty in rural India. *In*: *Climate Change and Critical Agrarian Studies*, 164–185. Routledge.
36. Zakirova, A., Klychova, G., Ostaev, G., Zugarova, E., Nigmetzyanov, A. and Zaharova, E. (2020). Organizational and methodological approach to managing financial flows of agricultural enterprises. *In*: *E3S Web of Conferences*, 164, 10009. EDP Sciences.
37. Jhariya, M.K., Banerjee, A. and Meena, R.S. (2022). Importance of natural resources conservation: Moving toward the sustainable world. *In*: *Natural Resources Conservation and Advances for Sustainability*, 3–27. Elsevier.
38. Lilian, I.U. (2024). Assessing the effectiveness of mobile apps in enhancing agricultural extension services delivery in Nigeria. *African Journal of Sustainable Agricultural Development*, 5(2): 13–31.
39. Bopp, C., Engler, A., Poortvliet, P.M. and Jara-Rojas, R. (2019). The role of farmers' intrinsic motivation in the effectiveness of policy incentives to promote sustainable agricultural practices. *Journal of Environmental Management*, 244: 320–327.

40. Benami, E. and Carter, M.R. (2021). Can digital technologies reshape rural microfinance? Implications for savings, credit, and insurance. *Applied Economic Perspectives and Policy*, 43(4): 1196–1220.
41. Shrestha, J., Subedi, S., Timsina, K.P., Chaudhary, A., Kandel, M. and Tripathi, S. (2020). Conservation agriculture as an approach towards sustainable crop production: A review. *Farming and Management*, 5(1): 7–15.
42. Glaros, A., Thomas, D., Nost, E., Nelson, E. and Schumilas, T. (2023). Digital technologies in local agri-food systems: Opportunities for a more interoperable digital farmgate sector. *Frontiers in Sustainability*, 4: 1073873.
43. Buitenhuis, Y., Candel, J., Feindt, P.H., Termeer, K., Mathijs, E., Bardají, I., ... Sorrentino, A. (2020). Improving the resilience-enabling capacity of the common agricultural policy: Policy recommendations for more resilient EU farming systems. *EuroChoices*, 19(2): 63–71.
44. Nyasimi, M., Kimeli, P., Sayula, G., Radeny, M., Kinyangi, J. and Mungai, C. (2017). Adoption and dissemination pathways for climate-smart agriculture technologies and practices for climate-resilient livelihoods in Lushoto, Northeast Tanzania. *Climate*, 5(3): 63.
45. Singh, A.K. and Burman, R.R. (2019). Agricultural extension reforms and institutional innovations for inclusive outreach in India. *In*: *Agricultural Extension Reforms in South Asia*, 289–315. Academic Press.
46. Gc, R.K. and Hall, R.P. (2020). The commercialization of smallholder farming: A case study from the rural western middle hills of Nepal. *Agriculture*, 10(5): 143.
47. Lewis, D., Kanji, N. and Themudo, N.S. (2020). *Non-governmental Organizations and Development*. Routledge.
48. Nuthalapati, C.S., Srinivas, K., Pandey, N. and Sharma, R. (2020). Startups with open innovation: Accelerating technological change and food value chain flows in India. *Indian Journal of Agricultural Economics*, 75(4): 415–437.
49. Wolfert, S., Ge, L., Verdouw, C. and Bogaardt, M.J. (2017). Big data in smart farming: A review. *Agricultural Systems*, 153: 69–80.
50. Duraiarasu E., Mahalakshmi S., Jose A.A., Manikandan T. and El Rhalibi, A. (2023). Precision farming for crop prediction. *In*: K. Kalaiselvi, Jose Anand A., Poonam Tanwar and Haider Raza (Eds.), *Advanced Technologies for Smart Agriculture*, 195–217. Denmark: River Publishers.
51. Nisha K., Augustince C., Jose A.A. and Kamesh M. (2023). Deep learning in smart agriculture. *In*: K. Kalaiselvi, Jose Anand A., Poonam Tanwar and Haider Raza (Eds.), *Advanced Technologies for Smart Agriculture*, 153–175. Denmark: River Publishers.
52. Jose A., Nirmala Devi, K., Vijayakumar, S. and Shiju C. Chacko. (2023). Decision-making support in smart farming. *In*: K. Kalaiselvi, Jose Anand A., Poonam Tanwar and Haider Raza (Eds.), *Advanced Technologies for Smart Agriculture*, 219–239. Denmark: River Publishers.
53. Augustine, C., Balaji, K., Dharanikumar, S.V. and Jose, A.A. (2023). Urban farming: Case study. *In*: K. Kalaiselvi, Jose Anand A., Poonam Tanwar and Haider Raza (Eds.), *Advanced Technologies for Smart Agriculture*, 321–338. Denmark: River Publishers.

8 | Climate Smart Agriculture (CSA) and Food Security in Rural Communities of Developing Countries: Qualitative Assessment

Edidiong Samuel Akpabio[1*], Kemi Funlayo Akeju[2] and Efehi Raymond Okoro[3]

[1] Department of Political Science, Trinity University, Lagos, Nigeria
[2] Department of Economics, Ekiti State University, Ado-Ekiti, Nigeria
[3] Department of Political Science, University of Benin, Nigeria

Climate change has posed serious challenges to food security globally. This sordid state of affairs has had more impact in the global south especially in the rural areas due to their inability to introduce and apply digital technologies to the agricultural sector. With most research focusing more on the urban settlements there is the need to interrogate the prospects and problems of climate smart agriculture in rural communities of developing societies with a view to appraising its impacts on food security. This explains the rationale for embarking on this study as it addresses the paucity of literature on the subject matter. The work observes that granted that climate smart agriculture was introduced to combat the untoward effects of climate change on the agricultural sector, there are a lot of hurdles to climb especially in the rural areas. Among the findings of this work is the unavailability of requisite support to this category of agriculturalist which is capable of preventing them from plugging into the benefits embedded in practicing climate smart agriculture. It is on the strength of this that we recommend that all stakeholders must work unitedly with those in the rural areas to ensure that all the challenges confronting climate smart agriculture are surmounted. It is only until this is done that the food security prospects of such societies can be enhanced and guaranteed.

*Corresponding author: esakpabio@yahoo.co.uk

1. Introduction

As the years draw near for the global goals, the prospects for achieving most of the goals continue to look bleak. This sordid state of affairs seems worse off in the developing parts of the globe especially in the global south. It remains a major challenge to the actualization of food security as it is fueled by a plethora of causal factors. Of particular interest is the dampened prospect of Sustainable Development Goal 2 (SDG2), i.e., eliminating hunger and malnutrition in all its manifestations, as it represents one of the biggest risks to humankind in this climate. It is equally glaring that it can herald more colossal consequences for a region already burdened by varied challenges. Also taking into cognizance the inability of the global south region to meet the MDGs (Millennium Developments Goals), the need to make concerted efforts to achieve the SDGs for them have become more expedient than ever. Transitioning to climate smart agriculture through the utilization of cutting-edge technologies [1], apart from being touted as the panacea to the food quagmire confronting the region, is also a viable strategy that may help lessen the untoward effects of climate change on the industry. It can also help expedite the attainment of the global goals.

It remains a truism that a multifaceted approach that takes into cognizance all the components that foster the issue of food security is needed to address hunger. With concentrated efforts and initiatives in the agricultural sector, food system sustainability, and vulnerable communities, we can move towards a world without hunger [2, 3]. Expanding access to food products, increasing output within the agricultural sector, strengthening existing social safety nets, encouraging and supporting the philosophy of sustainable agriculture, significant reduction of food waste, and along with deliberate efforts to nip the issue(s) that fuels poverty in our societies are a handful of ways these problems could be resolved efficiently. Internet of things (IoT)-based smart farming is much more efficient than traditional farming because urban smart farms use aeroponics to grow fresh, healthful food in a cleaner, easier, and more efficient way. Our state-of-the-art technology is the perfect choice for home or urban farming because it requires up to 90% less land and 98% less water [1].

It is not news that the agriculture industry globally suffers a plethora of encumbrances as a result of the untoward implications of climatic dysfunctionalities. The same situation is obtainable in other developing climes who have been contending with extraneous and intraneous challenges to their agricultural sector which has been exacerbated by the untoward effects of climatic changes. This explains the place of climate smart agriculture (CSA) as an adaptive measure and an instrument of resilience especially with respect to the attainment of SDG2. CSA is a method of growing food that tackles the problems of climate change (henceforth, CC) while also enhancing food security and farmers' livelihoods [4]. It does this by addressing hunger and CC at the same time. With the use of CSA, farmers may plant more drought-resistant crops and adjust to shifting rainfall patterns, among other practices and technology that can help them adapt to and lessen the effects of CC [5]. In addition to lowering greenhouse gas emissions and dependency on fossil fuels, farmers can increase soil health and increase their farms' resilience to CC by implementing renewable energy sources like solar power. By making investments in CSA, despite a changing climate, agriculture can continue to feed the globe's teeming populace.

That CSA can bolster the agricultural industry in any climate is not in doubt; the problem in rural communities of developing societies lies in its implementation. This challenge of execution which has persisted in this claim is premised on a plethora of conditionality such as financing, infrastructural availability, technical know-how among several other negatives. While agriculture ought to thrive more in rural settlements as most farming activities take place there, the vagaries the industry is exposed to have nullified this expectation. The reason for this inquiry is triggered by the degree of emphasis on urban centers with respect to the global goals and the neglect of the rural communities in developing climes which could be counter-productive due to the degree of volatility they experience and the vulnerabilities that could emanate from this state of affairs. More compelling was due to the centripetal role SDG2 plays and its capacity to facilitate the actualization of other goals.

The core objective of this paper, therefore, is to appraise the prospects of attaining SDG2 through the utilization of CSA with particular emphasis on the rural settlements in the global south due to the peculiarities of the region. We sought to achieve this objective by utilizing a qualitative methodology and engaging secondary literature sources such as peer-reviewed journal articles, published books (including e-copies), policy briefs, conference proceedings, and monographs among other outlets. We gleaned from extant studies that CSA encountered issues of acceptability and multifarious encumbrances in the rural areas when compared to the urban city centers. Among the issues it contends with are the problems of funding to carry out the practice, paucity of technical know-how, dearth of infrastructure among other issues. In a bid to remedy this situation, we advocated for massive investments in CSA, provision of soft loans and grants to farmers in such areas, training and re-training of agricultural personnel, aggressive and sustained sensitization of farmers among other actions which must be expeditiously taken. We posit that the prospects of fast-tracking SDG2 through the integration and adoption of CSA are manifold and attainable. What is required is the requisite political will to ensure that the necessary steps are taken to consolidate CSA and help developing societies attain the goal in no distant time.

2. Conceptual Frameworks on Climate Smart Agriculture

Climate change (CC) is one of the greatest issues bedeviling the 21st century international system [6], which has remained a source of concern to researchers [7]. Hitz and Smith described it as a kind of change laden with adverse impacts

[8], which can jeopardize progress towards achieving the SDGs [9]. The unique thing about this challenge is that CC affects developed and developing countries differently [10, 11]. Akpodiogaga-a and Odjugo opine that poor countries of the world are more susceptible to the adverse effects of CC [12], while according to Sarkodie and Strezov, lack of requisite capacity is a major explanation for the challenges of CC in developing countries when compared to their developed counterparts [13]. Across the world, the agricultural industry has been described as one of the major casualties of CC [14], with studies showing that CC could further worsen the challenge of food insecurity confronting the globe [15]. This is so because, as the climate changes, agricultural systems ought to change in response. Unfortunately, the reverse seems the case as CC adaptation faces a great deal of threat in developing countries [16]. Irrespective of the prevalent state of affairs, we must sound a caveat that the emergent challenges of CC which pose substantial threats to the agricultural sector has made CSA very expedient [17].

CSA is becoming increasingly popular both nationally and internationally as a means of tackling the problems traceable to agriculture as it pertains to CC towards improving gains in agricultural productivity and incomes [18, 19]. This practice can be described as modern agriculture as it discards the use of rudimentary farming activities and employs the usage of technology in carrying out farming activities [1]. In order to bring to the fore the connections between attaining food security and halting CC through agricultural development, the Food and Agriculture Organization (FAO) of the United Nations introduced the concept of Climate Smart Agriculture (CSA) in 2009 [20, 21]. CSA is defined by the Food and Agriculture Organization of the United Nations as "agriculture that improves national food security and development goals while reducing greenhouse gas emissions, improving resilience, and increasing productivity in a sustainable manner" [20]. CSA requires the engagement of pertinent strategies, tools, and procedures known as CSA technology and management practice operational at the micro- and the macro-level provision of enabling and supporting institutions, policies, and finance mechanisms [22].

To effectively promote development and provide food security in an unstable climate, the core tenet of the CSA concept is that developing nations should prioritize reducing poverty and ensuring food security by strengthening their agriculture and food industries transformation and reorientation in agricultural practices using appropriate technology [23]. Basically, CSA is a strategy to support agricultural system managers in effectively adapting to CC. CSA encompasses investments, institutions, regulations, technology, and actions that take place on and off the farm [18]. The operation of agricultural land, crops, farm animals, and aquaculture to balance critical food security and livelihood expectations with adaptation and mitigation priorities is one of the components of climate-smart agricultural systems [24].

A bottom-up prioritization process and consultative framework involving professionals, farmers, and key stakeholders are necessary to identify pertinent management practices as well as CSA technologies that have the most potential for positive effects on the economy, society, and environment. Three key initiatives are highlighted by the notion of CSA: (i) raising farmers' incomes and agricultural output in a sustainable manner; (ii) preparing for and enhancing resistance to CC; and (iii) cutting or eliminating greenhouse gas emissions in accordance with national development priorities [23, 25].

CSA is a method that incorporates several components that are anchored in local settings rather than a collection of activities that can be used everywhere. It encompasses investments, institutions, regulations, technology, and actions that take place on and off the farm [26]. The operation of agricultural land, crops, farm animals, and aquaculture to balance immediate core livelihood demands and food security, and with mitigation and adaptation priorities is one of the components of climate-smart agricultural systems. In order to preserve ecosystem services that are crucial for a broad food security architecture, mitigation, adaptation, and agricultural development in general, it entails managing the ecosystem and the landscape [27].

CSA has aided in raising awareness of the mutually reinforcing relationship between agriculture and CC, advancing the creation of increasingly advanced instruments for evaluating climate-smart technologies, and uniting a coalition of institutions and organizations committed to advancing CSA in a range of global contexts. To assist farmers increase productivity, develop resilience, and pursue lower greenhouse gas emissions, the World Bank Group, the Food and Agriculture Organization (FAO) of the United Nations, and the Consortium of International Agricultural Research Centers (CGIAR) are particularly dedicated to scaling up CSA [28, 29].

For millions of farmers to alter their practices, methods, and conduct in order to support CSA approaches and build capacity for their broader adoption requires acceptance of change [30]. The theory of change identifies the rationale for the desired change in a particular context. It showcases how result and progress at the early and intermediate stage of a course set motion for long-run results. It articulates the processes, strategies, and mechanisms through which change will occur [31]. Setting the theory of change in motion for the operation of CSA is necessary to help agriculturalists and producers to appreciate the impacts of CC and the choices at their disposal for switching to climate-smart practices [23]. Extension services serve an important role in helping farmers find new information and resources, as well as in promoting and supporting the behavioral adjustments required to increase the resilience of agricultural livelihoods and support sustainable development [32].

It is noteworthy that SDG2 specifically lists extension as one of the sectors in need of more funding to achieve the objectives of eradicating hunger, enhancing nutrition, and advancing agriculture. A significant proportion of poor people across the globe are mainly rural dwellers and their primary source of income is basically agriculture. Therefore, developing the prospects of increasing the productivity and revenues from small-scale farm owners, fish, livestock, as well as properly managed forest production systems will be the bedrock of state and non-state actors that genuinely seek to achieve global food sufficiency over the next 20 years [33]. The negative effects of CC are anticipated to be more severe in developing countries and such effects may manifest in the following forms: changes in precipitation patterns, higher temperatures, increased frequency of extreme weather events, and rising sea levels among others. It is indeed important to note that all of the above mentioned effects pose significant risks for farmers and their agricultural products including food and water supplies. With the above context in mind, resilience (that is, the ability to easily adjust or recover from the devastating effects of CC in rural communities) is therefore a predominant concern. In addition, agricultural practices are part of the key source of greenhouse gas emissions in communities globally. Mitigation can often be a significant co-benefit of actions to strengthen adaptation and enhance food security, and thus mitigation action compatible with national development priorities for agriculture is an important aspect of CSA [34].

3. Climate Change and SDG2 (Food Security): Assessing Intersections

The concept of food security has no singular definition [35]. This is because the concept and its definition have continued to evolve over the years [36], while attempts to provide a uniform definition for it have birthed a series of debates [37]. However, the FAO's definition of food security has gained currency and traction over the years. It defines food security as a situation when all individuals, irrespective of the time span, have access to food both physically, socially, and economically. The concept of food security has four components which are availability, access, utilization, and stability [38]. According to the global agency, the food must be safe, sufficient, and nutritious to meet their dietary needs and food preferences for a healthy and active life [39].

Food security which represents a core focus of sustainable development goal 2 remains a very crucial topic [40]. This is because it is a core requirement for individuals, communities and nations of the world [41], it is also a critical requirement for national security [42]. Unfortunately, this critical requirement has faced perennial threats which brings to the fore the need for governments across the world to take requisite steps to safeguard their societies from the threats of food insecurity [43]. The UN 1975 definition of food security which focuses on production speaks to our focus in this paper [44]. Certain conditions can predispose a society to food insecurity. For instance, wars, policy inconsistencies, climatic fluctuations among others act as catalysts to food insecurity [45]. The cacophony of issues plaguing food security has made it an issue of significant importance for governments and the international development community [46].

One major impediment to food security remains CC. To be very candid, it has been affirmed that CC portends a great deal of threat to SDG2 is not in doubt [47]. This threat which cuts across all the SDGs explains why SDG13 was enlisted to curb all threats emanating from climatic dysfunctionalities [48]. Research has shown that areas categorized as CC hotspots, face more threats of food insecurity when compared to other climes [49]. This explains why SDG2 which speaks exclusively to issues of food is threatened by the vagaries of CC [50], especially in developing countries who rely so much on agricultural production as a source of livelihood [47], and have a high degree of climate vulnerability [48]. Considering the centripetal role SDG2 plays on other goals, it becomes a source of worry if this goal remains threatened by climatic disruptions which can be contained with the right measures put in place. This situation is worsened by the failure of institutions to put effective adaptation policies in place [51], as CC comes with consequences and costs [52]. For rural communities in developing countries, one of the major consequences they experience is threats to their food sector.

In recent years, CSA is gaining relevance and increased acceptance [53]. It is an innovative system of agriculture that aims to transform the sector. Among the objectives of this agricultural system is to ensure food availability for the sprawling population of the world [54], little wonder why Akpabio et al, established a nexus between CSA and the actualization of SDG2 [55]. Apart from representing a resilient agricultural system to the vagaries of climatic dysfunctionalities [49], Akinsemolu et al. [56], described it as a solution-based agriculture. CSA is a climate adaptive measure to check the scourge of CC [57]. Taking into cognizance the potentials inherent in Africa's agricultural sector which have not been sufficiently tapped into, most African governments have promoted CSA as a response mechanism to tackle this menace [58].

4. Climate Smart Agriculture (CSA) and SDG2 Prospects in Rural Settings of Developing Countries

In modern scholarship, there are few topics that generate more public concern than CC. This may be an offshoot of the peculiarity of CC impacts—which may include, but not limited to, extreme temperatures (that is, excessive snowfall/cold or hot weather), erratic rainfall, and prolonged droughts in different geographic locations across the globe—that have been found to be a major challenge to farmers' productivity with far-reaching effect on every households' livelihoods [59]. It is indeed worthwhile to mention here that the impact of CC is usually severe in sub-regions and countries where their economic development is dependent on the climate [60]. This explains why CC is sometimes framed in literature by some scholars and policymakers as one of the popular forms of natural threat to human security [61].

Food security is one area where CC has proven to be a major security threat to humans. However, the long term and devastating threats posed by CC to food security, especially in developing countries have long been framed as myths and (in) visible by various means. But as more data emerge on the subject matter, states have begun to take steps that would enable them better understand the findings of relevant scientific-based studies with the sole purpose of crafting innovative ways to fully address the interlinked challenges of CC and food insecurity in their territorial spaces. To achieve this goal, the phenomenon of CSA—a modern strategy designed to enable countries—was introduced to facilitate state's attainment of the Sustainable Development Goal 2 in which member states of the United Nations are urged to collectively strive to create a world free of hunger by the year 2030 [62].

In many developed and developing countries today, different forms of mainstream CSA policies have been crafted and implemented at varying but appreciable degrees at the global and national levels. But at the rural (sometimes described as local in some literature) level, especially in developing countries, adaptation and mainstreaming of CSA remains a challenge. Hence, the need to x-ray the prospects of CSA mainstreaming and SDG2 at the rural level in such communities. In this study we relied on the use of three broad thematic subheadings: Readiness Mechanism, Services and Infrastructure, and Coordination Mechanism aligned with the triple wins of productivity, resilience, and mitigation to assess the prospects of CSA implementation in rural settings of developing countries [63].

Readiness Mechanism: Readiness mechanism is a broad theme which encompasses sub-indicators such as Agricultural adaptation policy, Agricultural mitigation policy, Economic readiness, Governance readiness, and Social readiness in countries of interest [63]. In the case of CSA and SDG2 in rural settings of developing countries, it could be argued that the possibility of fully implementing the policy in rural settings is slim for a number of reasons. For example, it has been established that in many developing countries their governments are not economically ready to shoulder the financial demands of the policy in their domains [64]. This is also the case of governance readiness in many developing countries. Many government officials are more concerned about how they could remain politically relevant and less concerned about issues of CC and SDG2 effective implementation and achieving results [13].

Services and Infrastructure: In much the same way, Service and Infrastructure also another broad theme with a number of key sub-indicators which include: Extension services, Agriculture Research and Development (R&D), Rural Access Index (RAI), Social safety nets, National GHG inventory system, National agricultural risk management systems, and Adaptive capacity [63]. Within the purview of this sub-theme, it can also be argued that the prospect of a fully implemented CSA and SDG2 in rural settings of developing countries is somewhat slim because of different factors, one of which is the lack of funding for agricultural R&D in many developing countries [65]. Poor RAI in many countries battling poor farm roads infrastructure is another major challenge to the successful implementation of the policy. The poor level of technological advancement in many rural settings in developing countries has also been identified as a key problem to the smooth and efficient actionability of the policy [59].

Coordination Mechanism: This is another major theme within CAS and SDG2 community. It has two major sub-indicators, namely, Disaster risk management coordination and Coordination mechanism [63]. Regarding the issue of disaster risk management coordination in rural communities of developing countries, it has been established that many low- and middle-income countries require some form of significant assistance to enhance their capacity to respond, manage, and reduce their susceptibility to natural occurrences and CC within their territorial spaces [66]. They also require some form of institutional and policy reform that would enable them to put in place or strengthen their disaster risk management capacity, especially in rural communities [66].

5. Suggestions for the Way Forward

Developing nations, in which agriculture which has long been the backbone of rural areas, are more susceptible to the negative effects of rising temperatures, variable rainfall, increased aridity, eroding soil, and drought-related precipitation. CSA techniques are being used by farmers in Africa and around the world to mitigate the effects of CC on crop productivity [4]. To ensure food and nutrition security, farmers can address the environmental, social, and economic aspects of sustainable development across agricultural landscapes with the use of the CSA strategy. Despite the fact that automated methods used in most modern agriculture squander important resources, CSA helps increase agricultural productivity and cut waste [67]. This strategy assists in bringing the goals and requirements of various stakeholders into line in order to create food systems that are more sustainable, egalitarian, and resilient.

Carbon-smart agriculture technologies boost resilience, lower or eliminate greenhouse gas emissions, and increase productivity [68]. Planning is necessary to handle trade-offs and synergies across the three CSA pillars of productivity, adaptation, and mitigation when implementing new technology. Thorough comprehension of the impacts of CC, the degree of agricultural sector vulnerability, and the creation of workable adaptation strategies are necessary for the proper use of the CSA strategy in rural communities [69]. In order to generate pertinent plans, strategies, investments, and coordination across processes and institutions in charge of agriculture, CC, food security, and land use, it entails embracing enabling policy frameworks and fortifying national and local institutions.

Numerous potential methods exist that could boost food security, sustain livelihoods, and increase the productivity of regional agricultural systems. Every program or policy must undergo a thorough, strategic evaluation to determine its possible advantages and disadvantages under various social, economic, and environmental scenarios. Evaluations can reveal how the local climate has changed and how it might change going forward, as well as how it affects the agricultural industries, food security, and livelihoods [70]. Additionally, they are able to assess if a certain set of measures is climate-smart or not. The process of implementing CSA varies depending on the setting and environment. There isn't a single approach that works in every circumstance.

To ensure that CSA is used effectively in rural communities, strong local institutions that support, encourage, and empower farmers must be established. The main keepers of information regarding their surroundings, agro-ecosystems, crops, livestock, and regional climate patterns are farmers. The knowledge needs and priorities of local farmers must be taken into consideration when adjusting to CSA. Farmers can find easily adoptable and implementable climate-smart choices with the help of local projects and institutions. In certain situations, efforts must also be taken to strengthen national policymakers' relationships with local government representatives and to increase their ability to participate in international policy fora on agriculture and CC. A crucial component of implementing CSA is the use of creative financing mechanisms that combine investments from the public and commercial sectors with financing related to agriculture and CC.

6. Conclusion

This chapter sought to interrogate the prospects of SDG2 through adopting CSA in rural communities of developing countries. To overcome obstacles and hurdles to food security both now and in the future, farmers must create creative and sustainable solutions for farming and agriculture. This is because the globe is experiencing food instability and famine. This became a necessity due to the neglect rural areas have suffered and the preference of researchers to focus on urban settlements. Arising from the discourse thus far, we observed that though there are huge prospects for these rural areas, it is imperative that the challenges they encounter be surmounted to enable them maximize the benefits embedded in this climate resilient and innovative agricultural system.

By supporting farmers, we may contribute to the reduction of poverty and the advancement of economic prosperity as farmers provide 70–80% of the world's food supply. CSA enables people to better survive the adverse consequences of CC by giving them tools, training, and financial support, which raises their income and productivity. Hence, we conclude by calling on the political leadership to provide requisite support that can help farmers and communities' members make accelerated progress towards the attainment of SDG2 within the shortest time possible.

References

1. Augustine, C., Balaji K., Dharanikumar S.V. and Jose Anand A. (2023). Urban farming: Case study. *In*: K. Kalaiselvi, J.A. Anand, P. Tanwar and H. Raza (Eds.),

Advanced Technologies for Smart Agriculture, 1st Edn. 321–338. Gistrup: River Publishers.

2. Blesh,, J., Hoey, L., Jones, A.D., Friedmann, H. and Perfecto, I. (2019). Development pathways toward "Zero Hunger". *World Development*, 118: 1–14.
3. Mustafa, M.A., Mabhaudhi, T., Avvari, M.V. and Massawe, F. (2021). Transition toward sustainable food systems: A holistic pathway toward sustainable development. *In*: C.M. Galanakis (Ed.), *Food Security and Nutrition*, 1st Edn. 33–56. Oxford: Academic Press.
4. Mutengwa, C.S., Mnkeni, P. and Kondwakwenda, A. (2023). Climate-smart agriculture and food security in Southern Africa: A review of the vulnerability of smallholder agriculture and food security to climate change. *Sustainability*, 15(4): 1–13.
5. Thottadi, B.P. and Singh, S.P. (2024). Climate-smart agriculture (CSA) adaptation, adaptation determinants and extension services synergies: A systematic review. *Mitigation and Adaptation Strategies for Global Change*, 29(3): 22.
6. Hamududu, B. and Killingtveit, A. (2012). Assessing climate change impacts on global hydropower. *Energies*, 5(2): 305–322.
7. Charles, A.O., Anetor, F.O. and Akpabio, E.S. (2023). The effect of climate change on the agricultural sub-sectors in Nigeria: An Autoregressive Distributed Lag (ArDL) approach. *The Journal of Developing Areas*, 57(2): 17–29.
8. Hitz, S. and Smith, J. (2004). Estimating global impacts from climate change. *Global Environmental Change*, 14(3): 201–218.
9. Ingutia, R. (2021).The impacts of Covid-19 and climate change on smallholders through the lens of SDGs; and ways to keep smallholders on 2030 agenda. *International Journal of Sustainable Development and World Ecology*, 28(8): 693–708.
10. Hakimi, A. and Inglesi-Lotz, R. (2020). Examining the differences in the impact of climate change on innovation between developed and developing countries: Evidence from a panel system GMM analysis. *Applied Economics*, 52(22): 2353–2365.
11. Maupin, A. (2016). The SDG13 to combat climate change: An opportunity for Africa to become a trailblazer. *African Geographical Review*, 36(2): 1–15.
12. Akpodiogaga-a, P. and Odjugo, O. (2010). General overview of climate change impacts in Nigeria. *Journal of Human Ecology*, 29(1): 47–55.
13. Sarkodie, S.A. and Strezov, V. (2019). Economic, social, and governance adaptation readiness for mitigation of climate change vulnerability: Evidence from 192 countries. *Science of The Total Environment*, 656: 150-–164.
14. Calzadilla, A., Rehdanz, K., Betts, R., Falloon, P., Wiltshire, A. and Tol, R.S. (2013). Climate change impacts on global agriculture. *Climatic Change*, 120: 357–374.
15. Wheeler, T. and Von Braun, J. (2013). Climate change impacts on global food security. *Science*, 341(6145): 508–513.
16. Khan, N., Ma, J., Zhang, H. and Zhang, S. (2023). Climate change impact on sustainable agricultural growth: Insights from rural areas. *Atmosphere*, 14(8): 1194.
17. Lipper, L., Thornton, P., Campbell, B.M., Baedeker, T., Braimoh, A., Bwalya, M. and Torquebiau, E.F. (2014). Climate-smart agriculture for food security. *Nature Climate Change*, 4(12): 1068–1072.
18. Mizik, T. (2021). Climate-smart agriculture on small-scale farms: A systematic literature review. *Agronomy*, 11(6): 1–16.
19. Zhao, J., Liu, D. and Huang, R. (2023). A review of climate-smart agriculture: Recent advancements, challenges, and future directions, *Sustainability*, 15(4): 1–15.

20. Saj, S. and Torquebiau, E. (2018). Climate-smart agriculture, agroecology, and soil carbon: Towards winning combinations. *47 Perspective the CIRAD Policy Brief.* Cirad, Agricultural, Research for Developed, Sep. 2018.
21. Lipper, L., McCarthy, N., Zilberman, D., Asfaw, S. and Branca, G. (Eds.). (2018). *Climate Smart Agriculture: Building Resilience to Climate Change*. Cham: Springer International Publishing.
22. Dinesh, D., Frid-Nielsen, S., Norman, J., Mutamba, M., Loboguerrero Rodriguez, A.M. and Campbell, B.M. (2015). Is Climate-Smart Agriculture effective? *A Review of Selected Cases, CCAFS Working Paper No. 129*. Copenhagen, Denmark: CGIAR Research Programme on Climate Change, Agriculture, and Food Security (CCAFS).
23. Matteoli, F., Schnetzer, J. and Jacobs, H. (2020). Climate-Smart Agriculture (CSA): An integrated approach for climate change management in the agriculture sector. *In*: J.M. Luetz and D. Ayal (Eds.), *Handbook of Climate Change Management: Research, Leadership, Transformation*, 409–437. Cham: Springer International Publishing.
24. Das, U. and Ansari, M.A. (2021). The nexus of climate change, sustainable agriculture, and farm livelihood: Contextualizing climate smart agriculture. *Climate Research*, 84: 23–40.
25. Ghosh, M. (2019). Climate-smart agriculture, productivity, and food security in India. *Journal of Development Policy and Practice*, 4(2): 166–187.
26. Woods, T., Ernst, M. and Tropp, D. (2017). *Community Supported Agriculture: New Models for Changing Markets.* United States Department of Agriculture, Agricultural Marketing Service, Transportation and Marketing Program Research Report 316239.
27. Venkatramanan, V. and Shah, S. (2019). Climate smart agriculture technologies for environmental management: The intersection of sustainability, resilience, well-being, and development. *In*: *Sustainable Green Technologies for Environmental Management*, Ch. 2, 29–51.
28. Zougmoré, R.B., Läderach, P. and Campbell, B.M. (2021). Transforming food systems in Africa under climate change pressure: Role of climate-smart agriculture. *Sustainability*, 13(8): 1–17.
29. Kirina, T., Groot, A., Shilomboleni, H., Ludwig, F. and Demissie, T. (2022). Scaling climate smart agriculture in East Africa: Experiences and lessons. *Agronomy*, 12(4), 820.
30. Partey, S.T., Zougmoré, R.B., Ouédraogo, M. and Campbell, B.M., (2018). Developing climate-smart agriculture to face climate variability in West Africa: Challenges and lessons learnt. *Journal of Cleaner Production*, 187: 285–295.
31. Reinholz, D.L. and Andrews, T.C. (2020). Change theory and theory of change: What's the difference anyway? *International Journal of STEM Education*, 7(1): 1–12.
32. Kalimba, U.B. and Culas, R.J. (2020). Climate change and farmers' adaptation: Extension and capacity building of smallholder farmers in Sub-Saharan Africa. *In*: *Global Climate Change and Environmental Policy: Agriculture Perspectives*, Ch. 13, 379–410. Springer Singapore.
33. Sekaran, U., Lai, L., Ussiri, D.A., Kumar, S. and Clay, S. (2021). Role of integrated crop-livestock systems in improving agriculture production and addressing food security: A review. *Journal of Agriculture and Food Research*, 5(100190): 1–10.
34. Smith, P., Arneth, A., Barnes, D.K., Ichii, K., Marquet, P.A., Popp, A., ... Ngo, H. (2022). How do we best synergize climate mitigation actions to co-benefit biodiversity? *Global Change Biology*, 28(8): 2555–2577.

35. Kantor, L.S. (2001). Community food security programs improve food access. *Food Review/National Food Review*, 24(1): 20–26.
36. Shetty, P. (2015). From food security to food and nutrition security: Role of agriculture and farming systems for nutrition. *Current Science*, 109(3): 456–461.
37. Alcock, R, (2009). *Speaking Food: A Discourse Analytic Study of Food Security*. School of Sociology, Politics, and International Studies. University of Bristol Working Paper (07–09).
38. Yu, B., You, L. and Fan, S. (2010). *Toward a Typology of Food Security in Developing Countries* (No. 945). International Food Policy Research Institute (IFPRI).
39. Food and Agriculture Organization. (1996). *Rome Declaration on World Food Security and World Food Summit Plan of Action*. World Food Summit 13–17 November 1996, Rome, Italy. FAO.
40. Burchi, F. and De Muro, P. (2016). From food availability to nutritional capabilities: Advancing food security analysis. *Food Policy*, 60: 10–19.
41. Tonukari, N.J. and Omotor, D.G. (2010). Biotechnology and food security in developing countries. *Biotechnology and Molecular Biology Reviews*, 5(1): 013–023.
42. Stringer, R. (2001). Food security in developing countries. *In*: B.N. Gosh (Ed.), *Contemporary Issues in Development Economics*, 1st Edn. Ch. 5, 70–104. New York: Routledge.
43. Abbade, E.B. (2017). Availability, access, and utilization: Identifying the main fragilities for promoting food security in developing countries. *World Journal of Science, Technology, and Sustainable Development*, 14(4): 322–335.
44. Clover, J. (2003). Food security in sub-Saharan Africa. *African Security Review*, 12(1), 5–15.
45. Samim, S.A. and Zhiquan, H. (2020). Assessment of food security situation in Afghanistan. *SVU-International Journal of Agricultural Sciences*, 2(2): 356–377.
46. Leroy, J.L., Ruel, M., Frongillo, E.A., Harris, J. and Ballard, T.J. (2015). Measuring the food access dimension of food security: A critical review and mapping of indicators. *Food and Nutrition Bulletin*, 36(2): 167–195.
47. Mugambiwa, S.S. and Tirivangasi, H.M. (2017). Climate change: A threat towards achieving "Sustainable Development Goal number two" (end hunger, achieve food security and improved nutrition, and promote sustainable agriculture) in South Africa. *Jàmbá: Journal of Disaster Risk Studies*, 9(1): 1–6.
48. Campbell, B.M., Hansen, J., Rioux, J., Stirling, C.M. and Twomlow, S. (2018). Urgent action to combat climate change and its impacts (SDG13): Transforming agriculture and food systems. *Current Opinion in Environmental Sustainability*, 34: 13–20.
49. Szabo, S., Hossain, M.S., Renaud, F., Traore, D., Hussain, A., Matczak, P. and Matthews, Z. (2018). Accelerating progress toward the zero hunger goal in cross-boundary climate change hotspots. *Environment: Science and Policy for Sustainable Development*, 60(3): 18–27.
50. Amanullah, A. and Khalid, S. (2020). Agronomy-food security-climate change and the sustainable development goals. *In*: A. Amanullah (Ed.), *Agronomy-Climate Change and Food Security*, Ch. 1, 1–8. London: IntechOpen.
51. Balasubramanian, M. (2018). Climate change, famine, and low-income communities challenge sustainable development goals. *The Lancet Planetary Health*, 2(10): e421–e422.
52. Koubi, V. (2019). Sustainable development impacts of climate change and natural disaster. United Nations *Background Paper Prepared for Sustainable Development*

Outlook. Available from: https://www.un.org/development/desa/dpad/wp-content/uploads/sites/45/publication/SDO_BP_Koubi.pdf.

53. Abegunde, V.O., Sibanda, M. and Obi, A. (2019). The dynamics of climate change adaptation in Sub-Saharan Africa: A review of climate-smart agriculture among small-scale farmers. *Climate*, 7(11): 132.
54. Taylor, M. (2018). Climate-smart agriculture: What is it good for? *The Journal of Peasant Studies*, 45(1): 89–107.
55. Akpabio, E.S., Akeju, K.F., Omotoso, K.O., Ohunakin, F. and Ogundokun, R.O. (2023). Climate smart agriculture and Nigeria's SDG 2 prospects: A review. *In*: *2023 International Conference on Science, Engineering, and Business for Sustainable Development Goals* (*SEB-SDG*), 1: 1–6.
56. Akinsemolu, A.A., Onyeaka, H.N. and Tamasiga, P. (2023). Climate-smart agriculture as a possible solution to mitigate climate change impact on food security in Sub-Saharan Africa. *Food and Energy Security*, 13(1): 1–6.
57. Barasa, P.M., Botai, C.M., Botai, J.O. and Mabhaudhi, T. (2021). A review of climate-smart agriculture research and applications in Africa. *Agronomy*, 11(6): 1–26.
58. Mrabet, R. and Moussadek, R. (2022). Development of climate smart agriculture in Africa. *In*: Mkomwa, S. and Kassam, A. (Eds.), *Conservation Agriculture in Africa: Climate Smart Agricultural Development*, Ch. 2, 17–65. Oxfordshire: CAB International.
59. Wakweya, R.B. (2023). Challenges and prospects of adopting climate-smart agricultural practices and technologies: Implications for food security. *Journal of Agriculture and Food Research*, 14(100698): 1–8.
60. Abidoye, B. and Odusola, A. (2015). Climate change and economic growth in Africa: An econometric analysis. *Journal of African Economies*, 24(2): 1–25.
61. Hough, P. (2018). *Understanding Global Security*. 5th Edn. New York: Routledge.
62. United Nations. (2024). *Sustainable Development Goals – Goal 2: Zero Hunger*. https://www.un.org/sustainabledevelopment/hunger/.
63. The World Bank. (2016). *Climate-Smart Agriculture Indicators*. The World Bank Group Report number 105162-GLB. Available from: https://documents1.worldbank.org/curated/en/187151469504088937/pdf/105162-WP-P132359-PUBLIC-CSA IndicatorsReportweb.pdf.
64. Georgieva, K., Gaspar, V. and Pazarbasioglu, C. (2022). Poor and vulnerable countries need support to adapt to climate change. International Monetary Fund (IMF) Blog. Available from: https://www.imf.org/en/Blogs/Articles/2022/03/23/blog032322-poor-and-vulnerable-countris-need-support-to-adapt-to-climate-change.
65. Alston, J.M., Pardey, P.G. and Piggott, R.R. (Eds.). (2006). Synthesis of themes and policy issues. *In*: *Agricultural R&D in the Developing World: Too Little, Too Late?* Ch. 13, 361–372. International Food Policy Research Institute.
66. GFDRR – Global Facility for Disaster Reduction and Recovery (2022). *GFDRR: Bringing Resilience to Scale*. Annual Report 22. Available from: https://documents1.worldbank.org/curated/en/099904102272326411/pdf/IDU0242fa9a607c8804e9a0b0d404a 431f9a65a9.pdf.
67. Anand, J.A., Nirmala Devi, K., Vijayakumar, S. and Shiju C. Chacko. (2023). Decision-making support in smart farming. *In*: K. Kalaiselvi, J.A. Anand, P. Tanwar and H. Raza (Eds.), *Advanced Technologies for Smart Agriculture*, Ch. 10, 219–239. New York: Routledge.

68. Yeboah, S., Owusu Danquah, E., Oteng-Darko, P., Agyeman, K. and Tetteh, E.N. (2021). Carbon smart strategies for enhanced food system resilience under a changing climate. *Frontiers in Sustainable Food Systems*, 5(715814): 1–16.
69. Everest, B. (2021). Farmers' adaptation to climate-smart agriculture (CSA) in NW Turkey. *Environment, Development, and Sustainability*, 23(3): 4215–4235.
70. Gunaratne, M.S., Radin Firdaus, R.B. and Rathnasooriya, S.I. (2021). Climate change and food security in Sri Lanka: Towards food sovereignty. *Humanities and Social Sciences Communications*, 8(1): 1–14.

9 | Smart Agriculture: Cultivating Sustainable Solutions for a Green Future

Boppudi Swanth and Sheela J.*
Department of School of Computer Science and Engineering, VIT-AP University, Amaravathi, Andhra Pradesh, India

The critical journey towards ending world hunger is discussed in this chapter. We examine the mutually beneficial connection between food security and sustainable innovative agricultural practices. We examine how technology and sustainable farming practices interact in the face of expanding global population, climate change, and resource shortages and we carefully examine the advantages for crop production, resource conservation, and environmental sustainability. Although there is a potential roadblock, there are ways to promote the wider adoption of sustainable smart agriculture. We emphasize the need for collaborative efforts among governments, communities, and industries to create an enabling environment for the integration of these practices. With an eye towards the future, the chapter examines cutting-edge discoveries, cooperative projects, and technologies that show promise for boosting sustainable smart agriculture capacity to ensure food security. In order to achieve the common objective of eradicating hunger, in this chapter we end with a strong call to action for stakeholders, decision-makers, and the international community to embrace and invest in sustainable smart agriculture as a group.

1. Introduction

The deliberate modification of a piece of the Earth's crust is the art and science of farming, also known as agriculture. This is accomplished through the raising of animals and the growing of plants and other crops for both commercial gain and the fulfilment of human nourishment and other basic necessities.

*Corresponding author: Sheela.j@vitap.ac.in

There is a direct correlation between the rapid population growth and the rise in food consumption. Food is becoming more and more necessary as the human population worldwide rises. According to the poll, about one in seven people globally do not have the ability to obtain sufficient food to sustain good health and maintain active lives [1]. Recent advances in remote sensing technologies have greatly eased the prediction of agricultural yields [2]. Generally speaking, in order to determine the effective agricultural yield, farmers must promptly predict the crop production in the future. Analysis may also be necessary for farmers to ensure that crop production is reaching its maximum potential. Nevertheless, due to urbanization and climate change, this kind of farmers' goal will be challenging. The weather's unpredictability has led to several risks and improbabilities in agriculture [3].

In addition to having effective dimension reduction skills, machine learning offers considerable learning capabilities for assessing nonlinear relationships between supplied information sets. By implementing the Internet of Things (IoT)—which entails dispersing a variety of sensors, actuators, cameras, and other devices throughout agricultural land—stands to help precision agriculture. The monitoring of crop growth, water levels, temperature, disease detection, soil moisture, and many other factors is made possible by these technologies [5].

One of the machine learning subgroups called "Deep Learning" uses different configurations of raw data to produce different outcomes. Deep learning uses a probability model built from a huge amount of field data to forecast crop performance based on a variety of meteorological circumstances [6]. To attain increased crop yields, it is crucial to develop cost-effective methods powered by artificial intelligence (AI) that provide nearly real-time intra-field phenology and crop condition assessments. Remote sensing emerges as an excellent tool for yield prediction, offering non-destructive, rapid, and affordable monitoring of the Earth's surface across expansive areas [7].

India excels as a significant producer of food crops, including rare varieties. Despite being a vulnerable component of the Indian economy, agriculture holds substantial importance as it contributes to all other sectors and exerts influence across diverse spheres. The implementation of automation in agriculture is essential, considering the technological progress seen in other industries. "Digital agriculture" involves the use of sophisticated computer systems to calculate different metrics, like crop verification, crop estimation, yield finding, crop quality, and using different machine learning approaches. Artificial neural networks, which have various advantages over traditional systems, are widely used in the agriculture sector. The smart agriculture system shows in Figure 1.

The primary advantage of neural networks lies in their ability for forecasting and prediction through parallel thinking. Instead of being entirely programmed, neural networks can learn from data. In the face of the complex challenges posed by climate change, ensuring food security has become more critical than ever [8]. The concept of "sustainable smart agriculture", which integrates cutting-edge

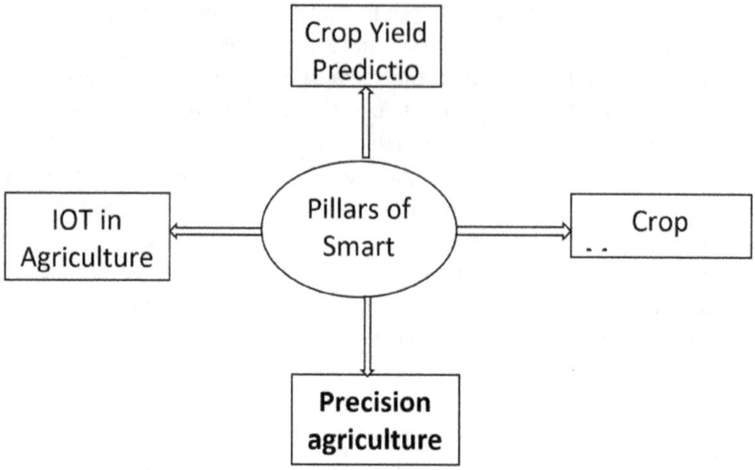

Figure 1: Smart agriculture.

technology with environmentally friendly methods, emerges as a transformative approach to address these challenges. Employing smart technologies is not merely a strategic choice but also a necessity for building resilient, efficient, and sustainable food production systems in this era characterized by unprecedented interconnectedness and innovation

The infusion of AI is triggering a significant revolution in agriculture, a cornerstone of human civilization. This chapter delves into the myriad applications of AI in agriculture, illustrating how state-of-the-art technologies are reshaping traditional agricultural approaches, optimizing resource utilization, and promoting the vision of a more productive and sustainable agrarian landscape. It explores the transformative impact of AI technologies like computer vision, machine learning, and others on the agricultural industry, scrutinizing the utilization of data, algorithms, and computing power in the development and implementation of AI applications in agriculture [10].

Elevasaran and Vincent (2021) [3] devised a smart agriculture system utilizing expert systems. In this system, the concept of the IoT was employed to relay data to the server, empowering field actuators to make informed decisions. To achieve this, the server must possess the capability to autonomously make decisions. Soil sensors, temperature, humidity, and leaf moisture makeup this system. It does not interfere with the watering operation; it merely provides information about the land. Cloud-based agricultural decision-making and assistance are expanding quickly these days. Through its website, the Agriculture Decision Support along with Automation System (DSAS) assists farmers in managing a variety of applications. Farmers can get real-time data with DSAS, which consists of numerous stages that can link and synchronize multiple devices at the same time [2].

The farmer plays a crucial role, utilizing software to operate each machine and monitor real-time information. For instance, a spray controller administers a specific quantity of insecticide in the field, a fertilizer controller manages the application of fertilizer, and an irrigation controller facilitates the supervision of irrigation. DSAS operates using data from multiple sensors, including soil moisture along with nitrogen sensors [1]. The system was divided into five parts: gathering images, where researchers took pictures of pomegranate leaves; and image pre-processing, which involved scaling, filtering, and processing the generated image in accordance with predetermined criteria. The next stage is colour image segmentation, where k-means clustering is used to separate the leaf segment affected by the disease from the leaf segment that is healthy. The adjusted image is then used to get the total leaf area.

The third segment is then utilized to assess the area affected by the leaf's infection. Ultimately, precise grading of the illness can be derived in the concluding section with the assistance of a fuzzy inference system. Researchers are starting to use IoT systems as well as AI techniques to the aquaculture industry in addition to agriculture [5].

The system devised by [7] employs advanced automation technologies to monitor water quality. While M2M technology is still in its early phases, it is advancing steadily. Yost and Hartemink (2020) [8] have developed a method facilitating self-communication among machines. Torky and Hassanein (2020) [4] have introduced a self-organizing, comprehensive sensor-based intense watering technique with upper and lower layers. Kamilaris and P-Boldú (2018) [5] attempted to showcase a prototype of a small-scale smart irrigation system. In their pursuit of an intelligent irrigation system, [2] focused exclusively on automation and the IoT.

Soil testing is the first step in soil preparation. It involves evaluating the present levels of nutrients and figuring out how much is right for a particular soil based on fertility and crop needs. Irrigation is a vital form of agriculture essential for soil and water conservation [19]. The utilization of sophisticated data in evaluating systems concerning crop, soil, water, and climate factors is key to maintaining irrigation efficiency and consistency [11].

Agriculture encompasses seven essential activities: land management, soil preparation, water monitoring, weed identification, pesticide recommendation, identification of diseased crops, and cost estimation. Land management involves monitoring physical aspects like weather patterns and geological features [8].

The technologies mentioned—IoT, Data Science, Machine Learning, Deep Learning, Blockchain—are primarily focused on data, offering valuable insights and comprehensive analyses. Consequently, these state-of-the-art technologies find application in various agricultural processes, including determining the most suitable crop for specific areas and identifying factors detrimental to crops, such as weeds, insects, and diseases. This aids in acquiring information about crop growth and facilitates decision-making

Smart agriculture significantly reduces hunger rates by boosting agricultural productivity and efficiency. By utilizing advanced technologies such as IoT sensors, drones, and data analytics, farmers can optimize growing conditions, monitor crop health, and detect diseases early. This precise management leads to higher yields and minimizes crop losses. Furthermore, precision farming techniques ensure that resources like water and fertilizers are used efficiently, decreasing waste and increasing food availability. Enhanced forecasting and climate adaptation strategies help farmers maintain stable food production despite changing weather patterns. By improving overall food production and reducing the risk of crop failures, smart agriculture contributes to a more reliable food supply, thereby helping to lower global hunger rates.

1.1. Benefits of Sustainable Smart Agriculture

1.1.1. Resource Efficiency

The goal of smart agriculture is to maximize productivity while reducing waste by utilizing cutting-edge technologies and data-driven techniques to optimize resource efficiency. Through accurate use of inputs like pesticides, fertilizers, and water, farmers may lower their environmental impact and increase crop yields. Precision irrigation systems, for example, minimize water wastage and save this essential resource by ensuring that crops receive precisely the right amount of water. In a similar vein, fertilizer application that is focused avoids runoff and misuse, which can cause soil erosion and water contamination. By utilizing these techniques, smart agriculture encourages environmentally conscious farming methods that protect biodiversity for future generations in addition to increasing agricultural productivity.

1.1.2. Climate Resilience

Climate change introduces significant uncertainty into traditional farming, disrupting weather patterns, increasing extreme weather events, and shifting growing seasons. Smart agriculture offers adaptive solutions through advanced technologies that monitor and respond to these changes. Utilizing IoT sensors, predictive analytics, and automation, farmers gain real-time insights into soil moisture, temperature, and crop health, enabling precise resource management and timely interventions. Climate-resilient crop varieties and precision farming techniques optimize water, fertilizer, and pesticide use, reducing environmental impact and enhancing sustainability. This technology-driven approach not only mitigates climate-related risks but also promotes long-term agricultural resilience and productivity.

1.1.3. Enhanced Productivity

Farmers that use technology in their agricultural techniques are able to keep a careful eye on the condition of their crops, spot diseases early on, and create the ideal environment for crop growth. Drones, satellite imaging, and IoT sensors are examples of cutting edge equipment that offer precise, real-time data on a variety of topics, such as moisture in the soil, levels of nutrients, and plant health. With the use of these technologies, diseases can be identified early enough to be treated quickly, reducing crop loss and stopping the transmission of infections. Furthermore, accurate management of growth circumstances, including fertilization, irrigation, and climate control, guarantees that crops get the best nutrients at the correct moment. This technological method helps meet the increasing worldwide demand for food by increasing production while also fostering a more stable and dependable food supply [10].

1.1.4. Supply Chain Enhancement

Beyond the farm, smart agriculture boosts the effectiveness of the whole food supply chain. Technology can ensure that more produce reaches consumers and prevent food waste in a variety of ways, from distribution and retail to logistics and storage. Making sure there is enough food has become a top priority in a world where both population expansion and climate change are formidable obstacles. In order to fulfil the global challenge of achieving zero hunger, responsible smart agriculture arises as a vital answer, combining state-of-the-art technology with ecologically conscious practices. The foundations, advantages, and difficulties associated with smart agriculture are examined in this chapter, along with how it can transform the systems for producing and distributing food [11].

1.1.5. Precision Agriculture

Precision farming, often known as precision agriculture, is an agricultural management technique that offers a thorough method to manage seasonal and regional variations in soil and crops. Effective management of these factors will maximize yield, boost profitability, and enhance overall production quality [1]. Precision agriculture can be used to increase crop productivity. A study of precision agriculture rates of adoption revealed that high-value enterprise farms were quicker to embrace the technology than low-value business farms. The speed at which precision agriculture gains traction depends on the country and particular areas. Those in the valley are more likely to use precision agriculture than those in the mountainous area [2]. The disparity in adoption rates may be due to the large investments required. Therefore, finding ways to lower the cost of machinery is essential if farms of all sizes are to implement precision agriculture. Modern technologies including GPS, indicators, and information analytics are used in precision farming to enhance irrigation, fertilization, and insect control methods. Precision farming reduces waste and maximizes efficiency by carefully managing resources, which raises agricultural yields and lessens the impact on the environment. Precision farming, also known as precision agriculture, is a cutting-edge technique that optimizes various farming activities through the use of cutting-edge technologies. Precision farming's primary goal is to increase the productivity, long-term viability, and efficiency of agricultural operations. This method precisely manages resources, reduces waste, and raises overall agricultural yields through the use of sensors, real-time data, and technology-driven decision-making [21].

2. Modern Technologies

2.1. Global Positioning System (GPS)

Because GPS technology provides accurate location information for field mapping as well as crop monitoring, it is essential to precision farming [8]. Precision navigation and control of agricultural machinery with GPS technology ensures precise planting, harvesting, and use of inputs like pesticides and fertilizers by farmers. Precision farming uses a variety of sensors, such as soil, moisture, and crop health sensors, to gather information on many elements of the farming environment [5]. Farmers may monitor soil conditions, evaluate crop health, and make well-informed decisions about irrigation, fertilization, and insect control with the use of sensor data. This focused strategy maximizes productivity while minimizing resource consumption. With the aid of GPS technology, farmers may precisely map their agricultural fields, drawing boundaries and comprehending the overall layout of their land [12].

2.2. Variable Rate Technology (VRT)

VRT allows for the variable application of inputs such as fertilizers, pesticides, and water within a field based on specific conditions. By tailoring input application to the specific needs of different parts of a field, VRT optimizes resource use, reduces environmental impact, and improves overall crop performance. Precision farming employs automated equipment, such harvesters and tractors that are fitted with GPS and electronic sensors to carry out operations with extreme accuracy. Improved total productivity results from automated machinery's increased operating efficiency, decreased need for laborious tasks, and assurance that jobs are completed precisely and consistently. Auto-steering systems, guided by GPS, automate the steering of tractors and other farm machinery [10]. Auto-steering enhances the accuracy of field operations by eliminating human errors associated with manual steering. This technology allows for straighter rows, consistent spacing, and optimized field coverage, contributing to increased efficiency and reduced input waste.

2.3. Data Management and Analysis

Precision farming relies heavily on data management and analysis to process large datasets generated by various sensors and technologies. Advanced analytics and machine learning algorithms help farmers make sense of complex data, providing insights into crop performance, disease detection, and yield predictions [3]. This data-driven decision-making enables proactive and targeted interventions.

2.4. Remote Sensing and Satellite Imagery

Remote sensing technologies, including satellite imagery and drones, provide valuable information for monitoring large agricultural areas. These technologies offer high-resolution imagery, allowing farmers to assess crop health, identify areas of stress, and monitor changes over time. This information is crucial for making timely and informed decisions.

2.5. Farm Management Software

Farm management software platforms integrate data from various sources and provide a centralized tool for planning, monitoring, and analyzing farm activities. These platforms enable farmers to create precise plans for planting, irrigation, and harvesting, monitor ongoing activities, and generate reports [21]. They facilitate efficient farm management and decision-making. GPS-enabled data logging devices capture detailed information about field activities, and this data is often integrated into precision farming software platforms. The development of comprehensive farm management plans is facilitated by the exact location-based data that GPS devices gather. This data can be analyzed by farmers to improve future planning, monitor the effectiveness of various field zones, and make well-informed decisions.

2.6. Benefits to the Economy and Environment

Precision farming aims to optimize resource use, reduce waste, and minimize environmental impact. By improving the efficiency of agricultural operations, precision farming contributes to cost savings for farmers and a more sustainable approach to farming. It helps lessen the environmental impact of agriculture, preserve water, and use less chemical inputs [11].

2.7. Yield Monitoring and Mapping

GPS is employed in yield monitoring systems, which track and record crop yields as the harvest progresses across the field. Yield mapping, coupled with GPS data, provides farmers with insights into spatial variations in crop performance. This information is valuable for decision-making in subsequent seasons, enabling farmers to adjust management practices based on historical yield data.

2.8. Guidance for Irrigation and Drainage

GPS technology aids in mapping and guiding irrigation and drainage systems in the field. Precision in irrigation and drainage is essential for water conservation and efficient resource use. GPS-guided systems help farmers design and implement irrigation and drainage strategies tailored to the specific needs of different areas within the field.

2.9. Monitoring and Documentation

GPS allows for the continuous monitoring and documentation of field activities and operations. Farmers can generate detailed records of planting dates, application rates, and harvest times with precise location information. This documentation supports compliance with regulations, facilitates traceability, and enhances the overall management of farm operations.

2.10. Future Developments and Connectivity

Ongoing developments in GPS technology involve increased precision, faster updates, and improved connectivity. Enhanced GPS capabilities contribute to more accurate and responsive precision farming systems. Improved connectivity, including integration with 5G technology, facilitates real-time data transfer and communication between farm machinery and centralized systems.

3. The Internet of Things (IoT)

In order to improve many facets of farming operations, the IoT for farming represents a revolutionary integration of digital technology, networking, and data analytics. The IoT in agriculture seeks to maximize resource utilization, boost productivity, and enhance overall efficiency through the use of sensors, devices, and connection solutions. Here is a comprehensive overview of the role of IoT in agriculture:

3.1. Smart Sensors

Smart sensors are used in fields and on equipment to collect data on a range of environmental conditions in real time. These sensors monitor temperature, humidity, and other factors. By providing farmers with the necessary information to make informed decisions about pest control, the fertilization process, and water supply, the data collected maximizes the utilization of available resources [12].

3.2. Precision Livestock Farming

IoT devices such as wearables and sensors are used to monitor the health, behaviour, and productivity of livestock. Farmers can track the well-being of individual animals, detect signs of illness, and optimize feeding schedules. This leads to improved animal welfare, increased productivity, and reduced losses. IoT-enabled devices are integrated into agricultural machinery and equipment to enable automation and remote monitoring. Automated systems, guided by real-time data, allow for precise planting, harvesting, and other tasks. Remote monitoring enables farmers to manage equipment and receive alerts for maintenance needs, reducing downtime.

3.3. Drones and UAVs (Unmanned Aerial Vehicles)

Drones equipped with sensors and cameras are used for aerial surveys and data collection over large agricultural areas [13]. Drones provide high-resolution imagery for crop monitoring, pest detection, and yield estimation. This data supports timely decision-making and allows farmers to identify issues early in the growing season. IoT-enabled irrigation systems monitor soil moisture levels and weather using sensors. These devices can dynamically adjust irrigation schedules based on real-time data to ensure crops receive the proper amount of water. This promotes efficient resource management and water conservation. Livestock movement and health are tracked and monitored using IoT devices such as ear tags and collars with GPS capabilities. Tracking livestock aids in preventing theft, finding lost animals, and keeping an eye on grazing patterns. Health monitoring makes it possible to identify illnesses early, which improves disease management techniques.

3.4. Supply Chain Visibility and Weather Monitoring

Throughout the supply chain, IoT is utilized to track the movement and condition of agricultural goods. IoT devices give real-time data on the position, heat, and moisture of crops from the field to the market. This visibility contributes to waste reduction, spoiling prevention, and the safety and quality of agricultural products. IoT devices gather meteorological data, such as humidity, temperature, and precipitation. With the use of current weather data, farmers can plan ahead for jobs like planting, harvesting, and other operations. Additionally, it helps in the early identification of weather-related hazards like frost or intense heat.

3.5. Data Analytics and Decision Support

IoT-generated data is analyzed using advanced analytics tools to extract meaningful insights. Data analytics allows farmers to make informed decisions about crop management, resource allocation, and overall farm strategy. It facilitates precision farming practices and contributes to increased efficiency. IoT devices are integrated into greenhouse systems to monitor and control environmental conditions. Smart greenhouses use sensors to regulate temperature, humidity, and light, creating optimal conditions for plant growth. This level of control enhances crop quality and yield. IoT devices, including sensors and imaging systems, are used to monitor for signs of pests and diseases. Early detection allows for targeted interventions, reducing the need for broad-spectrum pesticides. This approach contributes to sustainable pest management practices.

3.6. Smart Farm Management Platforms

IoT is integrated into centralized farm management platforms that aggregate data from various sources. These platforms provide farmers with a comprehensive view of their operations, allowing for efficient decision-making, resource allocation, and planning. They often enable remote monitoring and control of farm activities. In conclusion, the integration of IoT in agriculture represents a paradigm shift in how farming operations are conducted. By harnessing the power of connectivity and data-driven insights, IoT contributes to sustainable and efficient agricultural practices, ultimately playing a crucial role in addressing global challenges such as food security and resource optimization. Blockchain has quickly emerged as a crucial technology for many applications related to precision agriculture. Researchers are thinking of creating blockchain-based IoT systems for precision agriculture due to the need for intelligent peer-to-peer systems that can verify, secure, monitor, and analyse agricultural data. In order to store, organize, and distribute agricultural data in a digital format which is more credible, immutable, transparent, and decentralized, Blockchain technology is essential. A network of "smart farms" is created when Blockchain technology and IoT are combined with precision farming. This combination results in increased independence and adaptability [7].

3.7. Blockchain Technology in Agriculture

Transforming the Food Supply Chain: Blockchain, the decentralized and distributed ledger technology, has shown immense potential in revolutionizing various industries, and agriculture is no exception. Its application in agriculture brings transparency, traceability, and efficiency to the entire food supply chain. Here are key aspects where Blockchain technology is making an impact.

4. Impact of Blockchain Technology

4.1. Supply Chain Transparency

Inefficiencies and challenges in tracking down the source of products are frequently caused by a lack of openness in the supply chain of food. A transparent and unchangeable record of each transaction made along the supply chain can be created thanks to Blockchain technology. Farmers, processors, distributors, and retailers all contribute a block of data to the secure, traceable history. When there are problems with food safety, tracing the source of contaminated items can be difficult and time-consuming [7]. This makes it possible to follow the product's route from its source to the customer in real time. When a product is recalled, it is possible to locate the source fast, stopping the spread of dangerous goods.

4.2. Smart Contracts for Agreements

Traditional contracts can be cumbersome, leading to delays in payments and disputes. Smart contracts, self-executing contracts with the terms of the agreement directly written into code, automate processes like payments when predefined conditions are met. This guarantees fast and equitable payment to all stakeholders involved in the supply chain. Food fraud, including mislabelling and counterfeit products, is a significant concern. Block chain's transparency reduces the chances of fraud, as each step of the supply chain is recorded and verified. Consumers can trust that the information on the label accurately represents the product's origin and attributes.

4.3. Enhancing Market Access for Farmers

Smallholder farmers frequently have difficulty establishing the provenance and quality of their goods, which restricts their access to markets. Block chain helps small farmers build credibility and reach new markets by offering a transparent and safe record of farming methods, certifications, and transactions. This promotes inclusion throughout the agricultural value chain. Centralized data systems are susceptible to hacking and manipulation. Being decentralized, blockchain enhances data security. Each block is cryptographically linked to the previous one, making it extremely difficult for a single entity to alter the data. This ensures the integrity of the information stored. Delays in payment processes can affect farmers' financial stability. Through the use of crypto currencies and smart contracts, block chain facilitates faster and more transparent payment processes, ensuring that farmers receive fair compensation for their produce.

4.4. Environmental and Sustainable Practices

Proving adherence to sustainable and environmental practices is crucial for meeting consumer demands. Blockchains can record and verify data related to sustainable farming practices, organic certifications, and environmental impact assessments. This information is then readily accessible to consumers who prioritize sustainable products. Blockchain technology has the ability to build a more reliable, effective, and sustainable agriculture environment because of its decentralized and secure structure. Adoption will probably change how we grow, distribute, and eat food as it spreads. In the agricultural arena, the term "data-driven decision making" refers to the use of data to guide and inform various aspects of farming operations. Large volumes of agricultural data must be gathered, analyzed, and interpreted in order to produce insightful information that will help farmers make more strategic and knowledgeable decisions.

5. Farm Management Systems

5.1. Data Integration

Implementation of systems that consolidate data from various sources, including sensors, drones, and historical records, into a centralized platform. To facilitate effective data integration, the collected information is channelled into centralized repositories or farm management systems. These repositories act as hubs where data from disparate sources converges, creating a unified and easily accessible database. Centralization ensures that farmers and decision-makers have a singular platform from which they can draw insights and make informed choices regarding crop management, resource allocation, and overall farm strategy.

5.2. Decision Support Platforms

Utilization of advanced software tools that process and analyze integrated data, providing farmers with actionable insights for day-to-day and long-term decision-making. Decision Support Platforms (DSPs) represent a technological cornerstone in modern agriculture, seamlessly integrating data from diverse sources to empower farmers with informed and strategic decision-making capabilities. These platforms serve as comprehensive hubs where data, ranging from soil conditions and weather patterns to crop health and machinery performance, converges into a unified and accessible interface. Through the integration of advanced technologies such as AI and machine learning, DSPs analyze this amalgamated data, providing farmers with actionable insights and predictive analytics. In doing so, these platforms facilitate real-time monitoring of farm operations, enabling immediate responses to changing conditions. DSPs play a pivotal role in optimizing resource allocation, guiding planting schedules, and enhancing overall operational efficiency. The continuous evolution of DSPs holds the promise of revolutionizing agriculture by transforming vast datasets into practical knowledge, ultimately contributing to sustainable and resilient farming practices [14].

5.3. Real-time Monitoring

Real-time, ongoing observation of factors including crop health, weather, and soil quality enables quick reaction to changing conditions. A key component of agriculture's modernization is real-time monitoring, which gives farmers access to never-before-seen data and operational management [6]. This technique gathers and analyzes data continuously and instantly from a variety of sources, including sensors, drones, and IoT devices, to give a dynamic and current picture of the agricultural environment. Farmers may use modern technologies like AI and smart sensors to make rapid data-driven decisions in response to emerging issues in a proactive manner. This ability enhances efficiency as well as sustainability in the context of continuously changing circumstances by sharpening resource management skills and creating more resilient and adaptable farming methods. Real-time monitoring is ushering in a new era in agriculture where it is no longer merely a periodic endeavour but rather a continuously enhanced and responsive system [9].

5.4. Predictive Analytics

Predictive analysis in agriculture involves leveraging data and advanced analytics to foresee trends and outcomes related to crop growth, yield, and overall farm productivity. Predictive analytics assists farmers in making decisions by evaluating past data, present circumstances, and several affecting elements like weather trends, the condition of the soil, and pest prevalence. This proactive approach enables farmers to anticipate potential challenges, optimize resource utilization, and implement targeted strategies for improved crop management.

5.4.1. Crop Modelling

Utilizing statistical models to anticipate and simulate crop development while accounting for past data, weather patterns, and soil conditions. A sophisticated method of agricultural management is called "crop modelling", which uses computer and mathematical methods to simulate and forecast crop growth and development. This approach creates a dynamic model that replicates crop behaviour under diverse situations by integrating several data inputs, such as historical climatic data, soil properties, and crop-specific parameters. Modern technologies that help improve and evaluate these models include remote sensing and IoT devices, which provide useful real-time data. Crop modelling's predictive capabilities helps farmers proactively handle difficulties including climate unpredictability and market variations by reducing risk and improving resource efficiency. Crop modelling, a crucial aspect of precision agriculture, highlights how technology has the power to completely transform farming methods and secure a more robust and viable future for world agriculture.

5.4.2. Market Trends Analysis

Application of analytics tools to assess market trends, enabling farmers to anticipate demand, set optimal pricing strategies, and make informed choices about crop selection. Market trends analysis in agriculture involves the systematic examination and interpretation of various factors influencing the agricultural market, including consumer demand, pricing dynamics, and global trade patterns. Leveraging data analytics, machine learning, and historical market data, this practice enables farmers and stakeholders to identify patterns, anticipate shifts in market conditions, and make informed decisions regarding crop selection, pricing strategies, and market entry points. By scrutinizing supply and demand dynamics, as well as external factors such as weather events or geopolitical developments, market trends analysis empowers agricultural professionals to respond proactively to market fluctuations. This data-driven approach contributes to the optimization of production strategies, allowing farmers to align their practices with prevailing market conditions and enhance their competitiveness in the dynamic agricultural marketplace.

5.5. Resource Optimization

The goal of resource optimization in agriculture is to maximize the productivity of inputs like energy, fertilizers, and water to produce larger yields with the least amount of waste and negative environmental effects. By utilizing cutting-edge technologies such as precision agriculture and data analytics, farmers can adjust the use of resources to the unique requirements of the crops and fields [10]. Accurate irrigation management, tailored feeding, and frugal energy use are required for this. Farmers may help the agricultural industry achieve long-term financial and ecological benefits by optimizing resources, boosting overall productivity, reducing costs, and promoting sustainable farming practices [11].

5.5.1. Irrigation Management

Optimization of water usage through data analysis, ensuring that irrigation is applied based on actual needs and soil moisture conditions rather than fixed schedules. Irrigation management stands as a critical facet of precision agriculture, encompassing the strategic and efficient application of water resources to optimize crop growth. By combining sensor technology, data-driven systems for decision-making, and soil moisture monitoring, farmers can accurately choose when and the amount of water to administer, based on the demands of the crops. Advanced irrigation management involves real-time data analysis, enabling farmers to respond promptly to changing environmental conditions and plant requirements. Farmers can install controlled irrigation systems that adapt to the present climate trends, soil water levels, and crop growth phases by utilizing technologies such as IoT devices and weather predictions. An example of how technology-driven accuracy in water application supports resource-efficient and sustainable agricultural practices is irrigation management, which maintains ideal plant health while reducing environmental effect [16, 18, 19].

5.5.2. Fertilization Strategies

Applying fertilizer in a way that optimizes output while minimizing waste and the environment by taking into account crop requirements and soil nutrient levels. Fertilization strategies represent a pivotal component of precision agriculture, involving the precise application of nutrients to optimize crop growth and maximize yields. With data-driven insights from soil nutrient evaluations, prior crop performance information, and advanced analytics, farmers may customize fertilization procedures to meet the unique requirements of their crops with this method [18]. Farmers may instantly modify the types and rates of fertilizer applied by using real-time nutrient level monitoring provided by satellite imaging and soil sensors [3]. By minimizing over-application, this targeted and reactive strategy reduces environmental impact while simultaneously improving the accuracy of nutrient utilization. Fertilization strategies underscore how technology integration supports sustainable agriculture, ensuring that nutrient resources are applied judiciously to promote optimal plant health and contribute to overall farm productivity.

5.6. Risk Mitigation

Risk mitigation in agriculture involves the implementation of strategies to minimize the impact of uncertainties and potential threats to crop production.

Farmers employ various practices to address risks such as adverse weather conditions, pests, diseases, and market fluctuations. Diversification of crops, using resistant varieties, and adopting sustainable farming practices are common risk mitigation measures. Furthermore, weather forecasts, insurance programmes, and cutting-edge technologies aid in the early identification and control of possible hazards, empowering farmers to take well-informed decisions and protect their livelihoods. Farmers strive for a more secure and robust agricultural system by constantly addressing and managing risks.

5.6.1. Pest and Disease Control

Early detection of pests and diseases through data analysis, enables farmers to implement targeted interventions and minimize crop losses. Pest and disease control within agriculture involves a comprehensive and technology-driven approach to monitor, identify, and mitigate potential threats to crops. By utilizing cutting-edge technologies like sensors, picture recognition, and immediate information analytics, farmers are able to identify early indicators of disease outbreaks or insect infestations. Integrated pest management (IPM) strategies utilize this information to implement targeted interventions, including precision application of pesticides or biocontrol agents. By employing predictive modelling and historical data analysis, farmers can anticipate the likelihood of pest and disease occurrences, enabling proactive measures for prevention and control. This data-centric approach not only minimizes the environmental impact of pest control measures but also enhances overall crop health and productivity [6, 7]. The integration of technology in pest and disease control exemplifies a proactive and sustainable method, ensuring the resilience of agricultural systems against the challenges posed by pests and diseases.

5.6.2. Yield Prediction

Accurate prediction of crop yields using historical data and machine learning models, help farmers plan for harvest and market logistics. Precision agriculture's main component, yield prediction, uses cutting-edge data analytics and technology to estimate crop production with accuracy. By merging past yield data with current weather, soil quality, and crop growth phases, farmers can use neural networks and mathematical models to project yields in the future. This data-driven approach empowers farmers with insights into potential harvest outcomes, allowing for proactive decision-making in resource allocation, harvest planning, and market strategies [23]. By leveraging predictive analytics, farmers not only optimize their operations but also mitigate risks associated with fluctuations in climate and market conditions. In order to improve productivity and make wise decisions for long-term agricultural success, farmers can use yield prediction (Figure 2) as a valuable tool for sustainable and effective farming techniques.

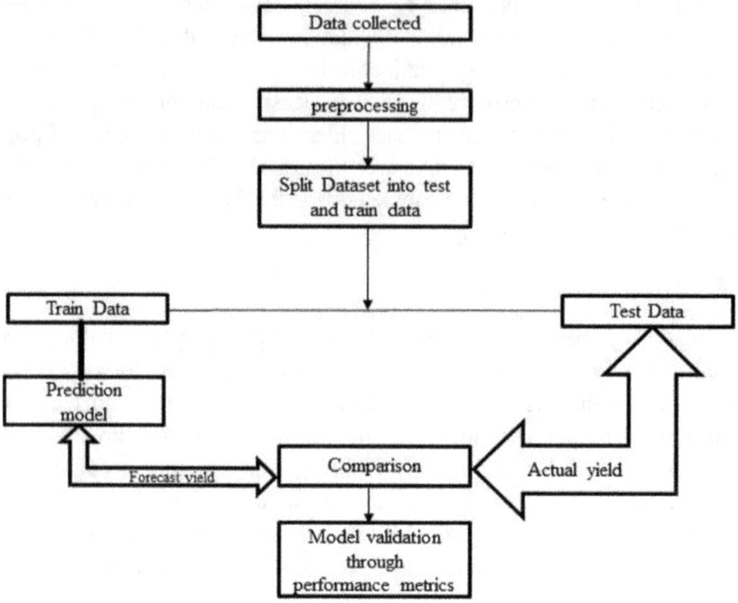

Figure 2: Yield prediction general architecture.

5.6.3. Operational Efficiency

Automated Equipment Scheduling: Applying AI algorithms to optimize the scheduling of automated equipment, including harvesters and tractors, to increase productivity and decrease downtime. In agriculture, operational efficiency refers to the smooth integration of cutting-edge technologies to optimize several areas of farming methods, leading to less environmental impact, higher productivity, and resource savings. Precision agriculture techniques, which employ automated machinery and GPS-guided machinery, can help farmers enhance their operational processes, including planting, harvesting, and irrigation. Simplified procedures are further enhanced by continuous tracking and based on data decision support systems, which minimize downtime and enable quick reactions to changing conditions [22]. By guaranteeing optimal resource allocation, the application of AI algorithms in supply chain management and machinery scheduling increases overall efficiency. In addition to being essential for cost-effectiveness, operational efficiency also helps to advance sustainable farming methods by coordinating farming operations with the concepts of accuracy and flexibility in order to create a more resilient and fruitful agricultural future.

5.6.4. Supply Chain Optimization

Utilizing data-driven insights to ensure timely delivery, cut waste, and optimize supply chain operations from seeding to distribution. Using data-driven solutions and technology strategically to improve the efficiency, sustainability, and transparency of the whole agricultural manufacturing and shipping process is known as supply chain optimization in agriculture. Optimization techniques make use of real-time data from automated machinery, IoT devices, and AI-powered analytics to maximize supply chain performance from the land to the customer. As a result, waste is decreased and logistics are optimized. Predictive analytics may be used by farmers to better plan planting times and match output to market demand, and supply chain traceability and transparency are guaranteed by technologies such as blockchain.

With the help of data insights, automation during harvesting and packaging minimizes delays and improves the freshness of product. In addition to making agriculture more economically viable by cutting costs and boosting market competitiveness, supply chain optimization promotes sustainability by lowering environmental effects and guaranteeing the timely delivery of high-quality goods to customers [20].

5.7. Continuous Improvement

In agriculture, continuous improvement refers to the process of continuously improving farming methods in order to increase production, sustainability, and efficiency. Farmers make an effort to use cutting-edge technologies, use best practices, and keep up with the most recent developments in the agricultural industry. This strategy promotes the combination of sustainable practices, precision agricultural methods, and data-driven decision-making. Farmers may minimize their influence on the environment, maximize resource utilization, and adjust to changing conditions like climatic variations by continuously evaluating and modifying their practices [19]. Agriculture needs to be continuously improved in order to promote resilience for long-term sustainability with the farming sector in addition to production increases.

5.7.1. Feedback Loops

Establishment of feedback loops that allow farmers to assess the impact of decisions and adjust strategies based on real-world outcomes. Feedback loops in agriculture represent a dynamic and iterative process where data-driven insights are continually used to refine and adapt farming strategies. These loops entail the ongoing gathering of data from numerous sources, including sensors, IoT devices, and previous performance information. The data is then evaluated to produce insights that can be put to use [3, 4]. These insights help farmers decide how best to manage their crops, allocate resources, and make operational changes. Crucially, the results of these choices are incorporated within the system, generating an ongoing cycle of development and enhancement [17]. The utilization of feedback loops enables farmers to fine-tune their practices based on real-world outcomes, fostering adaptability and resilience in the face of changing conditions. In addition to increasing production, this iterative method helps to promote sustainable agriculture by reducing resource waste and improving long-term success methods.

5.7.2. Adaptive Strategies

Utilization of data to iteratively refine farming practices, adopting adaptive strategies that respond to changing conditions and improve overall productivity and sustainability. In essence, data-driven decision making in agriculture is a dynamic and iterative process that empowers farmers to optimize resources, mitigate risks, and enhance overall farm performance [14].

6. Challenges and Future Developments

Precision farming faces challenges such as the initial cost of adopting technology, the need for education and training, and ensuring interoperability of different systems. Ongoing research and development are focused on enhancing the integration of AI, machine learning, and robotics in precision farming, making systems more intelligent, adaptive, and user-friendly. Precision farming represents a transformative shift in agriculture, aligning technological advancements with sustainable and efficient farming practices. As the world strives to achieve zero hunger, precision farming stands as a crucial tool in ensuring food security, optimizing resource utilization, and creating a more resilient and sustainable agricultural future [15]. To sum up, GPS technology is the foundation of precision farming since it offers the spatial intelligence required to maximize agricultural productivity. The use of GPS in conjunction with other cutting-edge technology has made traditional farming a data-driven, effective, and sustainable industry. The use of GPS in precision agriculture is growing as the agricultural industry changes, supporting international efforts to guarantee food security, reduce environmental impact, and achieve the aim of ending hunger [16]. Farmers may more precisely and effectively manage the intricacies of modern agriculture by utilizing the insights gained from thorough data analysis, which helps to promote resilient and sustainable farming methods.

7. Conclusion

In conclusion, the introduction of accuracy, efficiency, with data-driven decision-making by smart technology is a critical factor in the transformation of agriculture. Irrigation, fertilization, and pest control are just a few of the farming practices that have been completely transformed by the incorporation of cutting-edge technologies like GPS, sensors, and data analytics. These developments enhance agricultural productivity and sustainability while also optimizing resource usage. As time goes on, sustained advancements in intelligent technology will be necessary to handle global issues and guarantee the efficacy and robustness of contemporary farming methods.

References

1. Godfray, H.C.J., Beddington, J.R., Crute, I.R., Haddad, L., Lawrence, D., Muir, J.F., ... and Toulmin, C. (2010). Food security: The challenge of feeding 9 billion people. *Science*, 327(5967): 812–818.
2. Khaki, S. and Wang, L. (2019). Crop yield prediction using deep neural networks. *Frontiers in Plant Science*, 10: 621.
3. Elavarasan, D. and Vincent, P.D.R. (2021). A reinforced random forest model for enhanced crop yield prediction by integrating agrarian parameters. *Journal of Ambient Intelligence and Humanized Computing*, 1–14.
4. Torky, M. and Hassanein, A.E. (2020). Integrating blockchain and the internet of things in precision agriculture: Analysis, opportunities, and challenges. *Computers and Electronics in Agriculture*, 178: 105476.
5. Kamilaris, A. and Prenafeta-Boldú, F.X. (2018). Deep learning in agriculture: A survey. *Computers and Electronics in Agriculture*, 147: 70–90.
6. Becker-Reshef, I., Vermote, E., Lindeman, M. and Justice, C. (2010). A generalized regression-based model for forecasting winter wheat yields in Kansas and Ukraine using MODIS data. *Remote Sensing of Environment*, 114(6): 1312–1323.
7. Cramer, S., Kampouridis, M., Freitas, A.A. and Alexandridis, A.K. (2017). An extensive evaluation of seven machine learning methods for rainfall prediction in weather derivatives. *Expert Systems with Applications*, 85: 169–181.
8. Yost, J.L. and Hartemink, A.E. (2020). How deep is the soil studied: An analysis of four soil science journals. *Plant and Soil*, 452: 5–18.
9. Suchithra, M.S. and Pai, M.L. (2020). Improving the prediction accuracy of soil nutrient classification by optimizing extreme learning machine parameters. *Information Processing in Agriculture*, 7(1): 72–82.
10. Keswani, B., Mohapatra, A.G., Keswani, P., Khanna, A., Gupta, D. and Rodrigues, J. (2020). Improving weather dependent zone specific irrigation control scheme in IoT and big data enabled self-driven precision agriculture mechanism. *Enterprise Information Systems*, 14(9–10): 1494–1515.
11. Batchelor, W.D., McClendon, R.W., Adams, D.B. and Jones, J.W. (1989). Evaluation of SMARTSOY: An expert simulation system for insect pest management. *Agricultural Systems*, 31(1): 67–81.
12. Ferentinos, K.P. (2018). Deep learning models for plant disease detection and diagnosis. *Computers and Electronics in Agriculture*, 145: 311–318.
13. Irimia, M. (2016). Five ways agriculture could benefit from artificial intelligence. *AI for the Enterprise, IBM.* ISSN:0975-2927.
14. Banerjee, G., Sarkar, U. and Ghosh, I. (2017). A radial basis function network based classifier for detection of selected tea pests. *International Journal*, 7(5).
15. Bargoti, S. and Underwood, J. (2017, May). Deep fruit detection in orchards. *In*: *2017 IEEE International Conference on Robotics and Automation* (*ICRA*), 3626–3633. IEEE.
16. Anand, A.J., Devi, K.N., Vijayakumar, S. and Chacko, S.C. (2024). Decision-making support in smart farming. *In*: *Advanced Technologies for Smart Agriculture*, 219–239. Denmark: River Publishers.
17. Augustine, C., Balaji, K., Dharanikumar, S.V. and Anand, A.J. (2024). Urban farming: Case study. *In*: *Advanced Technologies for Smart Agriculture*, 321–338. Denmark: River Publishers.

18. Duraiarasu, E., Mahalakshmi, S., Anand, A.J., Manikandan, T. and El Rhalibi, A. (2024). Precision farming for crop prediction. *In*: *Advanced Technologies for Smart Agriculture*, 195–217. Denmark: River Publishers.
19. Nisha, K., Augustine, C., Anand, A.J. and Kamesh, M. (2024). Deep learning in smart agriculture. *In*: *Advanced Technologies for Smart Agriculture*, 153–175. Denmark: River Publishers.
20. Balamurugan, K., Rammya, B., Kumar, D., Kalanithy, K. and Mathan, J.N. (2023, Sep.). Implementation of effective rainfall forecast model using machine learning. *In*: *2023 4th International Conference on Smart Electronics and Communication* (*ICOSEC*), 1655–1660. IEEE.
21. Ahmed, L.J., Dhanasekar, S., Govindaraj, V. and Ezhilazhagan, C. (2024). Crop management system. *In*: *Advanced Technologies for Smart Agriculture*, 55–79. Denmark: River Publishers.
22. Sathish, R., Chand, T.P., Gummaluri, S.R.P., Kanmani, M.V. and Daniya, T. (2022, Nov.). A review on smart farming based on soil and environmental factors with deep learning techniques. *In*: *2022 International Conference on Augmented Intelligence and Sustainable Systems* (*ICAISS*), 222–230. IEEE.
23. Zineb, R., Samira, K. and Larbi, K.M. (2023). Applications of smart agriculture for environmental protection using deep learning techniques. *In*: *E3S Web of Conferences*, 412: 01083. EDP Sciences.

10 | Community Empowerment through Technology

Benjamin Kodai Kaje[1*], Ningchuiliu Gangmei[1] and K. Krelo Peter Kajeo[2]

[1] School of Education, Christ University, Bangalore, India

[2] Department of Political Science, Hill College Tadubi, Senapati District, Manipur

A UN report estimated that 10% of people on the planet will experience chronic hunger by 2022, with a large share of those people coming from rural origins. In Manipur, a Northeastern state of India characterized by hilly terrain and sparse population, predominantly inhabited by Nagas and Kukis tribes, urban migration threatens to exacerbate poverty. Although hunger is not frequent among Nagas at the moment, one cannot rule it out in future. When smart technologies ably support agriculture, it increases production in a sustainable way, thereby reducing hunger among the tribal populace. For this, the government must assist farmers in discouraging environmentally damaging activities like *Jhum* cultivation and cultivating poppies. For now, tribals have food to eat; nevertheless, future hunger problems could result from a lack of assistance. This research imagines a future where environmental sustainability is maintained while supplying for the requirements of the people of Manipur by giving technology-assisted farming prime importance. By using a participatory observation approach, the authors highlight how urgent it is to stop more ecological deterioration and protect the biodiversity of the area. Ultimately, a balanced ecological future for Manipur's rural and urban inhabitants depends on the adoption of sustainable smart agriculture.

1. Introduction

Manipur is a small state with big problems. There are challenges unique to every state, yet the situation in Manipur has also been gripping and alarming. The kind of social unrest since May 2023 has shown its dark side to the world. There are various reasons for the cause of social disparities which stir up different

*Corresponding author: benjamin.kaje@christuniversity.in

problems. One of the many problems is security, including food, land, and identity. Agriculture and its produce are vital factors that bind and give security to people everywhere. Hunger could drive people to go to any extent to commit anything without ruling out crime. Agriculture must feed the growing population with all its challenges and problems. To feed a growing population while preserving the environment and natural resources, agriculture must overcome difficulties that are getting worse [1]. The agricultural sector has a difficult task ahead of feeding the masses considering the ecology and preserving nature for the next generations.

One of the primary issues facing the Manipur hill districts is that there are no industries or institutions that could provide huge employment opportunities. For different reasons, Corporates and companies are not willing to venture into the hills to set up factories and institutions of importance. It is the rural nature of the industry, which has low productivity and operates in isolated, rural environments that are difficult for contemporary economic techniques to penetrate. When they compete in domestic and international markets, their lack of contemporary institutions and infrastructure puts them at a disadvantage [2]. Like many states of India, there has been unequal development in most sectors between the hills and valleys of Manipur. Improving agricultural produce is bound to improve the lives of the people in general. Yet, because of the undeniable fact of globalization, society has to live forward and cannot live backwards. At the same time, this cannot be done at the expense of destroying nature. Getting enough food is getting difficult due to climate change. It lowers agricultural productivity and jeopardizes natural resources like irrigation water availability. The agricultural industry confronts hurdles to profitability in addition to climate concerns because of increased input costs (such as fertiliser), variable commodity prices, and expanding regulatory requirements [3].

The present problem in Manipur could be largely due to the scarcity of land and resources. Tribals have the right to live in the capital city of Imphal, whereas the vast majority of Meiteis cannot live in the hills, thus creating some disquiet between the two communities. The hill districts that are occupied by the tribals are Senapati, Kangpokpi, Ukhrul, Kamjong, Tengnoupal, Chandel, Churachandpur, Pherzawl, Noney, and Tamenglong, and the remaining districts are dominated by the Meiteis who are the majority in the valley. Geographically, Manipur represents only 0.7% of the land and 0.24% of the people of India. The state has 28.56 lakh people living in it, or 128 people per square kilometre, according to the final results of the 2011 population census [4]. There is a trend among the tribals to head towards the towns and cities within the given limited geography of Manipur, which further intensifies the problem between the different communities. Like in other parts of the country, the propensity to migrate to towns and cities is an unfortunate reality. And this is for better job opportunities and scope for better education for their children.

A Green Revolution is the need of the hour to solve various complex issues plaguing the state. Researchers have developed various techniques and technologies to support agricultural production stability in the face of extreme seasonal changes. It seems more imperative than optional for farmers to adopt such robust techniques and technologies [5]. Digitalization has been crucial for enhancing agricultural output, rationalizing the use of natural resources, and promoting sustainable agriculture. The agricultural techniques that were previously empirical and labour-intensive have been transformed by internet of things (IoT)-driven agriculture [6]. Without a doubt, technological advancement will help people get rid of their poverty, and the menace of hunger will not arise among the tribals of Manipur.

People living in the hills were almost self-sustained in the past. However, people in the past may have had poor water systems, sanitation, and hygiene (WASH). Nevertheless, with modern education and style affecting the far and rural areas, the trend for better education and opportunities affected even the tribals of Manipur. Due to its hilly topography and inadequate infrastructure, North East India is not only geographically separated from the rest of the nation but also has a variety of sociocultural child-rearing customs because it is home to different tribal and ethnic communities. These native folks don't have access to basic facilities and live in traditional thatched roof huts. The high rate of childhood morbidities and malnutrition in this area can be attributed, either directly or indirectly, to the absence of essential utilities [7].

2. Methodology

The sources are mostly secondary as gleaned from various journals, books, and chapters. Besides, the primary authors, being natives of the place, have first-hand experience of the prevailing conditions and the challenges faced by the people. Though the hills of Manipur consist of about 90% of the total landmass of the state, the vast majority of the land is not fit for cultivation. Some even put the figure that 90% of the total hilly region of Manipur is not suitable for cultivation. Having traversed across the length and breadth of the hills of Manipur and seeing the quick depletion of the forest due to defective agricultural practices besides the widely cultivation method of *jhum*, smart agriculture will streamline the agricultural method in Manipur. The authors further employ the participation observation method and pooling from the rich online materials to conceptualize the paper's completion.

There is no pretension on the part of this chapter in its attempts to map the future of smart agriculture in the hills of Manipur from a sustainable perspective. It has to be conceptualized first to bring about any changes in the world. Aspects of smart agriculture can bring about tremendous transformation in the people's social, cultural, ecological, and even political well-being. Though one may say that the tribals, in general, do not go to bed hungry, all the same, if due care is not taken in the economic, social, and political milieu, one can foresee a grim future not far from now. After a brief introduction in the chapter, the background of the study is vital to get a bird's view of the geographical, social, and economic situation of Manipur, with special reference to the hills that the tribes inhabit. Thereafter, a short discussion is on the present scenario of hunger in the world, the cultural practice of the tribals in the past to sustain one another, and briefly dealt with the concept of smart agriculture. Additionally, the need for smart agriculture, with all its challenges in the hilly terrain of Manipur is discussed. The most advanced technologies of agriculture cannot possibly be implemented in the hills, and the reasons for this will be deliberated on in the later part of the chapter. The chapter progressively deliberates on the findings and discusses, wherein it concludes, that the need for smart agriculture is stressed for a sustainable future.

3. Background to the Study

The whole region of North East India lags behind the rest of India in most sectors. From infrastructural development to factories, educational hubs to sports amenities, advanced medical facilities to the latest mode of transportation, the people from the region are denied these opportunities. It is not because of a dearth of talent but a lack of opportunity. The people in this region have historically relied heavily on agriculture, but because effective participation and utilization of technical aids are lacking, the agro-based economy does not develop as it could [8]. Nevertheless, it has a rich reserve of biodiversity and natural resources. Almost half of the blooming plants on the Indian subcontinent are found in North East India, which is a relatively undisturbed region and part of the Indo-Burma biodiversity hotspot [9].

There is a constant and prolonged tension in Manipur between the three ethnic races, viz., Meitieis, Nagas, and Kukis. The Nagas and Kukis belong to the tribal communities, whereas the majority of the Meities belong to the general category. The state has a geographical area of 22,327 sq. km, of which 90% is hilly and occupied by the tribals, and the remaining 10% is the valley where the majority are Meiteis. Of this, the tribals could purchase land within the limited 10% of the land, whereas the Meities could not buy the land of the tribals. Hence, some tensions have been brewing over the years. The steady growth of the state's population over the decades has not helped the cause of the environment or peace in society. A growing trend in temperature, precipitation, and greenhouse gas emissions has been observed in Manipur, where the agriculture sector faces the effects of climate change. Approximately 80% of the state's population lives on agriculture [10]. With a population of more than 2.8 million, it had up to roughly 0.24% of the nation's total population according to the 2011 census, whereas its landmass comprised approximately 0.7% [4]. The state average is 103 people per square kilometre, although the population density in the hill areas is 44 people per square kilometre.

The size of the cultivated area is about 10.46% of the total geographical area. Agriculture is characterized by rugged terrain, wide variations in slopes and altitude, community land systems, and conventional cultivation practices [10]. It remains one of the main occupations in the state. In Manipur, agricultural labourers and cultivators comprise 52.81% of the workforce. Nonetheless, meteorological conditions and timely rainfall are the primary determinants of agriculture's performance in the state. While terrace cultivation is practised in some areas of the hills, where *jhuming* or shifting cultivation is commonly used in most hills, permanent cultivation is typically practised in the valley districts [4]. The fruits of science and technology have hardly been used, barring two-wheeled power tillers and mini tractors in the hills of Manipur. There are many tribes who depend mainly on *jhum* or shifting cultivation. When shifting cultivation is practised among the tribals, lots of trees are felled for the purpose of this cultivation.

Another menace that ails the hills of Manipur is the plantation of poppy cultivation. There are supposedly isolated highland regions throughout the state where poppies are grown. It is anticipated that in 2017–18, approximately 6,000 acres were under poppy cultivation spread throughout various hill districts in Manipur. Government agencies' annual eradication campaigns typically result in the destruction of about 10% of all farmed crops, a small amount [11]. Though poppies are cultivated in the hills, the primary objective is to sustain the family and assist in meeting the children's educational expenses. The motive of many struggling farmers may be ideal, yet in the process of entertaining poppy cultivation, it has harmed the ecology and, more specifically, cut short the future generations. The state government needs to give alternative and sustainable forms of agriculture to these cultivators to desist them from this type of adventure.

4. Situation of Hunger

The world has gotten closer due to globalization, better communications, and smart transportation like never before. As a result, serious issues in any nation always touch more than just that nation. According to the United Nations, approximately 735 million people, or 9.2% of the world's population, will experience chronic hunger by 2022, a startling increase from 2019. This data highlights the gravity of the issue and points to a developing crisis. Regarding sustainable agriculture, food security, and nutrition, the vast majority of sustainable development goals' (SDG) targets are not expected to be met by the globe on time [12]. Due to the severity of global weather extremism, the world has come to a belated realization to safeguard the fragile earth. The exploitation of natural resources and environmental damage by humans under human dominion on earth is no longer socially acceptable because it ultimately results in the planet Earth's self-destruction as a little 'spaceship' [13]. The internet can expand this field's possibilities because of the ongoing rise in global food demand. Agricultural pests are better managed, and yield is raised with the help of intelligent irrigation systems. As a result of these smart systems using less water during irrigation, soil erosion in agricultural fields is decreased. Thus, it is possible to monitor the field in order to measure key plant indicators like temperature, soil moisture content, and soil and water pH [14].

While the world is geared towards feeding people experiencing poverty and care is taken to ensure enough food production, another vital issue to be addressed is the issue of proper distribution and sound storage of food grains. The earth has produced much beyond the imagination of ordinary minds. Yet, people experiencing poverty continue to go hungry due to issues within and outside the state. The problem of the distribution of agricultural products is also essential. Even though there is enough food to feed everyone on the planet right now, 500 million people still suffer from malnutrition, and over 821 million go hungry [15].

4.1. Hunger among the Tribals

One of the chief reasons that there are no beggars among the tribals of Manipur and Nagaland is due to certain cultural practices. In the past, there were no filthy rich people among the Nagas. The Feast of Merit or Feast of Thanksgiving after a good harvest acts as a leveller to the gap between the rich and poor. Inadvertently, since the arrival of Christianity among the Nagas, the practice of the Feast of Merit has become redundant. With the adoption of Christian customs among the tribe, this tradition was eventually discontinued. It is commonly understood to refer to feasts that affluent people host as a gesture of hospitality and charity to elevate their social standing [16]. This could be one of the innate reasons why there was no extreme poverty and riches in the past. It is unknown among the Naga tribals that someone would go to bed hungry due to lack of food. Yet, with time, the gap between the rich and the poor is widening. The need to produce sufficiently to feed the growing population is a necessity and not an option any longer.

In a similar vein, Kukis observed *Chavang Kut* (autumn festival). It is observed on three different grounds. Firstly, it is a time for the hardworking agricultural community to take a break from their work. Secondly, it is a way for them to express their gratitude to the god they worship for an abundant harvest. Lastly, it serves as a channel for intercession between the community and the supernatural forces to benefit all [17]. As for the tribals among the Kukis of Manipur, the chiefs own the land, and ordinary citizens are mere occupants of the land. Lands have to be utilized in the best possible way to get maximum benefit out of it. If a person gets lucky with a consecutive surplus harvest, the family performs a *Chang-ai* (Paddy ritual). Someone who gathered far more grains than needed for their household for multiple years in a row performed this ceremonial dance. The villagers' young people would go out to the field to help carry the rice grains as part of the rite [17]. Hence, when a family hosts *Chang-ai*, the family feeds the villagers, and thereby, the surplus food grains are not hoarded for the future.

5. Agriculture in the Hills

Jhum cultivation was practised among the tribals of Nagas and Kukis. There was no issue with land some decades ago because fewer people and villages could afford to rotate places for *jhum* cultivation. Besides, not all the villages among the tribals have the luxury of wet fields and terrace cultivation due to the mountainous terrain. For many communities, this was the only way of life. And yet, with time and better education, people have come to understand the malady of shifting/*jhum* cultivation. The Mao Nagas, for instance, totally abandoned this type of cultivation. It has caused much harm to the ecology due to the massive felling of trees, thereby disrupting the floral and faunal world. In addition, landslides may occur occasionally, and soil erosion is a worry during the rainy season.

Regarding the *jhum* agricultural system's efficiency, research indicates that its peak efficiency was achieved over a 10-year cycle; when the cycle duration is shortened or lengthened, efficiency decreases [26]. The *jhum* system is typically used on hill slopes, where poor yield and efficiency are caused by blown-off, run-off, and the percolation of mineral nutrients, which are the main drivers of nutrient loss [26]. While the region's population density is substantially lower than that of the entire country (324 persons per km^2), the north-eastern region's decadal growth rate (31.18 in 1991–2001) is significantly higher than the national rate (21.35 persons per km^2) as recorded between 1991 and 2001 (as cited in Majumder, 2011) [26]. Hence, smart agriculture would be more than handy to tackle and bring about marked changes in the hills of Manipur. For the benefit of both people and the environment, smart agriculture should be used sooner rather than later.

6. Smart Agriculture

It has taken years for smart technology to catch on in the agricultural industry. It may be among the many industries where the adoption of modern technologies has happened least quickly. This sector's process and progress have been slow yet advanced for the betterment. The four distinct revolutions in agriculture are as follows: (i) the traditional age, characterized by the use of human and animal power; (ii) the mechanized age, characterized by rumbling sounds; (iii) the automated age, characterized by rapid development; and (iv) the smart agriculture age, characterized by emerging technologies [18]. For India to thrive and sustain its expanding population, the benefits of prosperity must spread to its rural areas [19]. Using technology such as sensors, robots, artificial intelligence, positioning systems, IoT, and robots on your farm is known as smart agriculture. The ultimate goal is to maximize the amount of human labour employed while simultaneously improving crop quality and yield. Using IoT-powered smart farming, farmers and producers can maximize output and minimize waste in everything from the amount of fertilizer used to the number of trips made by farm vehicles. Smart farming is a high-tech, capital-intensive method of producing sustainable, clean food for the general public [20]. Smart agriculture represents a sophisticated phase in the evolution of agricultural informatization, building upon digitalization, networking, and intelligence to create a new paradigm for agricultural production. This involves utilizing knowledge and information as the foundation and incorporating contemporary information technology, including big data, cloud computing, artificial intelligence, IoT, intelligent equipment, and so forth, to enable agricultural information perception, quantitative decision-making, intelligent control, precise input, and personalized service [21].

There are many terms, such as smart agriculture and smart farming, that all more or less denote the same thing. By providing each plant or animal with precisely what it needs to thrive in the best possible way, precision agriculture is a clever farming method that increases operational accuracy and maximizes overall performance while lowering waste, inputs, and pollution [22]. Climate-smart agriculture (CSA) and Smart Agriculture are closely connected. CSA aims to achieve three distinct objectives. The three main goals are to: (i) raise agricultural productivity to support higher incomes, food security, and development; (ii) build various layers of adaptive capacity (from farm to nation); and (iii) reduce greenhouse gas emissions and add to carbon sinks [23].

Being native to the place, one could see the vast barren land due to *jhum* cultivation and poppy plantations. If one travels from AH 1, from Senapati to Imphal, one can see vast stretches of almost barren land along the highway. The issue of water shortage has been a perennial problem in the state of Manipur and more so in the hilly districts of Manipur. The need for modern technology will assist in afforestation in the hills, which will improve the ecosystem and overall ambience of the atmosphere. Local CSOs, NGOs, and student bodies, along with the government machinery, could take the initiative to plant a variety of trees, vegetables, and fruits, which could be conducive to the hilly regions of the area.

GPS-enabled remote-controlled robots for weeding, spraying, measuring moisture, scaring birds and animals, being watchful, etc. are important technologies that will improve agricultural production. Second, it incorporates intelligent decision-making based on precise real-time field data and smart irrigation with smart control. Thirdly, intelligent warehouse management, encompassing temperature regulation, humidity control, and theft prevention within the warehouse [24]. As a result, the technology transfer approach needs to be reoriented. A higher emphasis is placed on sustainability, long-term agricultural development, environmental and social protection, and resource efficiency [5]. Global technology advancements in recent years have resulted in both higher production and far higher production efficiency. Agriculture-related tasks, including spraying, cutting, weeding, ploughing, and other tasks, can be accomplished using and developing agricultural machinery that can be handled manually and automatically [14]. Precision farming is all about getting the "*Right Product, Right Place, Right Time*" right. With more precise and controlled methods than conventional farming, precision farming eliminates agriculture's labour-intensive and repetitive aspects [25].

7. Need for Smart Agriculture

The truth of the growth of population is an undeniable fact for a developing country like India. The same truth holds for a state like Manipur. And so, with the growth of population comes the need to feed more people. North East India has a less developed economy, and because of the region's customary shifting (*jhum*) farming methods, woods are severely impacted by human activity [9]. However, correspondingly, the land remains the same, and the need to feed a greater number comes as a burden to the land. Progressive climate change exacerbates global difficulties that humanity faces, such as the need for higher agricultural production, population growth, urbanization, land degradation, and the transition to animal-based diets [27]. The majority of agricultural production systems are mono-cropping, rainfed, and intended for subsistence [10]. The farmers need technical help because they rely a lot on the rain. Agriculture experts can provide a step toward sustainable agriculture. Agriculture scientists must, however, be able to impart new knowledge to the civic society and to themselves using a systems paradigm that emphasizes relationships over the discrete elements of agricultural reality, as disciplinary approaches typically do [13].

Burgeoning IoT technology presents a plethora of fresh chances for growth and new solutions, especially for innovative ideas in the agricultural sector [28]. It is critically necessary to conduct research on the creation of crop varieties and cultivation techniques tailored to a given region, including the modernization of the *jhumming* system and broad technology dissemination initiatives, as well as improvements to irrigation, communication, marketing, and processing facilities, in order to raise farmers' living standards [29]. When rice growers overcome the challenges and limitations they face and improve their access to technology without compromising the environment, the rice production system can be considered sustainable [30]. Widespread variations in temperature and rainfall patterns endanger agricultural output and make those who depend on it for a living—the majority of the world's impoverished—more vulnerable [31]. Due to the remote locations, low population density, challenging topography, and expensive transportation of goods to more populated distant markets, there are few outlets for the vegetables grown in the hill regions. When it comes to competing with farmers in more accessible places in the domestic export markets, the hill people are significantly less advantaged, which means they do not get the same opportunities [2].

To suggest that Manipur, and especially the hill districts, would adopt a total smart agriculture would be unrealistic. A person who has lived and understands the challenges in the hills of Manipur will vouch for this reality. However, there are elements of smart agriculture that could be used to lower the Manipur tribal population's hunger index. Conservation agriculture, agroecology, ecosystem-based management, small-scale irrigation, agroforestry, soil/water conservation, and grazing land management are examples of low-cost sustainable agriculture techniques used for decades [32]. The small steps taken now will have a huge impact on the environment and society as a whole among the tribals of Manipur in the long run.

The need to adopt aspects of smart agriculture is vital and cannot be underrated or ignored. Every part of the world is feeling the effects of climate change, global warming, and erratic weather. The hill stations of Mao area in Senapati used to have pleasant summer temperatures in the recent past, yet the need to have fans at home has become a reality today. There were no mosquitoes in the cold regions of the hills of Manipur; however, with the rising temperature, bed nets have become a requirement today. In order to support people who are most vulnerable to the effects of climate variability, adaptation options must be identified as these occurrences become more frequent and severe [32].

8. Smart Agriculture and its Challenges in the Hills

The implementation of smart technology in Manipur, particularly in the hills, is not without its evident difficulties and barriers. In addition to poor transportation and unstable power supplies, there are significant problems with internet connectivity. There are issues related to storage and transportation, and there are barriers even to the selling of agricultural products in rural areas. As a result, during the season, seasonal fruits and vegetables are sold for a discount, and during the off-season, farmers must spend a lot of money on buying the basics, which occasionally includes buying their sold commodities. The financial support of rural economies is closely tied to agricultural policy in many nations. Increasingly, low-income nations have the ability to direct their production down more beneficial and sustainable paths [23]. Unless the rural (hills) areas are developed, Manipur cannot march forward in the agricultural field. For the state's progress, agriculture is one of the important areas that need attention. Senapati and Ukhrul were classified as moderately vulnerable, whereas Tamenglong, Churachandpur, and Chandel were classified as severely vulnerable. It is discovered that hill districts are more susceptible than valley districts [33].

8.1. Unpredictability of the Rain

One of the many hurdles the farmers in the hills of Manipur face is the overdependence on monsoon rain and the lack of an irrigation system. And farming has almost been a trial-and-error method, which has been practised for centuries. There has been hardly any scientific research put into use for the ordinary farmers of the day. The main focus of the hills development strategy is undoubtedly the requirement for progressive plans that are both environmentally and economically sound, as well as for their efficient implementation. The range of issues in the hill regions is enormous, even before development forces have had time to construct solid institutional foundations [2]. The processes of preparing the land, planting seeds, growing crops, and harvesting can all be harmed by a number of uncertainties. A source of income that is unpredictable due to all these risks is agriculture. Indian rural youth are moving to the country's cities due to the agriculture sector's erratic wealth [34]. Unless technology comes to assist the farming community, farmers will be at the mercy of the rain god.

8.2. Tendency to Go to Towns and Cities

Like many states of North East India, Manipur is faced with the problem of immigration from around the neighbouring states and the country. Moreover, it is a fact that many common people in the state want to move from their rural communities to towns and cities for multiple reasons. The region's worryingly high population growth rate is caused by both domestic and international immigration. The reasons could differ from better employment opportunities to access to quality education; the North East Region (NER) residents' access to arable land and the region's natural resources are both seriously threatened by immigration-induced population expansion [8]. There are lots of inconveniences living in the villages, like connectivity with networks, quality education, healthcare, conveyance, amenities at home, etc. Hence, a person who has experienced a city way of life would, in general, not want to bear up with those challenges in rural areas. In villages, there is a move toward non-farm businesses, and some villagers are leaving the villages for urban areas in search of better job opportunities [35].

9. Findings

Within a decade, the climatic conditions of Manipur as a whole have become comparatively warmer. Significant changes in Manipur's climate in recent years have affected the region's weather patterns, including altered monsoon patterns, heavier rainfall, and longer dry spells. These climate fluctuations have a significant effect on agricultural productivity, endangering food security as well as rural residents' income and way of life [33]. The irony of life is that, contradictorily, people who work primarily in the food and agriculture industries are also the most susceptible to hunger [36]. The apparent defects of *jhum* cultivation are well known to all. However, there are some aspects we need to consider. Although the system supports over 443 thousand *jhumia* homes, it is criticized for low productivity and environmental inefficiencies. Because of the system's diversity, *jhum* agriculture offers households both food and nutritional security [37].

The Imphal valley is comparatively better in terms of agricultural practices and the use of various technologies. Five metric tons (MT) of Chak-hao, or aromatic black rice, have been sent from Manipur to Kerala. This shipment, which came from Freshies Fresh, Nahakpam Food and Beverages in Chingmeirong, is evidence of the rising demand for this highly sought-after rice type. Additionally, preparations are in place to ship 20 MT of Chak-hao to Europe [38]. Yet, compared to the hills in Manipur, it is pathetic. People still use primitive methods of cultivation. There is hardly any modern technology put into use for cultivation. There are very few who use power tillers in the field during the agricultural season. Although rice is the principal grain grown in Manipur, farmers still have to contend with a number of obstacles when growing it, including financial, technological, institutional supply, irrigation facility, ecological, and marketing obstacles [30]. Not only is agriculture a victim of climate change, but it also produces greenhouse emissions. CSA is an approach that aims to boost agricultural output without contributing to greenhouse gas emissions from farming operations [39]. Nowadays, poppy plants can be seen growing just a few kilometres away from residential areas in a number of Manipur's hill districts. The poppy plant is widely grown in the state's interior hill regions despite the Narcotic Drugs and Psychoactive Act of 1985 listing it as a prohibited narcotic recognized for its psychoactive properties. Although illicit poppy farming for opium has been going on for about 10 years, production has skyrocketed in recent years. According to a study, the reasons behind Manipur's illegal opium production are poverty, food insecurity, and material demands [11].

10. Discussions

The world's population is predicted to reach nine billion by 2050. One cannot expect the United Nations to lead in providing food for all. Each country and each state should wake up and do their bit to alleviate hunger and poverty. Without a pronounced transformation in agriculture, the World Bank projects that there will be nine billion people on the planet by 2050. It is sometimes stated that humanity would simply struggle to feed everyone. The ideals of smart agriculture look appealing on paper, yet implementing these principles is challenging. It is the scientific community's responsibility to pave the way for a sustainable and environmentally friendly future, given that science has contributed to such amazing growth. To restore the ecological balance in today's world, science must lead the way [40].

Adopting every smart agriculture concept would not be feasible or palatable in Manipur's hills. Water management, irrigation management, soil management, weather management, precision farming, fertiliser management, waste management, tracking and tracing of livestock, and even crop management present a wealth of opportunities. If some modern scientific and technology-related characteristics could be applied in the Manipur highlands, it would bring about revolutionary changes and prevent blind migration to the state's towns and cities. Establishing a reliable storage system in the hills to hold seasonal produce during the year's lean season and even export it to distant locations is another crucial component of smart agriculture. Many farmers are forced to sell their produce at a low price since there aren't enough storage facilities, which prevents them from making enough money from their laborious efforts.

One should be aware of the dangers of blindly striving for higher production without taking care of the environment. By putting more focus on food production, it appears to be misattributing food insecurity to food scarcity rather than a lack of access to land, jobs, and social safety nets. This could hasten environmental deterioration [41]. To ensure that society, the economy, and the environment are protected for future generations, scientific groups worldwide must cooperate and exchange knowledge [42]. Along these lines, smart agriculture would not save Manipur or the world's future. The means to achieve zero hunger have to be sustainable. Mere high production of food grains, with the aid of modern technology, could harm more than help, if the goal is just for more production of food grains.

11. Conclusion

For any state to grow, various sectors of society have to grow together. Smart agricultural systems reduce waste, boost output, and enable the management of a greater range of resources thanks to remote sensing. When using traditional farming methods, the farmer would constantly be out in the field assessing the state of the farmland [28]. Smart agriculture will aid in bringing about higher production and reduce the chances of hunger among the tribals of Manipur. It will also ensure that the environment is taken care of for the better. The menace of *jhum* cultivation is highly prevalent among the many tribes of Manipur. Better-equipped technology will ensure that the farmers will make use of the best within the limited area. It means producing sufficient food for a growing population without compromising the ecology. This will ensure that the land is cared for and nurtured for future generations.

It would not be possible to put into practice every aspect of smart technologies given the harsh terrain of Manipur. Nevertheless, beginning somewhere to some degree would mitigate the pressure on the land, thus saving the ecology to some extent. There have been apprehensions that, given the lure of quick money, many poor farmers have resorted to the cultivation of poppies in the hills. Though the money earned is used for the education of children, the end does not justify the means of achieving such goals. It is, therefore, in the nation's best interest to make farming dependable and profitable, which may be done by giving farmers a chance to have a supplementary source of income that pays well enough to cover their daily needs [19]. Early introduction of smart agriculture in the hills among the tribals will encourage them to be engaged in the hills with higher yield and taking care of nature. Besides, it would bridge the gap between the majority of Meiteis who are advanced in multiple fields with better income and greener job opportunities within the hills of Manipur.

More research studies could be incorporated into the agricultural sector in the hills of Manipur. Studies on the different varieties of soil, climatic conditions, and temperature could greatly help farmers try out other crops in the area. There are hot and humid places and cold locations in the hills of Manipur. Central

Agricultural University Imphal can be handy in disseminating information and enhancing production for cultivators in the hills of Manipur. Increased production among the tribals in the hills will reduce hunger to the overall burden of the state and, thereby, fulfil the goal of SDG2 of a hunger-free world by 2030. Though the terrains in the hills are challenging, almost 90% of the land is with the tribals who occupy the hilly region of the state. Any improvement in using smart agricultural technologies will benefit the state's overall population and lessen the burden on the environment by not having recourse to defective and harmful *jhum* cultivation.

Acknowledgement

The authors, namely Benjamin Kodai Kaje and Ningchuiliu Gangmei, are grateful to NFST for granting PhD fellowship (SRF) to them, respectively.

Disclosure Statement

The authors reported no potential conflict of interest.

References

1. Morchid, A., El Alami, R., Raezah, A.A. and Sabbar, Y. (2024). Applications of Internet of Things (IoT) and sensors technology to increase food security and agricultural sustainability: Benefits and challenges. *Ain Shams Eng. J.*, 15(3): 102509. doi: 10.1016/j.asej.2023.102509.
2. Gonmei, G.H. (2013). Hills economy of Manipur: A structural change. *J. North East India Stud.*, 3(1): 61–73.
3. Gomstyn, A. and Jonker, A. (2023). What is smart farming? *IBM*. https://www.smart-akis.com/index.php/network/what-is-smart-farming/.
4. Government of Manipur. (2021). *Economic Survey, Manipur 2020–21*. [Online]. Available: http://desmanipur.gov.in/files/NewsFiles/20Sep2019051341Economic Survey, 2018–19.pdf.
5. Prasad, Y.G., Maheswari, M., Dixit, S., Srinivasarao, C., Sikka, A.K., Venkateswarlu, B. et al. (2014). *Smart Practices and Technologies for Climate Resilient Agriculture*. Central Research Institute for Dryland Agriculture. [Online]. Available: http://www.crida.in
6. Huo, D., Malik, A.W., Ravana, S.D., Rahman, A.U. and Ahmedy, I. (2024). Mapping smart farming: Addressing agricultural challenges in data-driven era. *Renew. Sustain. Energy Rev.*, 189(*PA*): 113858. doi: 10.1016/j.rser.2023.113858.
7. Singh, K.D., Alagarajan, M. and Ladusingh, L. (2015). What explains child malnutrition of indigenous people of Northeast India? *PLoS One*, 10(6): 1–15. doi: 10.1371/journal.pone.0130567.
8. Mandal, R.K. (2011). Changing agricultural scenario and its impact on food habits in North East states of India. *Food Biol.*, 1(1): 14–21. [Online]. Available: http://foodbiology.org/.
9. Tripathi, S.K., Roy, A., Kushwaha, D., Lalnunmawia, F., Lalnundanga, L.H., Lalnunzira, C. et al. (2016). Perspectives of forest biodiversity conservation in

Northeast India. *J. Biodiversity, Bioprospecting Dev.*, 03(02): 2–9. doi: 10.4172/2376-0214.1000157.

10. Roy, S.S., Ansari, M.A., Sharma, S.K., Sailo, B. and Devi, C.B. (2018). Climate resilient agriculture in Manipur: Status and strategies for sustainable development. *Curr. Sci.*, 115(7): 1342–1350. doi: 10.18520/cs/v115/i7/1342-1350.
11. Singh, S.J. (2021). The menace of poppy plantation in Manipur. *Imphal Free Press*, Dec. 11. [Online]. Available: https://www.ifp.co.in/opinion/the-menace-of-poppy-plantation-in-manipur.
12. FAO. (2019). *Tracking Progress on Food and Agriculture-Related SDG Indicators*, 42. [Online]. Available: http://www.fao.org/sdg-progress-report/en/#chapeau.
13. Caporali, F. (2011). Agroecology as a Transdisciplinary Science for a Sustainable Agriculture. *In*: *Biodiversity, Biofuels, Agroforestry, and Conservation Agriculture*, 1–72. Springer.
14. Badran, A. and Kashmoola, M. (2020). Smart agriculture using Internet of Things: A survey. *In*: *Proceedings of the 1st International Multi-Disciplinary Conference Theme: Sustainable Development and Smart Planning, IMDC-SDSP*, 1–8. doi: 10.4108/eai.28-6-2020.2298249.
15. Rettore de Araujo Zanella, A., da Silva, E. and Pessoa Albini, L.C. (2020). Security challenges to smart agriculture: Current state, key issues, and future directions. *Array*, *8*(July): 100048. doi: 10.1016/j.array.2020.100048.
16. Mayirnao, S. and Khayi, S. (2023). Decolonizing feasts of merit: Reasoning Marān Kasā from a Tangkhul Naga perspective. *Asian Ethn.*, 24(2): 258–277. doi: 10.1080/14631369.2022.2089093.
17. Sitlhou, H. (2018). The shifting "stages" of performance: A study of "Chavang Kut" festival in Manipur. *Asian Ethn.*, 19(4): 468–488. doi: 10.1080/14631369.2018.1476834.
18. Liu, Y., Ma, X., Shu, L., Hancke, G.P. and Abu-Mahfouz, A.M. (2021). From Industry 4.0 to Agriculture 4.0: Current status, enabling technologies, and research challenges. *IEEE Trans. Ind. Informatics*, 17(6): 4322–4334. doi: 10.1109/TII.2020.3003910.
19. Mahto, R., Sharma, D., John, R. and Putcha, C. (2021). Agrivoltaics: A climate-smart agriculture approach for Indian farmers. *Land*, 10(11): doi: 10.3390/LAND10111277.
20. Mat, I., Kassim, M.R.M., Harun, I.M. and Ahmad Nizar Yusoff. (2018). Smart agriculture using Internet of Things. *In*: *2018 IEEE Conference on Open Systems* (*ICOS*), 54–59. Langkawi, Malaysia, doi: 10.1109/ICOS.2018.8632817.
21. Chunjiang, Z. (2019). State-of-the-art and recommended developmental strategic objectivs of smart agriculture. *Smart Agric.*, 1(1): 1.
22. Friha, O., Ferrag, M.A., Shu, L., Maglaras, L. and Wang, X. (2021). Internet of Things for the future of smart agriculture: A comprehensive survey of emerging technologies. *IEEE/CAA J. Autom. Sin.*, 8(4): 718–752. doi: 10.1109/JAS.2021.1003925.
23. Campbell, B.M., Thornton, P., Zougmoré, R., van Asten, P. and Lipper, L. (2014). Sustainable intensification: What is its role in climate smart agriculture? *Curr. Opin. Environ. Sustain.*, 8: 39–43. doi: 10.1016/j.cosust.2014.07.002.
24. Gondchawar, N. and Kawitkar, R.S. (2016). IoT based smart agriculture. *Int. J. Adv. Res. Comput. Commun. Eng.*, 5(6): 838–842. doi: 10.1109/WiSPNET.2018.8538702.
25. Pathan, M., Patel, N., Yagnik, H. and Shah, M. (2020). Artificial cognition for applications in smart agriculture: A comprehensive review. *Artif. Intell. Agric.*, 4: 81–95. doi: 10.1016/j.aiia.2020.06.001.
26. Majumder, M., Shukla, A.K. and Arunachalam, A. (2011). Agricultural practices in Northeast India and options for sustainable management. *In*: E. Lichtfouse (Ed.), *Biodiversity, Biofuels, Agroforestry, and Conservation Agriculture*, 287–315. Springer.

27. Mwongera, C., Shikuku, K.M., Twyman, J., Läderach, P., Ampaire, E., Van Asten, P. et al. (2017). Climate smart agriculture rapid appraisal (CSA-RA): A tool for prioritizing context-specific climate smart agriculture technologies. *Agric. Syst.*, 151: 192–203. doi: 10.1016/j.agsy.2016.05.009.
28. Sinha, B.B. and Dhanalakshmi, R. (2022). Recent advancements and challenges of Internet of Things in smart agriculture: A survey. *Futur. Gener. Comput. Syst.*, 126: 169–184. doi: 10.1016/j.future.2021.08.006.
29. Singh, Y.S. and Bera, B.K. (2017). Present scenario of agricultural sector of Manipur. *Econ. Aff.*, 62(2): 225. doi: 10.5958/0976-4666.2017.00004.3.
30. Thangjam, B. and Jha, K.K. (2020). International Web-Conference on New Trends in Agriculture, Environmental & Biological Sciences for Inclusive Development Sustainable Rice production in Manipur: Analysis of constraints faced by farmers. *J. Pharmacogn. Phytochem.*, 6: 57–63. [Online]. Available: http://www.phytojournal.com.
31. Lipper, L., Thornton, P., Campbell, B.M., Baedeker, T., Braimoh, A., Bwalya, M. et al. (2014). Climate-smart agriculture for food security. *Nat. Clim. Chang.*, 4(12): 1068–1072. doi: 10.1038/nclimate2437.
32. Lasco, R.D., Delfino, R.J.P., Catacutan, D.C., Simelton, E.S. and Wilson, D.M. (2014). Climate risk adaptation by smallholder farmers: The roles of trees and agroforestry. *Curr. Opin. Environ. Sustain.*, 6(1): 83–88. doi: 10.1016/j.cosust.2013.11.013.
33. Feroze, S.M., Aheibam, M., Singh, R., Ray, L.I.P., Rai, M., Singh, K.J. et al. (2014). Assessment of agricultural vulnerability to climate change in Manipur: A district level analysis. *Indian J. Hill Farming*, 27(1): 22–29. [Online]. Available: www.kiran.nic.in.
34. Bisht, I.S., Rana, J.C. and Ahlawat, S.P. (2020). The future of smallholder farming in India: Some sustainability considerations. *Sustain.*, 12(9): 1–25. doi: 10.3390/su12093751.
35. Sitlhou, H. (2022). Agro-economy in the hills of Manipur: An interplay of peasants, middlemen and markets. *J. North*, 12(2): 70–85.
36. Laborde, D., Parent, M. and Smaller, C. (2020). *Ending Hunger, Increasing Incomes, and Protecting the Climate: What Would It Cost Donors?* [Online]. Available: https://ecommons.cornell.edu/handle/1813/72864.
37. Barah, B.C. (2006). Agricultural development in North-East India: Challenges and opportunities. *Policy Br.*: 1–4.
38. Desk, B.N. (2024). Manipur's Chak-hao rice dispatched to Kerala, Europe expansion underway. *Business North East.*
39. Devi, V.M., Singh, R., Devarani, L., Hemochandra, L., Singh, R. and Sethi, B. (2020). School of Social Science, College of Post Graduate Studies in Agricultural Sciences. *Indian Res. J. Ext. Edu.*, 20(3): 52–56.
40. Kaje, B.K., Katini, K., Krelo Peter Kajeo, K. and Andrew Thomas, K. (2023). Indigenous house construction of the Mao Nagas: A sustainable perspective. *Mater. Today Proc.*, June: 0–8. doi: 10.1016/j.matpr.2023.06.038.
41. Gonzalez, C.G. (2022). Goal 2 end hunger, achieve food security and improved nutrition, and promote sustainable agriculture. *Sustain. Dev. Goals*. doi: 10.5040/9781509934058.0008.
42. Kaje, B.K., Haokip, P.L., Kajeo, K.K.P. and Thomas, K. Andrew (2023). Role of education for sustainability among the Tribals of Manipur: A critique. *In*: D.A. Karras, S. Thakur and S.K. Oruganti (Eds.), *Integrating Advancements in Education, and Society for Achieving Sustainability: Research and Evidence-Based Strategies from the Developing World*, 143–147. Routledge.

11 | Leveraging Machine Learning for Paddy Disease Identification in Sustainable Agriculture

M.K. Vidhyalakshmi[1*], Shanmuga Sundar Dhanabalan[2], N. Nithiyanandam[1] and M. Rajesh[3]

[1] Department of Computing Technologies, Faculty of Engineering and Technology, SRM Institute of Science and Technology, Kattankulathur, Chengalpattu, India

[2] School of Engineering, RMIT University, Australia

[3] Department of Computer Science and Engineering, Sanjivani College of Engineering, Kopargaon, India

Sustainability in the field of agriculture requires creative solutions to the problems faced by the farmers. Disease management plays an essential part of sustainable agriculture. In case of paddy cultivation, disease management is a critical component which ensures longevity and productivity, and subsequently benefits the farmers. Early detection of the disease prevents the endangerment of food security and financial stability. Traditional disease detection techniques are a time-consuming process and may not provide early diagnosis, potentially leading to substantial losses for farmers. Involving deep learning-based computer vision techniques can help in early detection and diagnosis of the diseases in the paddy crop. In this chapter, we have introduced a novel paddy disease detection method using deep learning architecture EfficientNetV2B2. The accuracy obtained with the proposed model is 98%.

1. Introduction

1.1. SDG 2 for Achieving Zero Hunger

Zero Hunger is a commitment to ensure that every human being has access to sufficient, safe, and nutritious food. It is the commitment to end hunger, accomplishing food security and enhancing nutrition, and encouraging

*Corresponding author: vidhyalm1@srmist.edu.in

sustainable agriculture. This goal is working to resolve the underlying causes of hunger and malnutrition—poverty, inequality, and food insecurity—a world where no one goes to bed hungry. Ending hunger needs a mix of increasing agriculture productivity in a sustainable manner, supporting sustainable farming practices, reducing loss and food waste, increasing access to a nutritious diet, and bolstering climate resistance, among other measures. It also needs empowering smallholder farmers, investing in rural infrastructure, and spawning efficient, resilient food systems that profit everyone equally and involve everyone in the process. Thus, Zero Hunger is more than merely ensuring that enough food is available. It is also about ensuring that there is enough high-quality food for everyone, that this food is culturally significant, and that it is manufactured in a manner that is both environmentally sustainable and beneficial to the health and livelihoods of future generations of people.

1.2. Innovative Solutions in Agriculture to Achieve SDG2

The chapter's focus on paddy cultivation and disease control highlights the role of innovative technologies in supporting Sustainable Development Goal 2. In particular, by using deep learning-based computer vision solutions for early disease detection in the paddy crop. This is because these technologies enable a pre-emptive maintenance approach that allows farmers to detect and contain crop diseases long before their proliferation would otherwise be detected. The implications of this for farmers who grow paddy are numerous—in particular, as it enables them to prevent yield loss and promote the longevity and regular productivity of the crop, it directly contributes to increasing food availability and reducing hunger. Additionally, the sub-related implications of enhanced food crop use for improved resource use and minimal environmental impact also support the SDG2's more holistic aim of making resilient and environmentally sustainable food systems a reality for our time and that of our descendants.

1.3. Leaf Disease Prevention in Agriculture

Preventing the occurrence of leaf diseases in plants is critical in ensuring food security at the global level and sustainable agricultural activities. Leaf diseases, which can be caused by fungi, bacteria, and viruses, are responsible for much of the damage witnessed particularly in crop farms. The damage typically results in a substantial reduction in edible crop volumes leading to reduced income and access to food for millions of individuals across the globe. Aside from lowering direct crop yields, the occurrence of leaf diseases has an impact on nutritional quality, hence, the health and more significant well-being of populations whose major foodstuffs are derived from the affected crops. Moreover, the controlling of leaf diseases using various chemical substances that are harmful to human health, other living organisms, and ecological integration does more damage than mitigating the irresponsible application of leaf diseases. Leaf disease prevention, when focused on other strategies such as crop rotation, developing varieties resistant to such diseases, and integrated pest management, will not just ensure optimal conditions crop growth but also reduce the reliance on chemical applications that damage the environment and endanger human lives. Agricultural research, education, and extension services may better prepare farmers to pursue prevention as the ideal way of preventing leaf diseases and mitigate the limitations presented by uncertainties in the environment and pressure to feed a burgeoning global population. Generally, preventing leaf diseases is key to promoting food security, sustainable agriculture, and the ultimate success of other global development efforts such as the SDGs.

1.4. Tech-driven Solutions for Crop Disease Management

Paddy crops, primarily producing rice, hold immense global significance as they are the staple food for over half of the world's population, particularly in Asia. They provide employment to millions in rural areas and are a major commodity in global trade, significantly contributing to the economies of leading rice-exporting countries like Thailand, Vietnam, and India. Culturally, rice is central to many culinary traditions and rituals worldwide. Rice is a crucial staple food globally, providing up to 50% of caloric intake for millions in Asia [1]. Over 2 billion people across continents rely on rice for 60-70% of their food calories [2]. China and India are major producers, contributing to half of the world's rice production. The crop faces challenges like weed infestations, impacting yields in regions like eastern India [3]. Rice's historical significance is profound, deeply ingrained in the cultures of Asia and West Africa [4]. Farmers experience production loss and financial losses every year as a result of different pest and disease attacks on their crops. An estimated average annual loss for paddy farming is around 37% [5].

However, paddy crops are vulnerable to several leaf diseases, including rice blast, bacterial leaf blight, sheath blight, and leaf smut, which can drastically reduce yields and degrade grain quality, leading to economic losses for farmers. Controlling these diseases involves integrated pest management, the development of resistant varieties, and improved cultural practices. These measures are crucial for maintaining the global importance of paddy crops, ensuring food security, and sustaining economic and cultural practices associated with rice cultivation. Long-term agriculture requires constant disease detection and plant health monitoring. Typically, plant diseases are identified manually which involves long processing time, intensive labour, and knowledge of plant diseases. In the case of extensive farms, manual labour is highly expensive [6]. Also, it is crucial for farmers to identify diseases for timely intervention. Manual inspection by experts is particularly challenging when a disease suddenly outbreaks, and there aren't always enough experts to go around. So, there is always a great demand for a system that identifies plant diseases quickly, precisely, and inexpensively. The automatic identification of diseases in plants is one of the important study areas in the agricultural field [7]. The identification of plant diseases is a crucial phase in preventing reductions in the productivity and quantity of agricultural goods [8]. Many studies are being conducted to use technology to identify diseases through the analysis of the images of the plant. Involvement of the technology can reduce the need for expertise and save time by enabling faster approaches to identify the disease and increase in food production [9].

The computer vision-based plant disease identification technology replaced the manual identification of the diseases. Computer vision uses machine learning or deep learning algorithms for disease identification. While machine learning involves manual feature creation for classifiers for detecting plant diseases [10], deep learning algorithms can autonomously generate and learn features from the dataset. The deep learning models may indeed outperform regular machine learning models when it comes to predicting rice disease.

In this chapter we propose a novel deep learning architecture using EfficientNetV2B2 for paddy disease classification. The study proves that our method has better performance over the other cutting-edge methods. The paper is structured as follows: Section 2 outlines the literature survey, Section 3 describes the methodology adopted for paddy disease classification, and Section 4 discusses the results.

2. Literature Survey

2.1. Traditional Image Processing Techniques

The early works involve traditional feature extraction techniques followed by application of classifiers for disease detection. Kanesh et al. [11] combined SURF descriptors and HOG descriptors and used SVM as a classifier for detecting the pests in the paddy. A knowledge-based system for plant diseases detection was proposed by Watanee [12]. The authors used an ontology-based strategy for modelling plant diseases, by incorporating the domain knowledge about the signs of plant diseases from farmers. Iqbal [13] presents a detailed study of various pre-processing, feature extraction techniques and classifiers used for disease detection in citrus leaves. Johannes, A. [14] proposed a leaf disease algorithm for wheat plants with statistical analysis methods. Es-Saady [15] presented a method for automatic recognition of plant leaves diseases from images by using two SVM classifiers continuously one after another. Islam, M. [16] proposed a leaf disease detection in potato plants by employing 10 features including colour, Grey Level Co-Occurrence Matrix (GLCM) texture with an SVM as the classifier. Wang, H. [17] proposed an algorithm for leaf disease detection in grape and wheat plants using a combination of features like colour, texture and shape along with Back propagation networks. Gupta, T. [18] proposed an algorithm for leaf disease detection using a modified SVM classifier with Cuckoo Search (CS) for optimizing the classifier parameters. Singh, V. [19] proposed leaf disease detection algorithm for various plants by segmenting the diseased part of the leaves using a genetic algorithm and using an SVM for classification. A.K. Dey [20] proposed a segmentation algorithm by plotting the histograms of the hue and applying threshold to separate the disease affected part of the leaves in the betel vine plant.

While some of the existing techniques diagnose crop diseases with a reasonable level of accuracy, all of them depend on feature design. Deep learning algorithms tackle this challenge by extracting sophisticated characteristics that are considerably superior to manually designed features.

2.2. Deep Learning Techniques

Deep learning, as described by LeCun et al. [21], is a representation of a learning approach in which the algorithm determines the optimal way for representing data through a sequence of optimizations rather than semantic characteristics. The underlying concept of deep learning is the use of multiple layers to extract the features. These multiple layers are hidden layers of neurons. The neurons have connections with the layers. The initial layers capture the elementary features while the final layers integrate the elementary features and extract the sophisticated level features.

Convolutional neural network (CNN) is a popular model in deep learning. It comprises input, convolution, pooling, full connection, and output layers capable of extracting the complex features from the images. Y. Kawasaki [22] used a customized deep neural network to identify diseases in cucumber plant. The authors through their work inferred that training the model with data augmentation has more impact in the classification accuracy than training epochs. Guofeng Yang [23] proposed a convolutional rebalancing network for leaf disease and pest detection in the rice plant. The approach works well with an unbalanced image dataset gathered from a rice field. To improve the features extracted from unbalanced dataset the authors used uniform as well as reverse samples for feature extraction combined with augmented data. Deng, R. [24] proposed an ensemble model which integrates the sub-modules in order to increase the accuracy and decrease the probability of misclassification. The model is employed in a mobile phone, which is beneficial for directly identifying from the field. Rahman, C. [25] proposed a disease detection model using small CNN architectures using Nasnet Mobile which is suitable for mobile application.

Jain [26] proposed an automatic system via a smartphone application to detect diseases from paddy leaves using YOLO object detection. The authors observed that YOLOv4 tiny is more effective than YOLOv3 mini for smartphone applications. Althaf Shaik [27] highlights India's severe water wastage and the resulting agricultural decline, suggesting that implementing internet of things (IoT) and robotics in drainage basins could enhance sustainable farming and water conservation, thus supporting SDG2. Nisha, K. [28], in her paper, highlights that the application of deep learning, particularly when integrated with IoT systems, holds significant potential for enhancing agricultural productivity. This approach can effectively address critical challenges such as climate change, pesticide use, and water scarcity.

3. Methodology Adopted

In this study, we adopted an approach to paddy disease detection with EfficientNet architecture as the base model. The method uses the transfer learning approach for training. The EfficientNet model, originally pretrained on a large ImageNet dataset, has acquired knowledge during the training to effectively classify the different classes present in the ImageNet. The knowledge acquired during the training is transferred to the fine-tuned layers to perform the classification task for paddy diseases. Figure 1 shows the Proposed Model for Paddy Disease Classifier. We propose the one convolution layer, batch normalization layer, two global average pooling layers and dense layers along with leak Relu layers and dense layers as final fine-tuned layers. These fine-tuned layers contribute to comprehending the complex patterns and characteristics linked to various crop illnesses in a more sophisticated way.

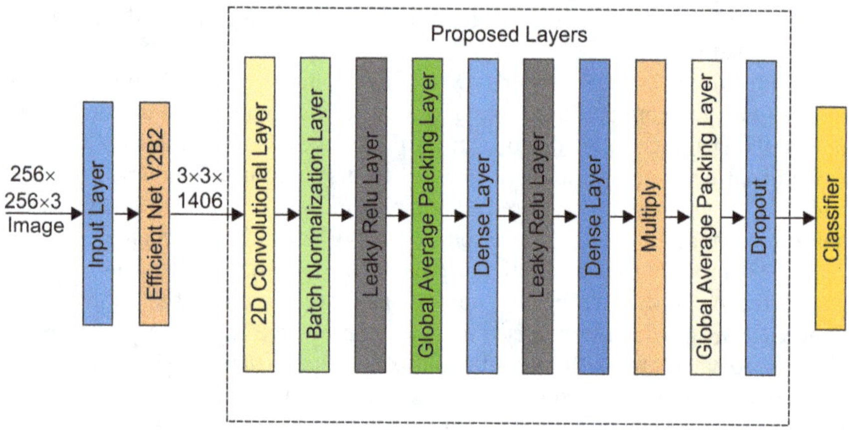

Figure 1: Proposed model for paddy disease classifier.

3.1. EfficientNet Model

In recent years, EfficientNet has gained its popularity for real-world applications due its balance between computational efficiency and good model performance. Mingxing Tan and Quoc V. Le [29] from the Google research team presented EfficientNet in 2019. While the regular convolutional networks use scaling in one-dimension, the EfficientNet model uses "compound scaling" which performs scaling in three dimensions namely depth, width, and resolution. This compound scaling makes EfficientNet incredibly efficient without needing high computational resources. With compound scaling, the model can achieve high efficiency with less resource requirements. The model showcases better efficiency in image classification tasks which is achieved through compound scaling.

The network's depth, breadth, and resolution are evenly scaled by the model using the compound coefficient (φ) as follows:

$$\left.\begin{aligned} &\text{Depth: } d = a^{\varphi} \\ &\text{Width: } w = b^{\varphi} \\ &\text{Resolution: } r = c^{\varphi} \\ &\text{Given that } a \,.\, b^2 \,.\, c^2 \approx 2 \\ &a \geq 1,\, b \geq 1,\, c \geq 1 \end{aligned}\right] \quad (1)$$

Here *a*, *b*, and *c* are constant coefficients that determine the network's depth, width, and image size and should be scaled while applying the compound scaling method. The model gets adjusted in a balanced way by changing these coefficients. Larger φ values result in larger models with more accuracy at the cost of more computational resources. Smaller φ values result in smaller models with fewer computational resources but lower accuracy. The EfficientNet model also employs several other techniques like inverted residuals, skip connections to optimize its architecture. The presence of inverted residuals with linear bottlenecks, effectively reduces the parameters and needs fewer calculations but increases the boosting accuracy. Skip connections assist with gradient flow in the training process, enhancing feature representation understanding. Moreover, depth-wise separable convolutions are commonly embraced in several layers. This approach slashes parameters and computations, as opposed to conventional convolutional layers, boosting effectiveness while keeping accuracy losses minimal.

4. Results along the Discussion

4.1. Dataset

A group of researchers collected the dataset known as "Paddy Doctor" [30] containing paddy leaf images captured in a village of Tamil Nadu, India during the timeframe between February and April 2021. The images of the paddy crop were captured between 40 and 80 days' old using a mobile phone. All the images uniformly cropped to the size of 1,080 × 1,440. Originally there were 30,000 JPEG images and after careful examination the repeated and poor-quality images were excluded. Ultimately, the dataset had 16,225 images with 13 different classes of diseases. The number of images for each class was not uniform. The distribution of the images across the different classes is shown in Figure 2. This dataset serves as a good resource for the study of paddy diseases.

4.2. Data Preprocessing

To train the proposed model we prepared the training, testing, and validation dataset. Stratified sampling is used to ensure that the same proportion of samples in each class is used in training, testing, and validation datasets. The stratified sampling prevents any potential bias by representing an adequate number of

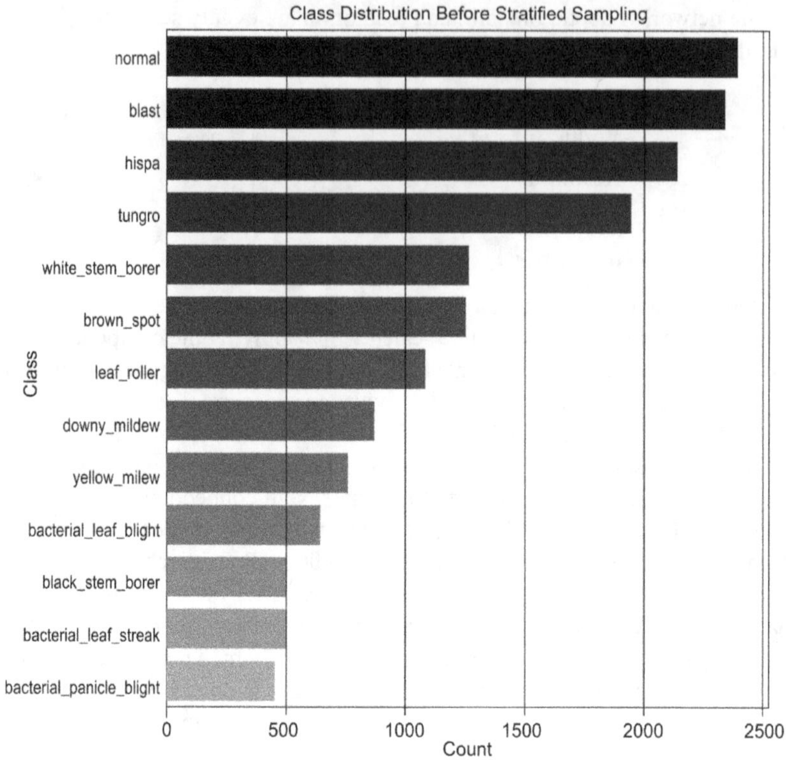

Figure 2: Distribution of the paddy diseases in the dataset before stratified sampling.

samples in class. Also, this sampling increases robustness and accuracy of the results. The class distribution in the original, training, and testing dataset is shown in Figure 3.

Data augmentation is adopted to enhance the diversity among the samples given during the training. These augmentations include random rotations of images up to 5 degrees, shearing with a factor of 0.2 to vary the distortion, zooming up to 20% to vary the scale and flipping to create mirror images. By adopting the augmentation, the model becomes more generalized in learning the features. Also, the dataset enriched with augmentation has the ability to create a wide range of inputs which results in improved accuracy. The sample images used for augmentation are shown in Figure 4.

4.3. Experimental Setup

Experimental setup for preparing the proposed paddy disease classification model involves importing the libraries like TensorFlow, and Keras. The EfficientNet model is loaded from the TensorFlow Hub. The dataset is also loaded, preprocessed and distributed as training, testing, and validation sets. The

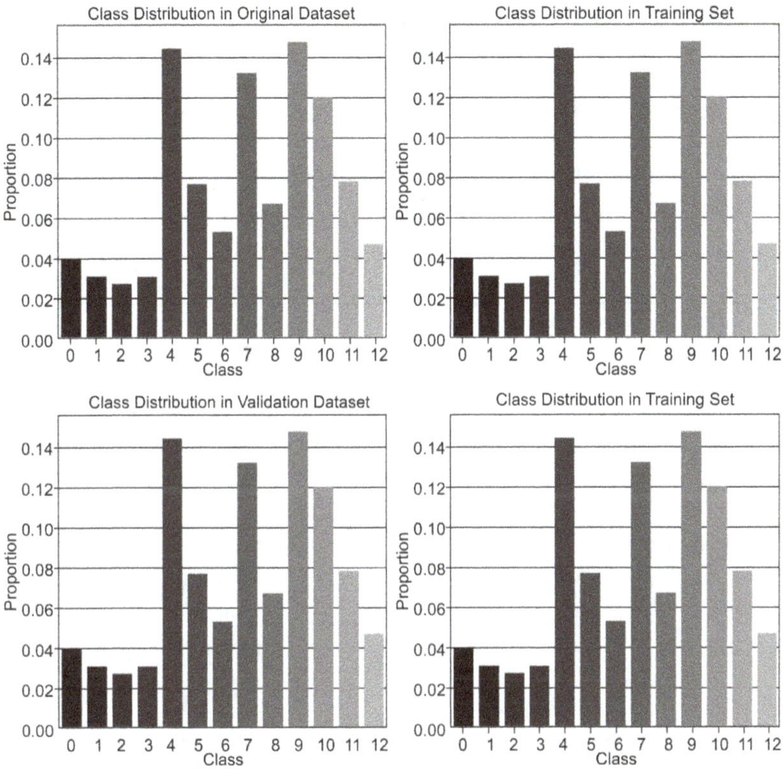

Figure 3: Class distribution in original, training, and testing dataset.

Figure 4: Images used for augmentation.

hyperparameters used for our model are listed in Table 1. The EfficientNetV2B2 model is used as a base model and the layers are pre-trained with a large dataset ImageNet. These pretrained layers play a key component in extracting the complex features. The pretrained layers are frozen and the final layers are fine-tuned especially for the paddy disease classification. Training is optimized using the Adam optimizer. The evaluation of the model is performed by conducting tests on the test dataset.

Table 1: Hyperparameters

Hyperparameter/Configuration	Value
SEED	21
DEBUG	FALSE
NUM_EPOCHS	75
IMG_HEIGHT	256
IMG_WIDTH	256
BATCH_SIZE	32
TARGET_SIZE_DIM	256
LEARNING_RATE	1.00E-04
LR_START	1.00E-08
LR_MIN	1.00E-08
LR_MAX	1.00E-04
LR_RAMPUP_EPOCHS	3.00E+00
LR_SUSTAIN_EPOCHS	0.00E+00
N_CYCLES	5.00E-01
NUM_EPOCHS	6.00E+01
Total parameters	9108203 (34.75 MB)
Trainable parameters	9025915 (34.43 MB)
Non-trainable parameters	82288 (321.44 KB)

To assess the performance of the plant disease classification model, accuracy and loss curves are plotted. The training accuracy curves serve as the graphical representation that illustrates the learning process from the training dataset. The validation accuracy curve provides the performance on the validation dataset and illustrates the model's ability to generalize the unseen data. Figure 5 shows accuracy and loss curves of our model. From Figure 5 it is evident that the curves are smooth and reach to near one value within 5 epochs. It is significant that the model learns quickly and effectively from the training set. Also, the sharp decrease in the loss curve indicates that it is minimizing the errors during the training.

Table 2 offers the overview of the performance of the model in different classes. The results show high values of precision, recall, and F1 score across various classes which indicate the model's efficiency in distinguishing the different disease classes. Especially for classes 2, 9, and 11, the precision, recall, and F1 scores are high. The overall accuracy of 97% reveals that the model is more robust to generalization. But the precision for classes 0, 3, and 4 are slightly lower than the other classes.

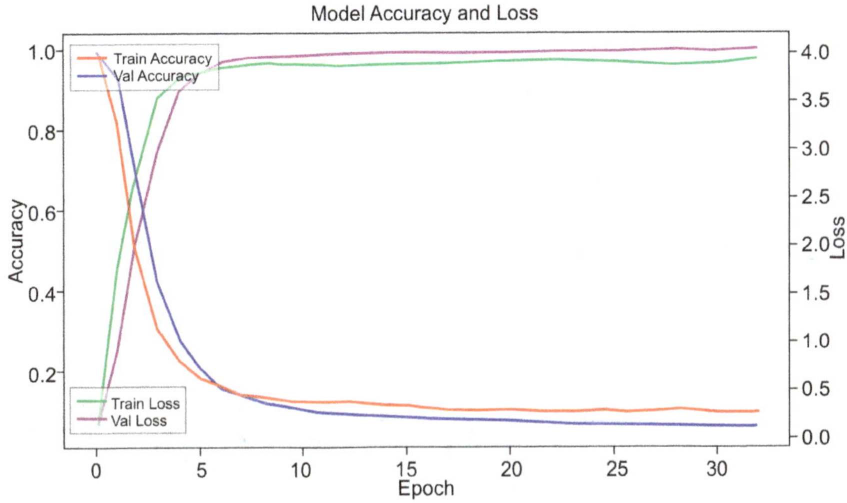

Figure 5: Model accuracy and loss curves.

Table 2: Class-wise performance metrics

Class	Precision in %	Recall in %	F1-Score in %	Support
0	96	94	95	130
1	97	96	97	101
2	99	99	99	90
3	93	99	96	101
4	98	95	96	470
5	97	98	97	251
6	96	95	96	174
7	96	97	97	430
8	97	96	96	219
9	98	99	98	481
10	96	99	98	390
11	99	97	98	255
12	98	97	98	153
Accuracy			98	3245
Macro avg	98	98	98	3245
Weighted avg	98	98	98	3245

Figure 6 shows the sample output images with true and predicted labels along with the confidence scores in percentage. From the true and predicted labels, direct comparison is possible whether the model accurately predicted the classes or not. The confidence score reveals the level of certainty of prediction of the class.

Figure 6: Sample output images showing actual and predicted class along with confidence of prediction.

Figure 7 presents the Receiver Operating Curve (ROC) of the model. It is the graphical representation to evaluate the model's performance. It represents the tradeoff between the true positive rate and false positive rate. It shows the model's capability to discriminate between positive and negative samples. Here, our model has shown a high superior discriminatory capability expressing high sensitivity and low false positive rates. The Area Under the Curve (AUC) which is derived from ROC is also very high which indicates the ability to balance the true positive and false positive.

We plotted the Confusion matrix to assess the performance of the paddy disease classifier by summarizing the predictions in tabular format. Both

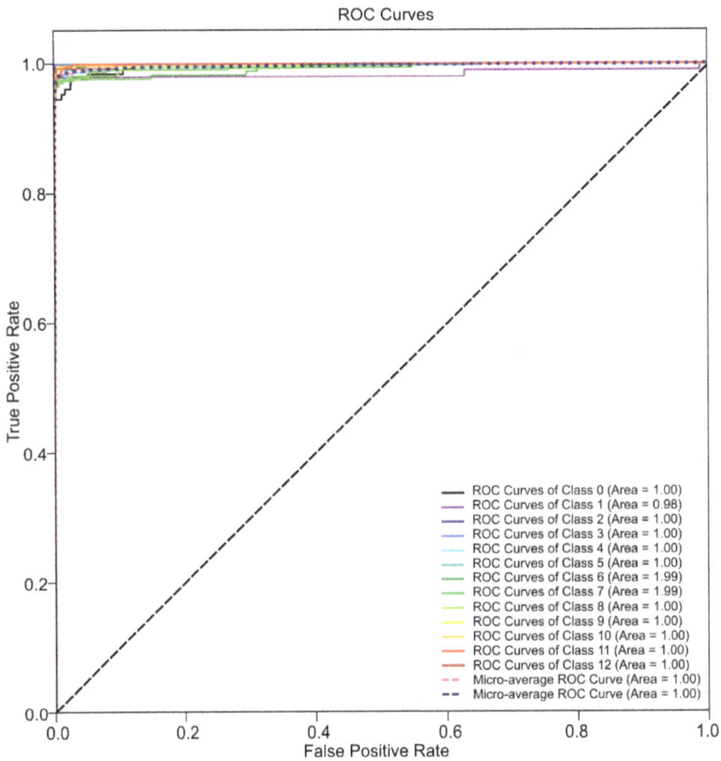

Figure 7: Receiver operating curve.

normalized and unnormalized confusion matrices are plotted as presented in Figures 8 and 9, respectively. The normalized confusion matrix gives the better comparison when the class sizes vary in different classes. It gives the understanding of the accuracy and error rates for each class. The un-normalized confusion matrix gives the approximate counts of true positive, true negative, false positive, and false negative predictions for each class. The normalized confusion matrix presents these values as percentage.

The model obtained an overall accuracy of 98%. This is compared with other existing classification models as shown in Table 3.

Table 3: Comparison with the state-of-art works on paddy disease classification

Method	Accuracy
Proposed EfficientV2B2 model	0.98
DenseNet169-MLP [28]	0.97
YOLOv4 tiny [31]	0.8537
EfficientNetB0+CBAM [32]	0.8689

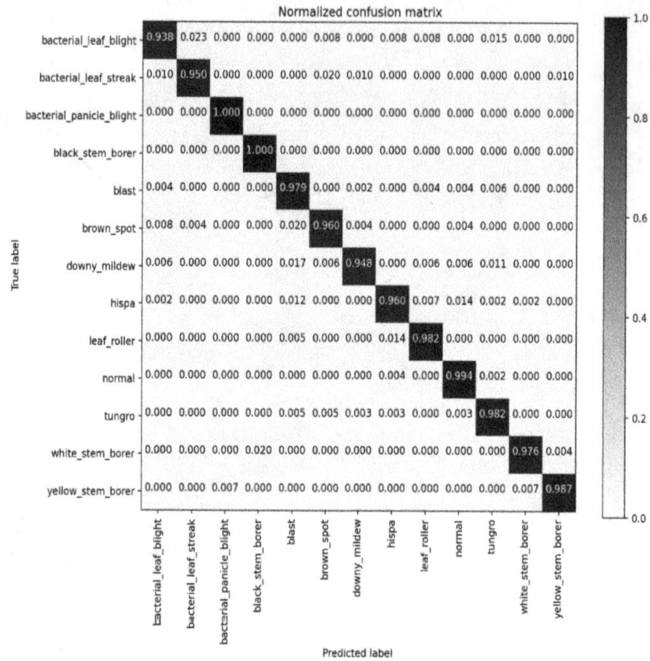

Figure 8: Normalised confusion matrix.

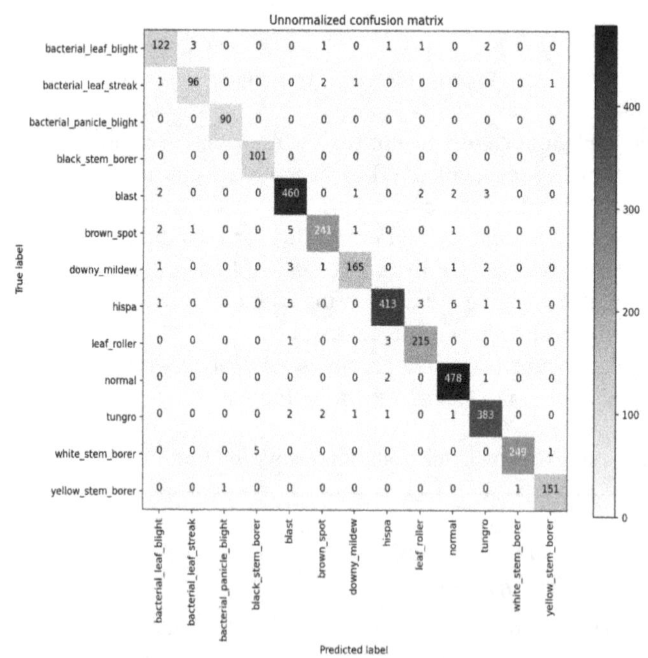

Figure 9: Un-normalized confusion matrix.

5. Conclusion

In conclusion, sustainability in agriculture demands innovative solutions to the challenges faced by farmers, with disease management standing out as a crucial component. Particularly in paddy cultivation, effective disease management ensures longevity and productivity, directly benefiting farmers by safeguarding their livelihoods. Early detection of diseases is paramount for preserving food security and financial stability, yet traditional methods often fall short in providing timely diagnosis, potentially leading to significant losses. Introducing deep learning-based computer vision techniques, as demonstrated in the chapter with an accuracy of 98% using the EfficientNetV2B2 architecture, offers a promising avenue for early disease detection in paddy crops. The commitment to Zero Hunger encompasses not only ensuring food availability but also promoting high-quality, culturally significant, and environmentally sustainable food production. By focusing on innovative technologies like deep learning for disease detection, the chapter underscores the role of technological advancements in supporting Sustainable Development Goal 2. Furthermore, preventing leaf diseases in plants is essential for global food security and sustainable agriculture. By adopting strategies such as crop rotation, resistant varieties, and integrated pest management, farmers can mitigate reliance on harmful chemical applications, contributing to healthier ecosystems and safer food production. Ultimately, preventing leaf diseases is fundamental to advancing food security, sustainable agriculture, and broader global development goals such as the SDGs.

References

1. Marton, Renee. (2014). *Rice: A Global History*. Reaktion Books.
2. Sardana, Virender, Gulshan Mahajan, Khawar Jabran and Bhagirath S. Chauhan. (2017). Role of competition in managing weeds: An introduction to the special issue. *Crop Protection*, 95: 1–7.
3. Muthayya, Sumithra, Jonathan D. Sugimoto, Scott Montgomery and Glen F. Maberly. (2014). An overview of global rice production, supply, trade, and consumption. *Annals of the New York Academy of Sciences*, 1324(1): 7–14.
4. Nayar, Narayanan Madhavan. (2014). *Origins and Phylogeny of Rices*. Elsevier.
5. Nidhis, A.D., Chandrapati Naga Venkata Pardhu, K. Charishma Reddy and K. Deepa. (2019). Cluster-based paddy leaf disease detection, classification, and diagnosis in crop health monitoring unit. *In*: *Computer-aided Intervention and Diagnostics in Clinical and Medical Images*. 281–291. Springer International Publishing.
6. Tiwari, Vaibhav, Rakesh Chandra Joshi and Malay Kishore Dutta. (2021). Dense convolutional neural networks based multiclass plant disease detection and classification using leaf images. *Ecological Informatics*, 63: 101289.
7. Astonkar, Shweta R. and Shandilya, V.K. (2018). Detection and analysis of plant diseases using image processing. *Int. Res. J. Eng. Technol.*, 5(4): 3191–3193.

8. Yusof, Munirah Mohd, Nur Fazliyana Rosli, Muhaini Othman, Rozlini Mohamed and Mohd Hafizul Afifi Abdullah. (2018). M-DCocoa: M-agriculture expert system for diagnosing cocoa plant diseases. *In*: *Recent Advances on Soft Computing and Data Mining: Proceedings of the Third International Conference on Soft Computing and Data Mining*, Johor, Malaysia, February 06–07, 2018. 363–371. Springer International Publishing.
9. Kim, Dae-Young, Avinash Kadam, Surendra Shinde, Rijuta Ganesh Saratale, Jayanta Patra and Gajanan Ghodake. (2018). Recent developments in nanotechnology transforming the agricultural sector: A transition replete with opportunities. *Journal of the Science of Food and Agriculture*, 98(3): 849–864.
10. El-kazzaz, Mohamed Kamal, Essa Ahmed Salem, Kamal Elsayed Ghoneim, Mohsen Mohamed Elsharkawy, Gabr Abd El-wanees, Nasr El-Kot and Zeinab Abd Elnaby Kalboush. (2015). Integrated control of rice kernel smut disease using plant extracts and salicylic acid. *Archives of Phytopathology and Plant Protection*, 48(8): 664–675.
11. Venugoban, Kanesh and Amirthalingam Ramanan. (2014). Image classification of paddy field insect pests using gradient-based features. *International Journal of Machine Learning and Computing*, 4(1): 1.
12. Hiranyawongkul, Watanee, Chutiporn Anutariya and Frederic Andres. (2018). An ontology-based approach to plant disease identification system. *In*: *Proceedings of the 10th International Conference on Advances in Information Technology*, 1–8.
13. Iqbal, Zahid, Muhammad Attique Khan, Muhammad Sharif, Jamal Hussain Shah, Muhammad Habib ur Rehman and Kashif Javed. (2018). An automated detection and classification of citrus plant diseases using image processing techniques: A review. *Computers and Electronics in Agriculture*, 153: 12–32.
14. Johannes, Alexander, Artzai Picon, Aitor Alvarez-Gila, Jone Echazarra, Sergio Rodriguez-Vaamonde, Ana Díez Navajas and Amaia Ortiz-Barredo. (2017). Automatic plant disease diagnosis using mobile capture devices, applied on a wheat use case. *Computers and Electronics in Agriculture*, 138: 200–209.
15. Es-saady, Youssef, Ismail El Massi, Mostafa El Yassa, Driss Mammass and Abdeslam Benazoun. (2016). Automatic recognition of plant leaves diseases based on serial combination of two SVM classifiers. *In*: *2016 International Conference on Electrical and Information Technologies*, 561–566. IEEE.
16. Islam, Monzurul, Anh Dinh, Khan Wahid and Pankaj Bhowmik. (2017). Detection of potato diseases using image segmentation and multiclass support vector machine. *In*: *2017 IEEE 30th Canadian Conference on Electrical and Computer Engineering*, 1–4. IEEE.
17. Wang, Haiguang, Guanlin Li, Zhanhong Ma and Xiaolong Li. (2012). Image recognition of plant diseases based on backpropagation networks. *In*: *2012 5th International Congress on Image and Signal Processing*, 894–900. IEEE.
18. Gupta, T. (2017). Plant leaf disease analysis using image processing technique with modified SVM-CS classifier. *Int. J. Eng. Manag. Technol.*, 5: 11–17.
19. Singh, Vijai and Misra, A.K. (2015). Detection of unhealthy region of plant leaves using image processing and genetic algorithm. *In*: *2015 International Conference on Advances in Computer Engineering and Applications*, 1028–1032. IEEE.
20. Dey, Amar Kumar, Manisha Sharma and Meshram, M.R. (2016). Image processing based leaf rot disease, detection of betel vine (*Piper Betle* L.). *Procedia Computer Science*, 85: 748–754.
21. LeCun, Yann, Yoshua Bengio and Geoffrey Hinton. (2015). Deep learning. *Nature*, 521(7553): 436–444.

22. Kawasaki, Yusuke, Hiroyuki Uga, Satoshi Kagiwada and Hitoshi Iyatomi. (2015). Basic study of automated diagnosis of viral plant diseases using convolutional neural networks. *In*: *Advances in Visual Computing: 11th International Symposium, 2015*, 638–645. Las Vegas, NV, USA, December 14–16, *Proceedings, Part II*. Springer International Publishing.
23. Yang, Guofeng, Guipeng Chen, Cong Li, Jiangfan Fu, Yang Guo and Hua Liang. (2021). Convolutional rebalancing network for the classification of large imbalanced rice pest and disease datasets in the field. *Frontiers in Plant Science*, 12: 671134.
24. Deng, Ruoling, Ming Tao, Hang Xing, Xiuli Yang, Chuang Liu, Kaifeng Liao and Long Qi. (2021). Automatic diagnosis of rice diseases using deep learning. *Frontiers in Plant Science*, 12: 701038.
25. Rahman, Chowdhury R., Preetom S. Arko, Mohammed E. Ali, Mohammad A. Iqbal Khan, Sajid H. Apon, Farzana Nowrin and Abu Wasif. (2020). Identification and recognition of rice diseases and pests using convolutional neural networks. *Biosystems Engineering*, 194: 112–120.
26. Jain, Siddhi, Rahul Sahni, Tuneer Khargonkar, Himanshu Gupta, Om Prakash Verma, Tarun Kumar Sharma, Tushar Bhardwaj, Saurabh Agarwal and Hyunsung Kim. (2022). Automatic rice disease detection and assistance framework using deep learning and a Chatbot. *Electronics*, 11(14): 2110.
27. Althaf Shaik, Nafisa Badruzaman, Devadharshini Gajendran, Geethalakshmi M. and Jose Anand. (2022). Organic farming in drainage system with advanced automation through robotics and IoT. *2022 International Conference on Data Science, Agents, and Artificial Intelligence*, 592–597. Chennai, India.
28. Nisha, K., Augustine, C., Jose Anand, A. and Kamesh, M. (2023). Deep learning in smart agriculture. *In*: K. Kalaiselvi, Jose Anand A., Poonam Tanwar and Haider Raza, *Advanced Technologies for Smart Agriculture*, 153–175. Denmark: River Publishers.
29. Tan, Mingxing and Quoc. V. Le. (2019). EfficientNet: Rethinking model scaling for convolutional neural networks. *arXiv Preprint* arXiv:1905.11946.
30. Petchiammal, Briskline Kiruba, Murugan and Pandarasamy Arjunan. (2023). Paddy doctor: A visual image dataset for automated paddy disease classification and benchmarking. *In*: *Proceedings of the 6th Joint International Conference on Data Science and Management of Data*, 203–207.
31. Narmadha, R.P., Sengottaiyan, N. and Kavitha, R.J. (2022). Deep transfer learning based rice plant disease detection model. *Intelligent Automation and Soft Computing*, 31(2).
32. Alirezazadeh, Pendar, Michael Schirrmann and Frieder Stolzenburg. (2023). Improving deep learning-based plant disease classification with attention mechanism. *Gesunde Pflanzen*, 75(1): 49–59.

12 | Innovative Technologies for Alleviating Hunger: Challenges and Opportunities

Shaziya Farhat* **and Sirajuddin Ahmed***
Department of Environmental Sciences, Jamia Millia Islamia, New Delhi

Global hunger is still a stark reality, despite significant progress. The Food and Agriculture Organization (FAO) estimates that in 2022, 750 million people were plagued by chronic hunger. This corresponds to approximately 9.2% of the world's population. Additionally, a hidden problem of micronutrient deficiencies plagues over 2 billion people globally. Technological advancements can offer a critical pathway to tackle this complex issue and have the potential to revolutionize food systems and ensure equitable access to nutritious food. This chapter examines several promising innovations, with a particular emphasis on Precision agriculture (PA), gene editing, vertical farming (VT), AI in agriculture, blockchain-based food traceability, and the Internet of Things (IoT). Each technology will be elaborated describing its concept, functioning, and potential benefits in alleviating hunger. Various issues and drawbacks of these technologies with regard to accessibility, affordability, as well as social inequalities, are also addressed. A diversified strategy is needed to overcome these obstacles. This chapter investigates how agricultural advancements can be used to significantly increase food production and ensure global food demands. It is expected that adopting these innovative technologies will pave the path for sustainable farming and alleviate global hunger.

1. Introduction

Hunger has been a problem for many years and still affects hundreds of millions of people worldwide. In 2022, an estimated 690–783 million people faced hunger worldwide. Approximately 122 million more people faced food insecurity due to the Covid-19 pandemic than there were prior to the outbreak, marking a tragic

*Corresponding authors: shaziyafarhat52@gmail.com; suahmed@jmi.ac.in

increase in hunger. In the Caribbean, Western Asia, and Africa, hunger rates are still rising. Africa has a far higher percentage of the population experiencing hunger than any other continent. Nevertheless, hunger isn't the only thing that worries us. In 2022, 2.4 billion people, a larger proportion of women and people in remote areas struggled to find adequate, safe, healthy food throughout the year. Millions of children below five still suffer from malnutrition [1]. According to the Global Hunger Index, the global hunger score for 2023 is 18.3, which is classified as moderate. Despite some country's efforts, this score indicates little progress in reducing hunger since 2015. Worryingly, 43 countries continue to experience alarming or severe hunger [2].

For the eradication of poverty and hunger, the international community committed to supporting the United Nations established 17 Sustainable Development Goals (SDGs) in 2015. There is no denying that reaching the 2030 SDG goal of zero hunger will be extremely difficult. According to projections, nearly 600 million people will still be hungry in 2030. We are falling short of the SDG2 targets due to the combination of increasing conflict, extreme weather, economic downturns and slowdowns, as well as the increased prices of nutritious foods and increasing inequality. Beyond just eradicating hunger, SDG Target 2.1 seeks to assure everyone has access to enough nutritious, and safe food throughout the year [1]. The most serious issue in developing nations is the loss of food, which results from infrastructural shortcomings, a shortage of technology, and inadequate investment in the agricultural sector, and occurs during the crop production, harvesting, post-harvesting, and subsequent processing stages [3]. Traditional farming approaches have some limitations, therefore to build a hunger-free future we will explore some technological advancements and the diverse opportunities presented by these technological innovations, followed by a discussion of the challenges that need to be overcome to make a significant impact on alleviating global hunger.

2. Technological Development for Hunger Alleviation

Technological advancements possess the ability to change food systems and make certain equitable access to nutritious foods. Advanced crop protection techniques, widespread mechanization, enhanced plant nutrition strategies, better crop varieties, and creative management techniques are just a few of the cutting-edge agricultural technologies that help to increase crop yields in a number of developing nations [4]. The agricultural industry is currently working towards modernization by utilizing modern technological advances to search for solutions for the efficient utilization of finite resources in order to meet the world's ever-increasing consumption needs.

This chapter examines several promising innovations, with a particular emphasis on Precision agriculture, gene editing, vertical farming, artificial intelligence (AI) in agriculture, blockchain-based food traceability, UAVs and the Internet of Things (IoT). Precision agriculture maximizes the use of resources by utilizing robotics and data-driven insights, which increases yields and decreases waste. Crop management is made more individualized by methods like soil analysis and sensor-based irrigation, which maximize yields while reducing environmental impact. Vertical farming, a controlled environment approach, utilizes stacked layers within urban areas for growing crops year-round under optimal conditions, resulting in increased productivity and less reliance on climate and land limitations. (AI) and machine learning (ML) models are being used extensively to meet a number of SDG2-related targets and to offer effective management strategies to combat hunger at the primary level. Blockchain technology offers enhanced traceability and transparency across the food supply chain. Through the immutable recording of data on a distributed ledger, it addresses food fraud, cultivates confidence, and enables customers to make knowledgeable decisions. IoT gives farmers access to real-time data collection and monitoring by connecting farm equipment and sensors to the internet. This gives farmers actionable insights. Another potent technology is gene editing, which precisely modifies plant genes to produce crops that are more resilient to drought, disease, as well as pests. Besides, this chapter highlights the application of additional technologies, which can greatly maximize crop yields, including Unmanned Aerial Vehicles (UAVs) for agricultural monitoring and beneficiary evaluations.

3. Artificial Intelligence (AI) in Agriculture

AI technology offers farmers useful information about the growth and health of their fields and crops, including determining when irrigation is necessary, applying pesticides and herbicides, using precision farming methods, and even forecasting the weather. Farmers can use AI to get more accurate data about their fields and crops, which will improve farming practices and productivity. With their growing popularity, robots and AI are now utilized for various tasks like picking fruit, pulling weeds, and detecting crops. In addition to labour input tasks, these machines are helpful in precision farming and the breeding of plants. Farmers can benefit from AI in a number of ways, including crop monitoring, irrigation, pest control, and data analytics through sensors, robots, IoT, cameras, drones, and widespread internet access. AI and ML serve as predictive analysts by evaluating past data, recommending perfect planting times, suggesting the best crops to grow based on conditions, and identifying the best hybrid seeds to maximize harvest [5].

3.1. Opportunities

AI is applied in agriculture in a variety of ways, including the management of water resources, weed control, crop growth enhancement, pest and disease detection, and nutrient provision. AI is widely used to track a variety of tasks, including fertilizer spraying and crop harvesting. Agri-bots, or AI robots, have been created to perform tasks similar to those of contemporary combine harvesters, boosting crop yields and working more quickly than human labour [6]. Farmers could identify and get rid of diseases with AI-based disease management. AI tools like ML, deep learning, and computer vision utilize multiple meteorological conditions to predict pest outbreaks. An efficient and automated way to manage weeds in and around the field is robotic weeding. Its goal is to lower the rate of dependency in agriculture as well as the use of herbicides. An intelligent spraying robot can use image processing and neural network techniques to identify the weeds in a crop and target them with liquid. This technique uses advanced image analysis to combat weeds. First, aerial photographs depict the crop field. These images are then fed into powerful AI algorithms, which can accurately distinguish between crops and weeds. With weeds identified, AI steps in again to estimate their size and maturity. This information is then used by high-precision robotic nozzles to deliver targeted herbicide doses, ensuring only the weeds get treated [7]. The conventional irrigation system has been replaced by an automated irrigation system. Once installed in the field, these systems don't need human intervention because they use sensors to analyze the moisture content of the soil. We can conserve water and increase crop yields with the help of this self-sufficient system [5]. Farmers use a range of sensors, such as air, pH meters, rain and frost, and soil moisture sensors, to design a variety of irrigation systems. Using historical data to prove effectiveness, these sensors track watering practices, real-time adjustments to watering schedules, and more. Soil moisture sensors are a vital component of an efficient irrigation system. These soil moisture sensors measure the water content of the soil to determine how much water is stored underground, helping farmers understand the processes occurring around the roots of crops. It gives information about the moisture content of the active root zone prior to the planned irrigation event. By utilizing AI, farmers are implementing smart irrigation systems. This innovative system integrates sprinklers and nozzles. It collects sensor data and transmits preset human commands to actuators that control the on-off functioning of sprinklers [7]. This technology can measure the water level and soil temperature without the need for human intervention. Successfully tackling several problems, including the rising costs of labour, crop cultivation, and pest infestation [5].

3.2. Challenges

Despite AI's emergence as a powerful tool, several factors hinder its adoption in developing nations. Financial constraints in developing countries severely impede AI adoption due to the high initial costs of software and hardware, the ongoing costs of managing these systems, and the inherent risks involved. Furthermore, uncertainties about the return on investment and potential risks discourage widespread implementation. Infrastructure challenges, including limited electricity access, unstable internet, hardware resources, and local support infrastructure, hinder AI adoption. Budget constraints and maintenance delays further impede the successful execution of AI-driven devices. Shortage of professionals in the agriculture sector, insufficient training, language barriers, and technical expertise in AI systems pose significant hurdles to AI adoption, as a result, customizing these systems to meet local requirements remains a challenge. Lack of historical data, data scarcity, and limited availability create significant obstacles for these technologies. AI adoption faces challenges in customization due to regional variations in languages, agricultural practices, and pests, making it difficult to develop a universal AI model. The absence of a robust regulatory framework for AI technology deployment in both developed and developing nations leads to uncertainties, ethical concerns, and regulatory gaps. Additionally, concerns about sharing of data, privacy, and labour market impacts, as well as compliance costs, pose significant challenges. Cultural norms, attitudes, and anxieties over potential disruptions or job losses could pose a barrier to AI adoption, particularly in developing regions. Developing countries encounter hurdles when it comes to integrating AI technology due to infrastructure issues, lack of market knowledge, digital divide, and inadequacy in key marketing and networking competencies, hindering farmers' decision-making and potential revenue generation [8].

4. Internet of Things (IoT)

IoT is transforming agriculture in several key areas, including precision farming, livestock management, and greenhouse optimization, they fall under various monitoring techniques. Wireless sensor networks (WSNs), collect real-time data that empower farmers and are used to monitor all of these applications using various IoT-based sensors and devices. Certain IoT systems use the service cloud for analysing and processing remote data, enabling researchers and agriculturists to make more informed decisions [9].

4.1. Sensors

Sensors are the backbone of IoT. Sensors serve as data collectors for IoT. The data they gather is then analysed to provide useful insights. Sensors play an important role in a variety of applications, including system monitoring and control, safety, alarm triggering, diagnosing issues, and in-depth analysis. Mobile sensors and apps are transforming agriculture by enabling remote field management, including water use optimization, real-time plant health monitoring, and targeted animal and pest protection. Sensors play an important role in agriculture by directly measuring soil parameters through tests like pH, nutrient content, humidity, moisture, and temperature. These results obtained by soil testing are critical for obtaining a high yield while maintaining quality. Ion selective electrode (ISE) and Ion-sensitive field effect transistor (ISEFT) sensors were additionally employed for tracking ion uptake by plants. Acoustic sensors, for instance, identify soil texture by measuring noise caused by tools interacting with soil particles. Airflow sensors delve deeper, determining air permeability, moisture content, structure, and even soil type. Finally, electrochemical sensors play a crucial role in measuring essential nutrients, salinity, and pH levels [10].

4.2. Opportunities

Farmers can leverage IoT to monitor soil health, air quality, temperature, water management, crop disease, location tracking, environmental conditions, pest, and fertilization strategies. By using inexpensive electronic devices and simple communication methods, the IoT allows us to interact with the physical world. Additionally, the user receives data about various environmental parameters through trigger alerts or recommendations sent to authorities via messages [11]. Cloud-based IoT control centres gather and analyse real-time data about the environment and crops in order to plant, fertilize, and harvest crops at the appropriate right time and for the right duration of time. In this way, IoT-embedded precision agriculture can be realized, improving agricultural production with regard to quality, quantity, sustainability, and cost-effectiveness [12]. The rise of IoT is revolutionizing disease and pest management. Farmers can monitor crop pests and diseases by using a variety of real-time IoT sensors that gather crucial data on location, crop development, and pest situation anywhere. After that, all raw data and images are transferred to cloud centres, where they are subsequently processed and examined using a variety of models and algorithms that are based on various illnesses and pests [9]. IoT-connected gadgets are able to continuously track not only the conditions within the field but also the surrounding environment. The information gathered by these devices feeds into a constantly updated online database. The user may utilize these saved data for additional analysis or to remotely monitor the field at all times from another location. This real-time monitoring environment facilitates the making of accurate decisions that will raise field productivity [13]. Resources like water, fertilizers, pesticides, and other materials can be used efficiently with careful tracking of the soil condition, farm, and environmental parameters. The agriculture field can be efficiently monitored with IoT. These technologies lessen the chance of intruders like wild animals damaging crops by keeping them out of the field. Monitoring nutrients and soil would streamline farming processes and help conserve water, which is a precious resource in many areas and could utilize smart irrigation systems [9].

4.3. Challenges

Several hurdles hinder integration of IoT in agriculture including hardware, interoperability, managing network connectivity and energy usage, safeguarding security and privacy, the costs of hardware and software, and need for education and training. Hardware challenges include harsh environmental conditions that can destroy devices, while interoperability challenges involve hardware, network protocols, syntax, semantics, and platform heterogeneity. Managing network and energy faces challenges related to low cost, flexible networking, and high scalability, but also include background noise and power drain. To improve data transmission reliability, power management and renewable energy sources like solar power and wind can be used. Blockchain technology offers potential for supply chain improvements but has challenges like scalability and energy-intensive mining processes. Security measures for IoT require careful consideration of device limitations. This includes deploying encryption algorithms, intrusion detection systems, secure key distribution methods, and robust routing protocols, all while keeping in mind the specific characteristics of these end devices. Hardware and software costs are also a major objective. Education challenges arise from uneducated farmers in developing countries, who may struggle to utilize information effectively [14]. The widespread adoption of IoT in agriculture, creating a smart farming environment, introduces potential security and privacy vulnerabilities. These concerns include disruptions to network communication, hacking of collected agricultural data, and compromise of sensitive farm properties. Farmer acceptance of IoT in agriculture could face hurdles due to both attitudes and knowledge gaps. Some farmers may prefer traditional methods, while others might lack the skills and knowledge necessary to adopt this new technology. IoT devices are typically deployed in open fields across diverse agricultural regions, they're often exposed to extreme weather events, the design of IoT hardware needs to prioritize resilience against these harsh climate conditions. For widespread adoption of IoT in smart agriculture, both reliability and scalability are crucial factors. Smart agriculture powered by IoT relies heavily on data processing to inform effective decisions. This necessitates a substantial network infrastructure, including numerous sensor nodes, gateways, communication protocols, and supporting middleware for device management [15]. Poor telecommunication infrastructure, particularly through mobile phones and tablets, makes data transmission unreliable, especially for remote operations. Farmers also face challenges in organizing and manipulating data from sensors, which can lead to incorrect calculations [16].

5. Blockchain Based Food Traceability

Blockchain constitutes a distributed database system for sharing and recording digital events, such as transactional data among participants in the value chain. It offers a new approach for ensuring food traceability, transparency, safety, and security. It achieves this by leveraging a combination of established computer technologies, including public/private key cryptography, cryptographic hash functions, and database technologies for efficient record-keeping. It eliminates the need for manual monitoring and tracking systems based on papers, preventing inaccurate effects on the traditional value chain [17]. Because of its numerous applications, including immutable databases, traceability, peer-to-peer networks, and smart contract procedures, blockchain technology can enhance long-term food security. This technology was created to record all of the financial transactions that have occurred online. Every computer, or node, on the blockchain maintains a history of all previous transactions. Furthermore, no one entity controls the blockchain because it is decentralized. Rather, every transaction on the blockchain is transparent and verifiable by any node in the network [18].

5.1. Opportunities

Blockchain technology enables product tracking from manufacturing to delivery, providing trustworthy information for all supply chain participants. Blockchain can enable transparent and traceable supply chains in agriculture, allowing consumers to verify the origin and quality of agricultural products. This transparency reduces fraud, ensures food safety, and builds trust between producers and consumers. Blockchain provides an immutable record of the entire journey of agricultural products, including information about seed quality, farming practices, use of fertilizers and pesticides, transportation, and storage conditions. In case of food safety issues or contamination, blockchain can quickly trace back the source, facilitating targeted recalls and minimizing consumer impact. Blockchain-based smart contracts automate and enforce agreements between farmers and buyers, enabling direct peer-to-peer transactions. This allows farmers to receive fair prices for their produce while buyers have confidence in the quality and authenticity of the products. It supports sustainability efforts by enabling certification and verification of eco-friendly practices, organic production, and fair trade. Blockchain also facilitates the collection and management of large amounts of agricultural data, enabling informed decision-making and resource allocation [19]. By integrating blockchain technology with IoT devices, small payments in farms, factories, or warehouses can be tracked and information shared across the supply chain [20].

5.2. Challenges

Blockchain technology faces challenges in accessibility due to its complexity and integration with various components such as IoT, RFID (radio frequency identification), sensors, actuators, robots, biometric data, and big data. There is a lack of skills and awareness on blockchain technology, and training platforms are also scarce. Policy development and regulation are both necessary and a major hurdle for the wider adoption of blockchain practices. Technical challenges and design decisions affect existing blockchains and those under development, such as permissioned, permission-less, open, or closed systems. Existing blockchain protocols face serious scalability obstacles, and privacy issues remain a challenge. The digital gap between developed and developing nations is evident in the adoption of blockchain technology. It requires high computing equipment, which is difficult to find in developing countries [19].

The accuracy of data is often influenced by the data provider, and in certain supply chains, there may be untrustworthy providers. This makes blockchain ineffective in preventing food fraud unless all data is thoroughly examined [20]. Another obstacle to successful blockchain implementation is lack of technical knowledge among participants, while unfamiliarity in business supply chains is also a major barrier. This technology addresses challenges faced by developing and underdeveloped nations in agriculture, such as non-uniform crop prices, excessive middlemen in the supply chain, farmer exploitation, crop wastage, and exclusion of small-scale farmers from government subsidies [21].

6. Gene Editing

Genome editing a new plant breeding that can unlock the potential for crops that are high-yielding, high-quality, and resilient to climate challenges, while minimizing environmental impact [22]. Through utilizing gene editing techniques, DNA can be added, changed, removed, or substituted at specific sites within a living organism's genome [23]. In plant research, sequence-specific nucleases (SSNs)-based genome editing techniques have gained popularity. These tools are applied to increase the crop's nutritional content, increase crop adaptability to environmental stresses, and develop crops with high yields. There are currently four SSNs in use: meganucleases, transcription activator-like effector nucleases (TALENs), zinc finger nucleases (ZFNs), and protein systems that are associated with CRISPR/CRISPR-like repeats (CRISPR/Cas) [24].

6.1. Opportunities

Utilizing techniques of plant gene editing, it is now common practice to create crop varieties that are more resilient to harsh environmental conditions or possess superior traits [24]. CRISPR-Cas technology, based on clustered regularly interspaced short palindromic repeats, is extensively employed in plant gene editing and has the potential to be a tool for achieving zero hunger [22]. Crop improvement programmes have made extensive use of the CRISPR-Cas genome-editing tool due to its simple design, cost-effectiveness to use, higher efficiency and reliable results, and capacity to rapidly develop new and improved crops. By editing crop genes, researchers are achieving significant increases in desirable qualities like starch, protein, and vitamins. Some crops have been genetically engineered to fight off stress and disease, making them more resilient. It is currently accepted that CRISPR/Cas technology holds immense potential to enhance crops in a number of ways, including expanding shelf life of food products, improving quality of crops and production, integrating abiotic stress-resistant characteristics (like resistance to drought, herbicides, and insecticides) and eliminating the need for antibiotic-resistant markers, leading to a safer food chain [24]. Important characteristics which have been created through both non-transgenic and transgenic gene editing techniques include resistance to virus, fungus, and bacteria in a variety of crops, such as several cereals, cassava, bananas, papayas, and various vegetables. Crops resistant to diseases and pests can lower the need for chemical pesticide treatments and boost yields by minimizing crop losses [23].

6.2. Challenges

When it comes to these technologies' efficient and useful implementation in the improvement of crops, there are still hurdles in the way. Off-target effect of these technologies, which depends on employing SSNs for deliberate alteration, addition, or replacement of particular loci, is one of their main limitations [24]. The strict regulation and comprehensive risk assessments on genetically modified crops present a significant obstacle to the application of genome editing technologies to develop improved crops. Genome editing faces hurdles when implementing certain edits, such as large deletions of genetic materials, changes in single-nucleotide polymorphisms, insertions of new genes, complete gene replacements, and inversions. Additional challenges include the lengthy process of backcrossing, which can take years, the labour-intensive crossing protocols, the ongoing cost of genotyping, and the potential labour and financial burden of relocating these plants to different environments. The number of unlinked gene edits increases, which strains the capabilities of many breeding platforms. Moreover, introducing multiple gene edits simultaneously hikes the risk of disrupting a beneficial genetic region. This can cause unintended negative effects on the plant's overall performance when bred with high-performing varieties, thereby hindering breeding progress [25]. Another hurdle lies in the limited gene pool for desirable traits, which is crucial for gene availability and improving major crops. Additionally, challenges persist with inefficient methods of transformation and complex, time-consuming processes required to regenerate plants. Biosafety concerns further complicate its widespread adoption. The large size of the CRISPR/Cas9 system hinders its efficiency for delivering gene edits to somatic tissues. Problems in the commercialization of transgenic crops arise due to the high costs and stringent regulations associated with releasing these organisms into the environment [26].

7. Vertical Farming (VF)

In VF systems, agricultural produce is grown vertically in multiple layers stacked on top of one another, making efficient use of available space. Sunlight and soil are not required in VF technology because the environment is completely under control. VF uses indoor, climate-controlled buildings in cities to dramatically increase food production and reduce environmental impact. The idea behind VF is straightforward: use less space to grow more food. It is not just about growing crops efficiently and sustainably, it also has the potential to boost economic growth, minimize environmental impact by reducing pollution, lower energy and water consumption, create new employment prospects, replenish ecosystems, and enhance accessibility to wholesome food. Crops grown in regulated environments are less susceptible to pesticides, dust, polluted water runoff, crop rotation, nutrient cycles, and climatic fluctuations [27].

7.1. Opportunities

It is possible to produce crops year-round in an air-conditioned facility and greater yield per space unit through VF, which also eliminates the need for transportation, increases control over food safety and biosecurity, and uses considerably less water, fertilizer, pesticides, and herbicides. Indoor farming operates independent of weather, enabling year-round cultivation. This leads to potentially higher yields and a more consistent income for farmers. Year round produce availability, reduced crop failures, managed agricultural runoff, and nutrient recovery are some of the direct advantages of such systems [28]. Compared to traditional farming, VF offers a number of benefits, such as increased efficiency, flexibility, and environmental advantages. There is no waste or pollution in VF. This approach fosters optimal growing conditions for each plant, leading to increased crop yields while minimizing the need for herbicides, pesticides, and nitrogen and phosphorus fertilizers. VF can help cool down urban environments, CO_2 sequestration, mitigate the urban heat island effect, and fight climate change. They can absorb high frequency noise generated by traffic and machineries [27].

7.2. Challenges

Limited scientific research on capabilities of VF, like yield potential, crop quality, and energy efficiency, hinders its full assessment and potential [29]. Barriers that exist in implementing vertical farms include social resistance and high start-up costs, lack of expertise, and the need for a controlled growing system. Urban locations tend to be expensive, and the energy consumption required for artificial lighting and climate control can significantly increase operational costs. Additionally, crops requiring insect pollination may need to be manually pollinated, requiring staff time and labour. Other challenges include limited fresh vegetable production, lack of clear USDA vegetable production method identification, and elusive profits. Indoor vertical farms compete against conventional farms, which still enjoy a cost advantage, limiting market penetration. Energy and labour are major contributors to these high costs. Other challenges of VF include social resistance, high start-up costs, and affordability issues in city centres with expensive land and spaces. Low-value agricultural commodities like wheat are economically unviable, making vertical farms' products confined in scope and quantities [27]. There has been relatively little scientific research into VF to date. Important factors like system design, cost-effectiveness, and production method optimization are constantly being assessed to maximize and improve the potential of VF [30]. Though it is still in its early stage, currently VF cannot compete with traditional farming methods in terms of cost-effectiveness and scalability.

8. Role of Drones in Agriculture

Drones will provide farmers with more optimized planting decisions and strategic crop rotation as well as comprehensive daily monitoring of crop health across the entire crop field. Drones of the current generation, which utilizes sensors like Normalized Difference Vegetative Index (NDVI) or near-infrared (NIR), are able to cover greater areas in less time and capture data that the human eye cannot see. Farmers and agronomists can differentiate healthy crop growing areas from weedy areas easily, using NDVI data and post-flight image processing [31]. Drones, also referred to as UAVs, are controlling and optimizing agricultural activities. Its entire operation is based on flight. Precise state estimation (knowing its position) and control are crucial for drone flight. The use of drones for agricultural tasks like watering, monitoring, identification, disaster relief, and photo collection, processing, and analysis has been made feasible by AI. Drones can scan fields using ML to determine which areas require irrigation and which have diseases, pests, and weeds [7].

8.1. Opportunities

UAVs offer several advantages, including limited physical constraints, shorter travel paths, and the ability to fly even in darkness or near zero visibility. UAVs can also save time and labour by reducing the need for additional labourers and spraying crops faster than human labourers. UAV empowers farmers to rapidly acquire aerial imagery of their entire farm or targeted areas of interest, which allow them to monitor crop development, from initial fruit sprouting to identifying pests and weeds problems that threaten crop health [32]. Utilizing drones equipped with advanced sensors allows for soil analysis. This analysis provides valuable data for irrigation planning, planting decisions, and monitoring nitrogen levels in the soil. Drones can also carry suitable reservoirs for fertilizers, herbicides, or pesticides, making crop spraying safer and cost-effective. They can also identify areas with moisture deficits using multispectral indices, planning timely irrigation. It can also help farmers control weeds, insects, pests, and diseases by detecting and informing them about field areas inflicted by these pests. Tree/crop biomass estimation can be done by utilizing ultra-compact LiDAR sensors mounted on drones, enabling estimation of tree/crop biomass change [33]. Drones equipped with scanning technology can precisely target areas for liquid application, optimizing efficiency, and minimizing chemical infiltration into groundwater. Mounted on drones, specialized devices can distinguish variations in how plants reflect green and NIR light. This information translates into multispectral images, allowing farmers to monitor plant health by tracking changes in reflectance [34].

8.2. Challenges

UAVs application in agriculture presents several challenges, including system integration, ethics, privacy, limited flight time, limited payload size, autonomy, data processing, empowering farmers, safety, and availability. Technical challenges include system integration, high-performance data analytics, and the need for net-centric infrastructure. The ethical implications include disruptions caused by GPS jamming and potential security breaches due to weaknesses in control systems. Limited flight time is a significant issue, as UAVs are largely dependent on battery capacity, it makes them less efficient and costlier compared to larger ground-based crop spraying equipment. Autonomy of UAVs also pose a challenge, as most are operated manually by humans, limiting their use to clear visibility. UAVs are still in the early stage, and effective techniques are needed to develop for data collection, transmission, and data conversion into usable information. Cost is another challenge, as the ideal UAV for agriculture purposes requires a balance between durability, extended flight times, operational stability, and autonomous flight. A major safety concern with UAVs is their limited ability to detect and avoid other flying objects, raising the risk of mid-air collisions [32]. Drones for agricultural use have limitations, including short flight durations of 20–60 minutes, high initial costs, and strict national laws. Connectivity is often unavailable in arable farms, requiring farmers to invest in local data storage or buy a drone with local data storage capability. Drones are weather dependent, making them difficult to fly in windy or rainy conditions. Farmers need specialized skills and knowledge to analyse drone images, and misuse can infringe privacy and illegally transfer information [33]. The rapid adoption of drones is hindered by gaps in laws and regulations, as they can be operated in public spaces and disrupt privacy expectations. Several factors hinder effective enforcement of UAV regulations. These include a lack of educational and training resources, challenges in communication with drone manufacturers and retailers, and the need for qualified professionals to evaluate risks and implement appropriate safety measures [35].

9. Remote Sensing

Agricultural remote sensing measures electromagnetic radiation interactions with soil and plant materials, using tractors, handheld sensors, satellites, and aircraft without having contact with them. Agricultural remote sensing has various applications including crop nutrient and water stress, crop yield and biomass, detecting insect and plant disease threats, and assessing various soil properties like organic matter content, moisture levels, clay composition, pH, and salinity. The NDVI is commonly used with satellite and aerial imagery for analysing spatial patterns in crop biomass across a field and predicting potential crop yield. Proximal remote sensing utilizes sensors directly mounted on farm machinery like sprayers, spreaders, tractors, or irrigation booms. Proximal sensing provides real-time data in the field, allowing for on-the-spot management of irrigation pesticides and fertilizers. Hyperspectral remote sensing offers a detailed picture by measuring reflectance across a broad spectrum divided into small intervals [36]. It is an effective tool for assessing several indicators of soil and crop health, such as moisture content, organic matter levels, nutrient availability, carotenoid and chlorophyll presence, cellulose content, crop biomass, and leaf area index [37].

9.1. Opportunities

Remote sensing technology that analyses images can be used to detect weeds and create targeted weed management plans, allowing farmers to apply site-specific herbicides based on spectral reflectance properties between weeds and crops. It can also be utilized to monitor and measure the stress placed on crops by biotic and abiotic factors. This technology allows for monitoring and assessment of insect defoliation, changes in spectral reflectance caused by chlorosis, and progressive reductions in leaf cover. Remote sensing is crucial for managing nitrogen fertilizer at specific sites and for monitoring atmospheric dynamics, such as cloud cover, wind patterns, humidity, and temperature. By combining remote sensing with GIS, farmers can identify areas with nutrient deficiencies in their crops. This targeted approach, known as site-specific nutrient management, reduces cultivation costs and improves fertilizer efficiency [38]. Remote Sensing provides a rapid and non-destructive method for detecting, measuring, and mapping crop stress, this valuable information guides site-specific management decisions regarding nutrient application and insecticide use. Crop disease monitoring is another area where RS has potential. Pathogens often lead to leaf loss, reduced shoot growth, or discolouration of leaf due to a reduction in photosynthesis. Thermal imagery has been shown to be more powerful in early detection of diseases. Remote sensing images demonstrate promising applications in creating detailed yield maps, allowing farmers to analyse variations in crop productivity within a single field [39]. Remote sensing is a useful tool for agricultural land use monitoring, as it allows for the mapping of the agro-ecological landscape. Crop yield forecasting is crucial for various stakeholders, including farmers, national governments, and private entities such as crop insurance providers and commodity traders [40].

9.2. Challenges

Remote sensing has the potential to revolutionize agriculture; however, many challenges need to be addressed. Cloud cover frequently restricts the accessibility of remote sensing imagery from aerial platforms and satellites. This limitation has a less significant impact on ground-based remote sensing [36]. Moran et al. [37] state that Satellite imagery in visible and NIR bands requires cloud-free days and provides the most reliable data when irradiance shows minimal variations over time. At the same time radar imagery collected from airplanes

Table 1: Potential benefits and barriers for the use of innovative technologies in agriculture

S. No.	Technology	Opportunities	Challenges
1.	Artificial Intelligence	Accurate fields and crop data, enhance productivity and quality	Limited electricity, unstable internet, Software and hardware cost, limited data availability
		Predictive analysts	Algorithmic bias risks
		Crop monitoring, nutrient provision, weather forecasting	Lack of market knowledge, insufficient skills, training, technical expertise
		Pest identification and control, disease management	Cultural norms, attitudes, fear of negative impacts, digital divide
		Sensors to analyse soil moisture, automated irrigation system	Customization challenges due to regional languages variations, agricultural practices, pests
		Track fertilizer spraying and crop harvesting, robotic weeding and spraying	Lack of regulatory framework, ethical concern, data sharing, and privacy concern
2.	Internet of Things (IoT)	Sensors enabled remote field management include water use optimization, real-time plant health monitoring, harvesting and animal and pest protection	Interoperability challenges: hardware, network protocols, syntax, semantics, and platform heterogeneity; Education challenges
		Disease and pest management	Knowledge and attitude of farmers towards tech adoption
		Improving agricultural production with regard to cost-effectiveness, quality, quantity	Security and privacy concerns, such as network and data hacking, regulatory gaps
		Monitor soil health, air quality, temperature, location tracking, environmental conditions, and fertilization strategies	Network and energy management challenges: low cost, flexible networking, high scalability, and poor infrastructure
		Greenhouse optimization	Hardware and software costs

3.	Blockchain Based Food Traceability	Collection and management of large data, enabling informed decision-making	Complexity, lack of awareness and skills, limited training platforms, technical challenges
		Food traceability, transparency, safety, security	High cost of accessing data, need for large data collection
		Eliminates manual monitoring and paper tracking systems	Digital gap between developed and developing countries
		Immutable databases, decentralization	Requires high computing equipment
		Verify the order of events and confirm transactions details	Serious scalability obstacles, and privacy issues
		Direct peer-to-peer transactions, reduces fraud	Data accuracy influenced by untrustworthy data provider
4.	Gene Editing	Cost-effective, higher efficiency, rapidly develop new and improved crops	Backcrossing time, cost of continual genotyping, cost and labour of relocating plants
		High-yielding, superior traits, harsh environment and climate resilient, fight off stress and disease, increase nutrition	Challenges in implementing large deletions, single-nucleotide polymorphisms, gene insertions, gene replacements
		Resistance to virus, fungus, bacteria	Complex and time-consuming plant regeneration processes
		More precise and controlled method for crop modification, targeted gene editing	Limiting larger scale genome editing, commercialization problem, biosafety issues
		Free from ethical concerns associated with transgenic crops	Strict regulation and comprehensive risk assessments, unlinked gene edits increase
		Expanding food shelf-life, abiotic stress-resistant characteristics and eliminating need for antibiotic-resistant markers, safer food	Off-target effect of these technologies, which depends on employing SSNs for deliberate alteration, addition, or replacement of particular loci

Contd...

Contd...

S. No.	Technology	Opportunities	Challenges
5.	Vertical Farming	Farms situated near consumers, eliminates transportation need	Limited scientific research, lack of expertise
		Sunlight and soil not required; weather independent	Energy consumption for artificial lighting and climate control
		Uses less water, fertilizer, pesticides; absorbs traffic noise	Manually pollinated
		Sequestering CO_2, reducing Urban Heat Island effect	Limited fresh vegetable production, lack of clear USDA
		Reduced crop failures; managed agricultural runoff, no waste or pollution	High start-up costs, affordability issues in central cities with expensive land and space
		More food in less space, produce crops year-round, increases food safety	System design, cost-effectiveness, and production method optimization
6.	Drones	Shorter travel paths, zero visibility ability to fly	Limited payload size, high initial costs
		Save time and labour, spraying crops faster, control weeds, insect, pest, and diseases	Gaps in laws and regulations, misuse, Ethical concerns: GPS jamming, and hacking
		Detailed crop field observation, informed decision-making	Limited flight time, specialized skills and knowledge
		Soil analysis for irrigation, planting planning, nitrogen levels, strategic crop rotation	Technical challenges: Need of net-centric infrastructure, system integration, data analytics
		Identify moisture-deficit areas, planning timely irrigation	Weather dependent, restricts flying in wind and rain
		Monitoring, photo collection, processing, and analysis, Tree/crop biomass estimation	Improper enforcement, lack of education and training, safety risks

7.	Remote Sensing	Assess crop yield, soil properties, and biomass	Lack of fine spatial resolution data, revisit frequency, spectral data, and processing capacity
		Management of irrigation, pesticides	Require expert knowledge, software
		Site-specific management of nutrient, weeds, and diseases	Limited availability and training, and interoperability with data and tools from various sources
		Monitor insect defoliation	Cloud cover restricts accessibility
		Monitoring agricultural land use and atmospheric dynamics	Lack of established techniques and frameworks applicable under a wide variety of conditions
		Detecting, quantifying, and mapping crop-related stresses	Costly high resolution satellite data

and satellites offers all weather solutions. Additional challenges include calibration of raw digital data for accurate interpretation, correcting imagery for atmospheric interferences or off-nadir view angles, and ensuring accurate image positioning using GPS ground control points (geo-rectifying pixels) [36]. Limited knowledge about the effectiveness of different sensors and platforms, a lack of readily available and well-trained decision-support tools based on RS (remote sensing) data, and difficulties in integrating data and tools from various sources also pose challenges [39]. The key challenges hindering high-precision land use monitoring with remote sensing are: a lack of data with high spatial resolution, infrequent data acquisition (limited updates), restricted spectral information, and the computational limitations in processing these complex datasets [40]. The high cost of acquiring high-resolution satellite data can be a significant hurdle, particularly for smaller farms. The absence of well-proven techniques and frameworks that consistently deliver accurate results across diverse climatic conditions, soil types, crops, and farm management practices hinder wider adoption in precision agriculture [41].

10. Conclusion

Hundreds of millions of people worldwide are affected by the ongoing crisis of hunger despite tremendous efforts to combat it. Unfortunately, it appears that the situation is getting worse, necessitating attention and creative solutions for ensuring food security. This chapter examines a number of ground-breaking inventions such as gene editing, precision agriculture, VF, AI, blockchain technology, UAVs, and the IoT. These innovations offer tremendous potential for alleviating hunger. Advancements in technology development and digitization can rise significantly and minimize the environmental impact of agriculture, strengthen crops' resistance to stress, improve soil health, and ultimately, make farming more cost-effective. Despite the potential benefits, the integration of these novel farming technologies lags behind expectations and is unevenly distributed worldwide. Several interconnected challenges hinder international efforts to eradicate hunger and ensure global food security, they are poor infrastructure and limited internet connectivity, restricted access to technology and the digital literacy skills required to use it effectively, insufficient funding, concerns about data privacy and security, a lack of comprehensive regulations in some countries, and farmer resistance to adopting new agricultural practices. In the face of daunting challenges, the key lies in a multifaceted approach. Establishing a collaborative environment where farmers, cooperatives, extension agents, scientists, and private entities can effectively work together and jointly develop knowledge will significantly enhance the pace of innovation in agriculture.

References

1. FAO; IFAD; UNICEF; WFP; WHO. (2023). *The State of Food Security and Nutrition in the World.*
2. *2023 Global Hunger Index: The Power of Youth in Shaping Food Systems.* (2023).
3. Fan, S. (2017). Supporting country-driven innovations and agri-food value chains for poverty and hunger reduction. *Global Food Security and Wellness*, 1–13.
4. Gunasena, H. (2003). *Food and Poverty: Technologies for Poverty Alleviation.* ResearchGate.
5. Pujitha, A.S., Anusha, B. and Prusty, A.K. (2023). *Artificial Intelligence in Agriculture: A review.* New York: SP Publishing.
6. Nawaz, A., Nadaf, H., Kareem Maniyar, A. and Hanumanthappa, N. (2020). Application of Artificial Intelligence in agriculture - Pros and cons. *Vigyan Varta*, 8.
7. Hossen, I., Fahad, N., Sarkar, R. and Rabbi, M.R. (2023). Artificial Intelligence in agriculture: A systematic literature review. *Turkish J. Comp. and Math. Educ.*, 14(1): 137–146.
8. Ahmad, A., Liew, A.X.W., Venturini, F., Kalogeras, A., Candiani, A., Di Benedetto, G., Ajibola, S., Cartujo, P. et al. (2024). AI can empower agriculture for global food security: Challenges and prospects in developing nations. *Frontiers in Artificial Intelligence*, 7.
9. Farooq, M.S., Riaz, S., Abid, A., Umer, T. and Zikria, Y.B. (2020). Role of IoT technology in agriculture: A systematic literature review. *Electronics*, 9(2): Article 2.
10. Navulur, S., S.C.S. Sastry, A. and N. Giri Prasad, M. (2017). Agricultural management through wireless sensors and Internet of Things. *International Journal of Electrical and Computer Engineering* (*IJECE*), 7(6): 3492.
11. Raut, R., Varma, H., Mulla, C. and Pawar, V.R. (2018). Soil monitoring, fertigation, and irrigation system using IoT for agricultural application. *In*: Y.-C. Hu, S. Tiwari, K.K. Mishra and M.C. Trivedi (Eds.), *Intelligent Communication and Computational Technologies* (Vol. 19), 67–73. Singapore: Springer.
12. Zhang, L., Dabipi, I.K. and Brown Jr., W.L. (2018). Internet of Things Applications for Agriculture. *In*: *Internet of Things A to Z*, 507–528. John Wiley & Sons, Ltd.
13. Srivastava, A. and Das, D.K. (2022). A comprehensive review on the application of Internet of Things (IoT) in smart agriculture. *Wireless Personal Communications*, 122(2): 1807–1837.
14. Friha, O., Ferrag, M.A., Shu, L., Maglaras, L. and Wang, X. (2021). Internet of Things for the future of smart agriculture: A comprehensive survey of emerging technologies. *IEEE/CAA Journal of Automatica Sinica*, 8(4): 718–752.
15. Thilakarathne, N.N., Yassin, H., Bakar, M.S.A. and Abas, P.-E. (2021). Internet of Things in Smart Agriculture: Challenges, Opportunities, and Future Directions. *In*: *2021 IEEE Asia-Pacific Conference on Computer Science and Data Engineering* (*CSDE*), pp. 1–9.
16. Dhanaraju, M., Chenniappan, P., Ramalingam, K., Pazhanivelan, S. and Kaliaperumal, R. (2022). Smart farming: Internet of Things (IoT)-based sustainable agriculture. *Agriculture*, 12(10): Article 10.
17. Zhao, G., Liu, S., Lopez, C., Lu, H., Elgueta, S., Chen, H. and Boshkoska, B.M. (2019). Blockchain technology in agri-food value chain management: A synthesis of applications, challenges, and future research directions. *Computers in Industry*, 109: 83–99.

18. Arora, S., Nabi, T. and Bhardwaj, M. (2023). Blockchain technology: A double-edged sword when it comes to sustainable food security. *The Empirical Economics Letters*, 22: 67–88.
19. Kamilaris, A., Fonts, A. and Prenafeta-Boldú, F.X. (2019). The rise of blockchain technology in agriculture and food supply chains. *Trends in Food Science and Technology*, 91: 640–652.
20. Sandeep Kumar, M., Maheshwari, V., Prabhu, J., Prasanna, M. and Jothikumar, R. (2021). Applying blockchain in agriculture: A study on blockchain technology, benefits, and challenges. *In*: A. Suresh and S. Paiva (Eds.), *Deep Learning and Edge Computing Solutions for High Performance Computing*, 167–181. Springer International Publishing.
21. Meena, S. (2024). Blockchain technology an emerging mechanism in agriculture. *In*: *Agri Business: A Managerial Perspective*, 46–53. New Delhi: Scientific Publishers.
22. Ahmad, S., Tang, L., Shahzad, R., Mawia, A.M., Rao, G.S., Jamil, S., Wei, C., Sheng, Z. et al. (2021). CRISPR-based crop improvements: A way forward to achieve zero hunger. *Journal of Agricultural and Food Chemistry*, 69(30): 8307–8323.
23. Qaim, M. (2020). Role of new plant breeding technologies for food security and sustainable agricultural development. *Applied Economic Perspectives and Policy*, 42(2): 129–150.
24. Hamdan, M.F., Karlson, C.K.S., Teoh, E.Y., Lau, S.-E. and Tan, B.C. (2022). Genome editing for sustainable crop improvement and mitigation of biotic and abiotic stresses. *Plants*, 11(19): Article 19.
25. Tian, Z., Wang, J.-W., Li, J. and Han, B. (2021). Designing future crops: Challenges and strategies for sustainable agriculture. *The Plant Journal*, 105(5): 1165–1178.
26. Rasheed, A., Gill, R.A., Hassan, M.U., Mahmood, A., Qari, S., Zaman, Q.U., Ilyas, M., Aamer, M. et al. (2021). A critical review: Recent advancements in the use of CRISPR/Cas9 technology to enhance crops and alleviate global food crises. *Current Issues in Molecular Biology*, 43(3): Article 3.
27. Al-Kodmany, K. (2018). The vertical farm: A review of developments and implications for the vertical city. *Buildings*, 8(2): 24.
28. Despommier, D.D. (2010). *The Vertical Farm: Feeding the World in the 21st Century*. Macmillan.
29. Beacham, A.M., Vickers, L.H. and Monaghan, J.M. (2019). Vertical farming: A summary of approaches to growing skywards. *The Journal of Horticultural Science and Biotechnology*, 94(3): 277–283.
30. Roberts, J.M., Bruce, T.J.A., Monaghan, J.M., Pope, T.W., Leather, S.R. and Beacham, A.M. (2020). Vertical farming systems bring new considerations for pest and disease management. *Annals of Applied Biology*, 176(3): 226–232.
31. Verougstraete, F. (2015). The rise of the drones in agriculture. *EC Agriculture*, 2(2), 325–327.
32. Yinka-Banjo, C., Ajayi, O., Yinka-Banjo, C. and Ajayi, O. (2019). Sky-farmers: Applications of unmanned aerial vehicles (UAV) in agriculture. *In*: *Unmanned Aerial Vehicles*. IntechOpen.
33. Pathak, H. (n.d.). *Use of Drones in Agriculture: Potentials, Problems, and Policy Needs*. ICAR-National Institute of Stress Management, Maharashtra, India.
34. Delavarpour, N., Koparan, C., Nowatzki, J., Bajwa, S. and Sun, X. (2021). A technical study on UAV characteristics for precision agriculture applications and associated practical challenges. *Remote Sensing*, 13(6): 1204.

35. Ayamga, M., Tekinerdogan, B. and Kassahun, A. (2021). Exploring the challenges posed by regulations for the use of drones in agriculture in the African context. *Land*, 10(2): 164.
36. Mulla, D.J. (2013). Twenty-five years of remote sensing in precision agriculture: Key advances and remaining knowledge gaps. *Biosystems Engineering*, 114(4): 358–371.
37. Moran, M.S., Inoue, Y. and Barnes, E.M. (1997). Opportunities and limitations for image-based remote sensing in precision crop management. *Remote Sensing of Environment*, 61(3): 319–346.
38. Adhikary, S., Biswas, B., Kumar Naskar, M., Mukherjee, B., Pratap Singh, A. and Atta, K. (2023). Remote sensing for agricultural applications. *In*: M. Eyvaz, A. Albahnasawi, E. Gürbulak and M. Tekbaş (Eds.), *Arid Environment: Perspectives, Challenges, and Management*. IntechOpen.
39. Khanal, S., Kc, K., Fulton, J.P., Shearer, S. and Ozkan, E. (2020). Remote sensing in agriculture: Accomplishments, limitations, and opportunities. *Remote Sensing*, 12(22): 3783.
40. Weiss, M., Jacob, F. and Duveiller, G. (2020). Remote sensing for agricultural applications: A meta-review. *Remote Sensing of Environment*, 236: 111402.
41. Sishodia, R.P., Ray, R.L. and Singh, S.K. (2020). Applications of remote sensing in precision agriculture: A review. *Remote Sensing*, 12(19): Article 19.

13 | Multidimensional Aspects of Zero Hunger in Sub-Saharan Africa

Oluwatosin O. Adesina and Omolara Campbell*

Department of Economics, Lead City University, Ibadan, Nigeria

This article explores the multifaceted challenges and strategies for achieving Sustainable Development Goal 2 (SDG 2)—zero hunger—in Sub-Saharan Africa (SSA). Despite global efforts, hunger has worsened since 2015 due to Covid-19, regional conflicts, and climate change, significantly impacting SSA. We discuss the current state of hunger, undernourishment and food insecurity in SSA, highlighting regional disparities and exacerbating factors like climate variability, conflict, economic instability, and health issues. A detailed examination of East, West, and Southern African countries reveals fluctuations in agricultural orientation indices, stored genetic resources, and food price anomalies, underscoring the need for strategic investments and robust policies. We emphasize integrating food security, nutrition, sustainable agriculture, social equity, economic development, and environmental conservation. Key strategies include investments in agriculture, social protection programs, climate adaptation, nutrition interventions, market access, capacity building, and women's empowerment. The transformative potential of smart technologies offers innovative solutions to improve agricultural productivity, sustainability, and resilience including precision agriculture, climate-smart practices, blockchain for market access, mobile applications for nutrition, and early warning systems. In conclusion, addressing hunger and the insecurity of food in SSA requires country specific, regional, and global coordinated actions, sustained commitments, strategic investments, and leveraging smart technologies to achieve zero hunger in SSA.

1. Introduction

To sustain life and well-being, food is essential as it is said that "a hungry

*Corresponding author: campbell.omolara@lcu.edu.ng

man is an angry man". Anyone lacking food is subjected to uncomfortable circumstances suffering both in the brain and body. For children specifically, it retards growth, learning, and reaching the best of their potential. In extreme cases, when hunger is severe, lives are taken. Factors that have caused hunger to increase over time include inequality, climate change, conflict, disasters and emergencies, poverty, to mention a few. Upon this basis and more were the SDGs drawn as a sequel to the Millennium Development Goals (MDGs) by the United Nations which have 17 distinct goals and several targets. The focus of Goal 2 of the SDGs is on eradicating hunger in all its dimensions.

The aim of the Sustainable Development Goal 2 (SDG 2): "Zero Hunger" is to eradicate global hunger by 2030. However, since 2015, the problem has worsened due to various factors including the 2019/20 pandemic, various regional conflicts, climate change, and so on. In 2019 for instance, an estimated 690 million people were undernourished. Nearly 60 million people more were recorded in 2020; between 79 and 161 million persons were experiencing hunger than in 2019 [1]. On an average, both moderate and severe food insecurity has increased. Available statistics indicate that food insecurity rose from 26.6% in 2019 to 30.4% in 2020, with negative repercussions for health and well-being [2-5]. According to the United Nations' Report for 2023, the amount of chronically hungry people surged to 735 million (representing 9.2% of the world's population) in 2022, which is a significant increase from 2019. Concurrently, an estimated 2.4 billion individuals faced moderate to severe food insecurity, marking a distressing rise of 391 million compared to 2019. This escalating hunger crisis underscores the urgent need for coordinated global efforts. Moreover, as hunger and food insecurity increased, existing inequalities worsened and hampered sustainable development. Approximately 2 billion people lack regular access to safe, nutritious, as well as food that is sufficient, perpetuating a cycle of poverty and malnutrition. In 2022 alone, about 148 million children experienced stunted growth, while children under 5 years, amounting to 45 million, have been observed to suffer from wasting, highlighting the profound impact on the most vulnerable populations [6]. Thus, addressing these critical challenges requires immediate and concerted action on multiple fronts to alleviate global hunger, promote food security, and create a more equitable and sustainable future for all, especially for Sub-Saharan African (SSA) countries.

"Zero Hunger" is a pivotal mandate to eliminate global hunger as well as malnutrition, nonetheless, as data reveals that global hunger persists (and still rising in SSA countries), thereby necessitating a comprehensive analysis of the current state of the goal. Existing body of literature often tends to pay attention to food security, coordination between SDG2 and other goals, amongst others. However, the condition of SSA countries, as compared with other regions concerning this goal, lacks evidence. According to data, the region has put up an abysmal performance in attaining this goal, zero hunger, specifically after the global pandemic in 2020. While numerous studies have delved into the issue of zero hunger for achieving SDGs, critical gaps persist in synthesizing a comprehensive probe into the multifaceted features of the goal, challenges around the current situation and the opportunities to be maximized by applying smart technologies to tackle such challenges. Although perspectives from authors have called for food systems for realizing SDG2, highlighting limitations, while others have underscored the positive role of institutions without detailing specific mechanisms; these have constituted challenges to policymakers due to insufficient empirical evidence.

As such, this chapter is clearly set out to bridge these gaps; first, we explored the multidimensional aspects of zero hunger, delving into the existing nature of the goal in SSA countries. Secondly, the challenges faced in the present circumstances in the region were explored while proffering solutions to them. Furthermore, opportunities in these circumstances were examined following smart technologies in enhancing the attainment of the goal. These will shed light on the urgency and importance of achieving the goal while considering the wider perspective of sustainable development. Based on data obtained between 2015 and 2021 from United Nations' Reports, and Food and Agriculture Organization, descriptive statistical techniques such as graphs were employed to analyze these data. Section 2 describes the multidimensional measures of the zero hunger goal and the extent of its realization in SSA countries. Section 3 exposes and discusses the challenges faced in achieving the targets of this goal. Section 4 suggests solutions required to overcome the challenges, while Section 4 presents the opportunities in the existing situation using smart technologies in achieving the SDG2. Section 5 concludes this chapter.

2. Multidimensional Aspects of Zero Hunger in Sub-Saharan African Countries

The world's common race is targeted towards the attainment of sustainable development. This race started in 2015 with some specified goals of which SDG2: Zero Hunger, is among them. This specifies what indicators to check out for a zero hunger world among countries and regions. As outlined in this goal, Zero Hunger extends beyond mere eradication of hunger to encompass broader issues of security of food, nutrition, and sustainable development [7]. The 14 indicators state that by 2030:

1. Hunger will be put to an end to ensure that all year round there will be access to food that is safe, nutritious, and sufficient, thus reducing the prevalence of:
   1.1. Undernourishment among the population;
   1.2. Moderate or severe food insecurity.
2. End All Forms of Malnutrition, thus reducing the prevalence of:
   2.1. Children under 5 years with a height-for-age less than –2 standard deviations from the WHO Child Growth Standards median (Stunting);
   2.2. Children under 5 years with a weight-for-height greater than +2 or less

than –2 standard deviations from the WHO Child Growth Standards median (Childhood Malnutrition - Wasting and Overweight);

2.3. Anemia in women aged 15 to 49 years, categorized by pregnancy status.

3. Double Agricultural Productivity and Incomes of Small-Scale Food Producers:
   3.1. Volume of production per labor unit by farm size;
   3.2. Average income of small-scale food producers.
4. Ensure Sustainable Food Production Systems:
   4.1. Agricultural area under productive and sustainable agriculture practices.
5. Maintain Genetic Diversity in Agriculture:
   5.1. Plant and animal genetic resources for food and agriculture that are available in conservation facilities;
   5.2. Local Breeds at Risk of Extinction.
6. Increase Investment in Agriculture:
   6.1. Agriculture Orientation Index;
   6.2. Official development assistance and other official flows into agriculture.
7. Correct and Prevent Trade Restrictions in Agricultural Markets:
   7.1. Status and reduction of agricultural export subsidies.
8. Ensure Proper Functioning of Food Commodity Markets:
   8.1. Monitor and manage extreme food price volatility.

Thus, the multidimensional aspects of zero hunger, and the present condition of these indicators in SSA countries are being examined.

3. Challenges Faced in Achieving the Targets of SDG2

Prevalence of undernourishment measures the share of the population that is undernourished, reflecting the insufficiency of food intake and the amount of hunger within a population. *Prevalence of moderate or severe food insecurity in the population* is the absence of reliable access to enough food for an active, healthy life which encompasses both measurable and immeasurable aspects of food availability, access, and utilization. *Prevalence of stunting among children under 5 years of age* indicates chronic malnutrition and inadequate nutrition during the early years of life. It reflects long-term insufficient nutrient intake and is related to developmental issues. *Prevalence of malnutrition among children under 5 years of age, by type* encompasses both wasting (low weight for height) and overweight (high weight for height). This involves all forms of malnutrition, including both undernutrition and overnutrition. *Prevalence of anemia in women aged 15 to 49 years, by pregnancy status (percentage)* is often linked to inadequate iron intake and poor nutrition. *Volume of production per labor unit* relates to agricultural productivity and efficiency, which are crucial for ensuring food security leading to sustainable agriculture.

Average income of small-scale food producers highlights the socioeconomic status of small-scale food producers, who play an important part in food production, particularly in developing countries. *Proportion of agricultural area under productive and sustainable agriculture* involves preserving natural resources, mitigating environmental degradation, and ensuring the long-term viability of the system of food production. *Number of (a) plant and (b) animal genetic resources for food and agriculture secured in either medium- or long-term conservation facilities* is associated with maintaining resilient and productive food systems, particularly against environmental challenges and changing climatic conditions. *Proportion of local breeds categorized as being at risk of extinction* includes the amount of indigenous breeds possessing unique genetic traits adapted to specific environmental conditions, a loss leading to reduction in genetic diversity and resilience in agricultural systems.

Agriculture orientation index measures the extent to which government expenditures are directed towards agriculture and rural development. *Total official development assistance and other official flows into the agriculture sector* represent financial support from governments and international organizations for agricultural development initiatives. S*ubsidies on agricultural export* are funds provided by the government to pay part of the amount paid for exporting agricultural produce to other countries. They have implications for world-wide security of food and market dynamics. They can distort trade patterns, affect the price of food, and impact the incomes of farmers, particularly in developing countries. *Food price anomalies* such as sudden spikes or fluctuations in the prices of food, can have significant implications for food access and affordability, particularly for vulnerable populations.

Summarily, achieving zero hunger requires an all-inclusive and combined approach that addresses the interconnected dimensions of security of food, nutrition, agricultural sustainability, social equity, economic development, environmental conservation, policy and governance, and community engagement. Food security defines the condition in which all people have access to sufficient, safe, and nourishing food to achieve their dietary needs. It is a fundamental prerequisite for achieving zero hunger. Guaranteeing food availability, accessibility, affordability, and utilization are key components of food security strategies. In addition to addressing hunger, efforts to address zero hunger must prioritize improving nutrition outcomes for all individuals. This entails combating malnutrition in its various forms, including undernutrition, micronutrient deficiencies, and overnutrition [8]. Promoting diverse and nutritious diets is essential for optimal health and well-being.

Also, sustainable agriculture practices are critical for ensuring durable food security and environmental sustainability [9]. Investments in sustainable farming methods, soil conservation, water management, and biodiversity conservation are necessary to enhance agricultural productivity while minimizing negative environmental impacts. Achieving SDG2 requires addressing underlying social inequalities and disparities that perpetuate insecurity of food and malnutrition

[10]. Promoting social equity involves ensuring equitable access to food, education, healthcare, and economic opportunities for all individuals, particularly marginalized and vulnerable groups. In addition, economic development plays a critical part in reducing hunger by creating opportunities for income generation, poverty reduction, and sustainable livelihoods [11]. Rural infrastructure investments, productivity in agriculture, as well as smallholder farmers can stimulate economic growth as well as increasing the security of food. Likewise, environmental conservation is essential for maintaining the ecological balance and resilience of food systems [12]. Sustainable land use practices, conservation agriculture, and mitigation of the effect of changes in climate are necessary to safeguard natural resources and biodiversity for future generations.

In the same vein, effective governance and policy interventions are crucial for addressing the foundation of hunger and malnutrition [13]. Policies such as social safety nets, agricultural subsidies, and nutrition education programs can help improve the security of food and nutrition outcomes at the country specific and international levels. Lastly, empowering domestic communities and encouraging participatory approaches are vital for the security of food [14]. Engaging communities in decision-making processes and supporting grassroots initiatives can engender inclusive and resilient food systems. Turning to the conditions prevalent around available data for SSA countries, six indicators were considered including the occurrence of moderate or severe insecurity of food among the total population, the occurrence of undernourishment, Credit to Agriculture, plant genetic resources accessions stored *ex situ*, Forestry and Fishing (Agriculture orientation index US$), Total disbursements of official flows for agriculture, by receiving countries (millions of constant 2021 United States dollars), and Indicator of Food Price Anomalies (IFPA), by Consumer Food Price Index [27].

Data reveals that countries in East Africa, including Ethiopia, Kenya, Somalia, South Sudan, Tanzania, and Uganda, have witnessed steady increases in the incidence of moderate or severe insecurity of food between 2015 and 2021. These trends suggest persistent problems in realizing the security of food in the region. For West African nations, such as Sierra Leone, Burkina Faso, Cameroon, Guinea, Côte d'Ivoire, Gambia, Nigeria, Ghana, Guinea-Bissau, Mali, Benin, Mauritania, Niger, Senegal and Togo, varying degrees of food insecurity have been experienced. While some countries have shown increasing trends, others have witnessed fluctuations or relatively stable prevalence rates. Southern African countries, including Zimbabwe, Botswana, Madagascar, Angola, Malawi, Mozambique, Lesotho, Namibia, and Zambia, have also experienced rising levels of food insecurity. Likely factors have exacerbated conditions such as the variability in climate, conflict, economic, agricultural productivity, poverty levels, social protection programs, drought, economic inequality, and HIV/AIDS to mention a few.

Comparing this with the North African countries like Algeria, Libya, Mauritania, and Sudan, unlike the SSA countries, they have seen steady increases in insecurity of food between 2015 and 2021. However, other countries in Africa, such as Egypt, Morocco, and Tunisia, may have more steady or decreasing trends caused by political instability, economic challenges, and changes in climate impacts as presented in Figure 1. Thus, the increasing frequency of moderate or severe insecurity in food across SSA countries underscores the urgent need for comprehensive strategies to address this issue. Policymakers must prioritize investments in agriculture, social protection programs, and resilience-building initiatives to ensure security of food for all populations. Efforts to lessen the effect of climate change, conflict, and economic instability on the production of food and distribution are crucial for long-term sustainability and resilience. Collaborative approaches are critical to addressing the complex challenges of insecurity of food in Africa to achieve the goal of ending hunger and malnutrition.

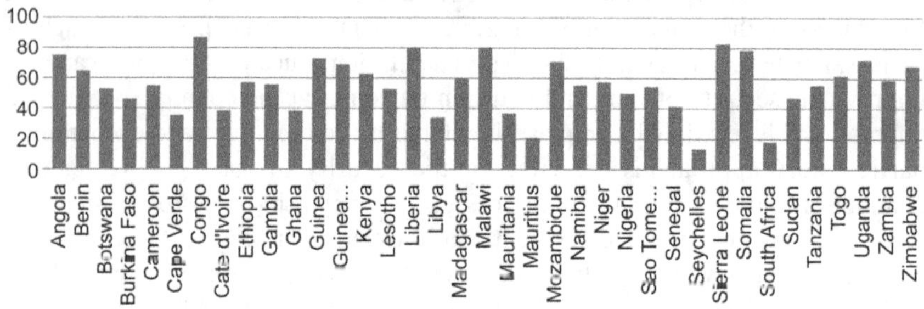

Figure 1: Prevalence of moderate or severe food insecurity in the total population (%). *Source*: Our World in Data Team, 2023.

Countries in East Africa, such as Ethiopia, Kenya, and Rwanda, have experienced fluctuations in the frequency of undernourishment (see Figure 2), with some observing slight decreases in current years. However, these fluctuations indicate ongoing challenges in attaining SDG2 in the region. Also, West African nations, including Ghana, Senegal, and Nigeria, have experienced fluctuations in the occurrence of undernourishment, with minor overall changes observed. While some countries have maintained relatively stable rates, others, like Mali and Nigeria, have witnessed notable increases in recent years. In Southern Africa, countries like Lesotho and Zimbabwe have experienced notable fluctuations in the incidence of undernourishment, with slight decreases noted in later years. However, challenges persist, as seen in countries like Angola, where undernourishment has increased significantly. Thus, it is revealed that complex challenges are faced by African countries in combating undernourishment and realizing SDG2. Sustainable development strategies, including investments in agriculture, social protection programs, and poverty alleviation efforts, are crucial for addressing the foundation of hunger and malnutrition. Comprehensive and coordinated actions at domestic, regional, and global levels are essential to guarantee the well-being of all individuals across the continent and achieve the goal of zero hunger in Africa.

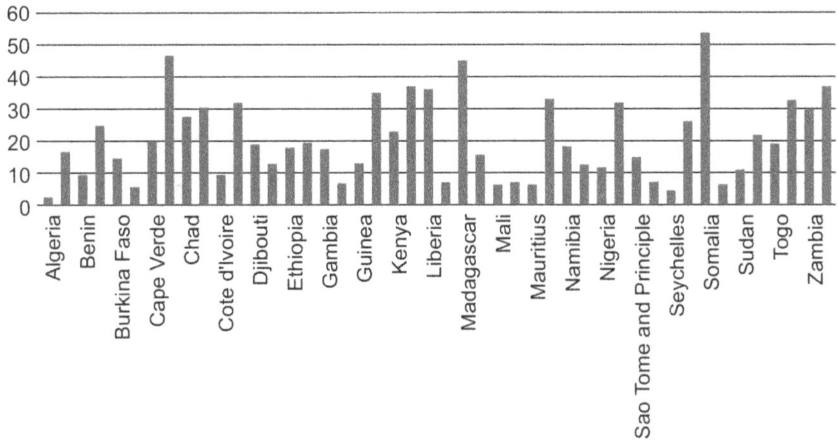

Figure 2: Prevalence of undernourishment (%). *Source*: Our World in Data Team, 2023.

Countries like Ethiopia and Kenya in East Africa have made significant strides in attaining SDG2, which aims to end hunger, attain the security of food and improved nutrition, and promote agricultural sustainability. Both nations have maintained steady or slightly increasing numbers of stored genetic resources over the analyzed period in Figure 3. This indicates a commitment to preserving genetic diversity for the security of food and agricultural development. Tanzania also showed notable increases in stored genetic resources in 2016, reflecting efforts towards attaining SDG2. In West Africa, Ghana and Senegal have demonstrated consistency in the quantity of stored genetic resources with minor fluctuations. While there may not be drastic increases, the stability suggests efforts towards maintaining genetic diversity for the security of food. Nigeria exhibited fluctuations but showed a notable increase in 2018, indicating efforts to address challenges related to genetic resource conservation in alignment with SDG2. In Southern Africa, countries like South Africa exhibited fluctuations in stored genetic resources, reflecting challenges in realizing the targets of SDG2. However, Botswana maintained a relatively steady number with a slight rise noted in 2020, suggesting efforts towards genetic resource conservation for agricultural sustainability. Namibia and Zimbabwe also experienced fluctuations, which may pose challenges to realizing SDG2 targets in these countries. Overall, the data highlights diverse trends in genetic resource management across African regions, with some countries making significant advancement towards attaining SDG2, while others face challenges. Fluctuations in stored genetic resources underscore the dynamic nature of conservation efforts, emphasizing the necessity for continuous monitoring and collaboration among countries and international organizations. Sustained commitments and collaborative actions are necessary to address challenges and promote biodiversity conservation and agricultural sustainability, ultimately contributing to the attainment of SDG2 in Africa.

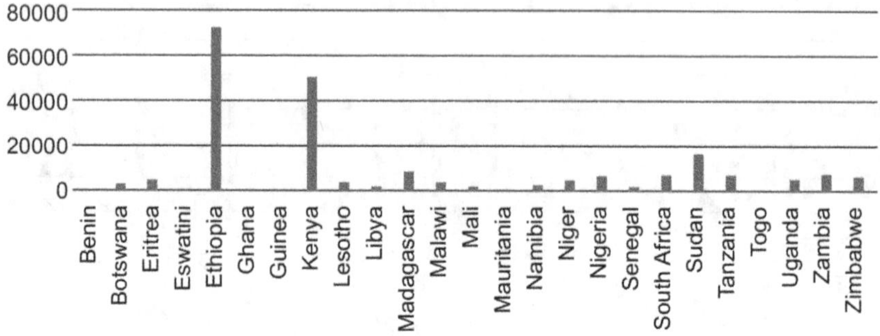

Figure 3: Plant genetic resources accessions stored *ex situ* (number).
Source: Our World in Data Team, 2023.

As depicted in Figure 4, countries in East Africa, such as Ethiopia, Kenya, Rwanda, and Uganda, have experienced fluctuations in the agriculture orientation index, with minor overall changes observed. While some countries experienced slight decreases, others have maintained relatively stable rates. However, the fluctuations underscore the necessity for targeted policies and investments to promote sustainable growth in the agriculture, forestry, and fishing sectors in the SSA. West African nations, including Ghana, Nigeria, and Senegal, have also witnessed fluctuations in the index, with minor overall changes observed. While some countries have seen improvements or stable rates, others like Burkina Faso and Mali, have experienced slight decreases in later years. This highlights the complexity of agricultural development in the SSA and the significance of addressing challenges such as access to credit and market access. In Southern Africa, countries like Zambia have experienced significant fluctuations in the index, with notable increases observed in later years. However, countries like Botswana and Zimbabwe have seen minor overall changes. The disparities highlight the prominence of addressing issues such as access to credit and

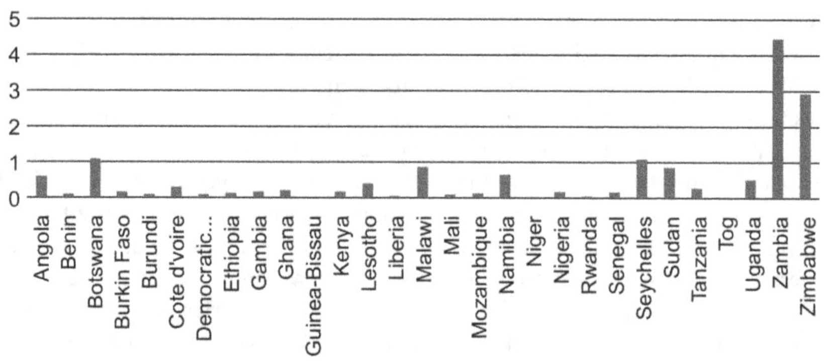

Figure 4: Credit to agriculture, forestry, and fishing (Agriculture orientation index) US$.
Source: Our World in Data Team, 2023.

technological innovation to drive growth in the economy and enhance the security of food in SSA. The fluctuations in the agriculture orientation index underscore the challenges and opportunities in the agriculture, forestry, and fishing sectors across Africa. Policymakers and stakeholders need to address issues such as access to credit, technological innovation, market access, and climate resilience to guarantee long-term sustainability and development of these vital sectors. Continued efforts and collaborations are indispensable to harness the full potential of the agriculture, forestry, and fishing sectors for the overall development of the continent.

Countries in East Africa, such as Somalia, Ethiopia, Rwanda, Uganda, South Sudan, Tanzania, and Kenya, have experienced fluctuations in total official flows for agriculture (Figure 5), while some countries have seen notable increases in disbursements in recent years, whereas have witnessed fluctuations or slight decreases. These variations underscore the need for strategic investments and robust funding mechanisms to support sustainable agricultural development in the region. West African nations, including Senegal, Benin, Burkina Faso, Mali, Ghana, Guinea, Nigeria, Togo, and Sierra Leone have also witnessed fluctuations in total official flows for agriculture. Some countries have experienced notable increases in disbursements, while others have seen fluctuations or slight decreases. This highlights the significance of addressing issues such as aid effectiveness and alignment with national development priorities to maximize the effect of foreign assistance to agriculture. In Southern Africa, countries like Angola, Botswana, Mozambique, Namibia, South Africa, Zambia, and Zimbabwe have witnessed fluctuations in total official flows for agriculture. While some countries have experienced overall increasing trends in disbursements, others have seen fluctuations or decreasing trends. This highlights the complexity of foreign aid and investment in the agriculture sector and the importance of addressing

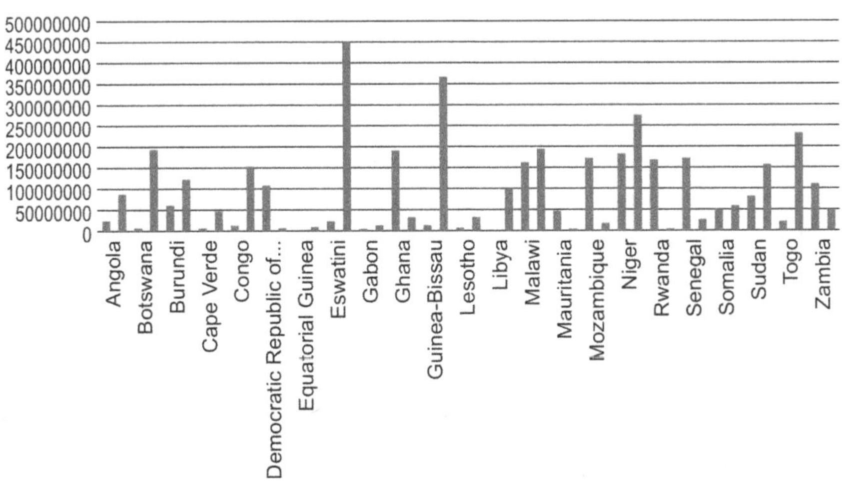

Figure 5: Total official flows (disbursements) for agriculture, by recipient countries (millions of constant 2021 United States dollars). *Source*: Our World in Data Team, 2023.

issues such as aid effectiveness and project sustainability. The fluctuations in total official flows for agriculture emphasize the significance of robust funding mechanisms and strategic investments to aid agricultural development in African countries. Policymakers and stakeholders need to address issues such as aid effectiveness, project sustainability, and alignment with national development priorities to maximize the impact of foreign assistance in the agriculture sector. Effective resource allocation and targeted interventions are essential to address the diverse challenges and opportunities facing the agriculture sector across the continent. Continued collaboration and innovative approaches are crucial to ensure sustainable agricultural development and food security in Africa.

The outlook of the indicator for food price anomalies is present in Figure 6; it depicts that countries that have experienced fluctuations in food price anomalies in East Africa, which include South Sudan, Rwanda, Somalia, Ethiopia, Kenya, Tanzania, and Uganda, while some have seen notable increases in anomalies in recent years, others have witnessed fluctuations or slight decreases. These variations highlight the complexity of food markets and economic conditions in SSA, underscoring the necessity for proactive measures and effective policies to address the volatility of price and ensure economic stability. West African nations, including Ghana, Guinea, Mali, Burkina Faso, Senegal, Benin, Sierra Leone, Nigeria, and Togo, have also witnessed fluctuations in food price anomalies. Some countries faced significant increases in anomalies, while others have seen fluctuations or slight decreases. This underscores the significance of monitoring food markets closely and implementing appropriate measures to mitigate the impact on susceptible populations and ensure economic resilience in West Africa. Southern African countries, such as Angola, Zambia, Mozambique, Namibia, Botswana, South Africa, and Zimbabwe, have witnessed fluctuations in food price anomalies. Some have experienced overall increasing trends in anomalies, others

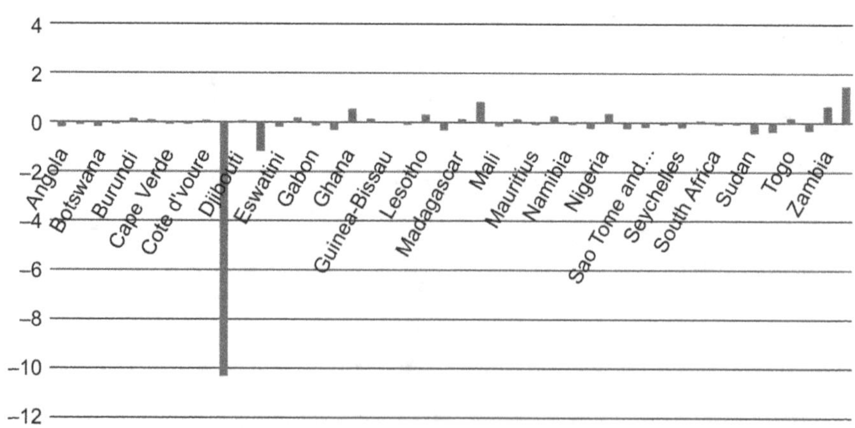

Figure 6: Indicator of Food Price Anomalies (IFPA), by consumer food price index.
Source: Our World in Data Team, 2023.

have seen fluctuations or decreasing trends. This highlights the complexity of food markets and economic conditions in Southern Africa and the prominence of targeted interventions to address volatility of price and ensure economic stability. Fluctuations in the price of food can have significant implications for the security of food, inflation rates, and economic stability. Policymakers and stakeholders need to monitor these trends closely and implement appropriate measures to moderate the effect on susceptible populations and ensure economic resilience. Proactive measures and effective policies are essential to address food price anomalies and promote stability in African markets. Collaborative efforts and targeted interventions are crucial to fight against the effect on the security of food and livelihoods, particularly in regions prone to volatility and economic shocks.

The challenges facing SSA countries in achieving SDG2 aimed at ending hunger, attaining food security, expanding nutrition, and encouraging sustainable agriculture are multifaceted and complex. These challenges are divided into two: first, challenges relating to the targets of the goal, and, secondly, challenges outside the targets of the goal. Recent evidence on the first division of the challenges indicates persistent issues in ensuring food security across SSA, with steady increases in food insecurity prevalence observed in countries like South Sudan, Kenya, Somalia, Tanzania, Ethiopia, and Uganda in East Africa between 2015 and 2021. Similarly, West African nations like Benin, Burkina Faso, Cameroon, Ghana, Nigeria, and Senegal have experienced varying degrees of insecurity in food production, with some countries showing increasing trends while others witnessing fluctuations or relatively stable prevalence rates. Also, Southern African countries like Angola, Mozambique, Zambia, and Zimbabwe have faced rising levels of insecurity in food production, further complicating efforts to achieve SDG2.

Despite efforts to address insecurity in food production in this region, the second division of challenges including climate variability, conflict, economic and political instability, low agricultural productivity, poverty, inadequate social protection programs, drought, economic inequality, and diseases such as HIV/AIDS persist, thus exacerbating the situation [15]. These challenges interact in complex ways, making it problematic in addressing food insecurity comprehensively.

4. Solving Challenges in the Current Circumstance of Multidimensional Aspects of Zero Hunger

There are several initiatives underway in SSA to achieve SDG2, which aims to end hunger and guarantee food security. These efforts include increasing investments in agriculture, implementing social protection programs, promoting climate adaptation, enhancing nutrition interventions, market access, capacity building, and prioritized empowering women. These challenges require holistic approaches such as integrating social, economic, and environmental dimensions. Additionally, strategies including fostering innovation, addressing gender inequality, and building resilience to shocks from outside the economy are essential. Further, strengthening governance, supporting smallholder farmers, and encouraging food systems that are sustainable remain crucial. Overall, collaborative efforts and sustained commitments are central to overcome these challenges and create a future where all individuals have adequate, harmless, and nutritious food [15].

Also, there are innovative solutions that utilize Smart Technologies. Smart technologies have been identified by researchers as possessing immense capabilities for transforming agriculture and addressing food security challenges in SSA. Smart technologies offer innovative solutions to overcome challenges currently facing SSA countries in attaining SDG2 and improving agricultural productivity, sustainability, and resilience for zero hunger. Smart technologies have the ability to provide real-time data and insights that empower farmers to make informed decisions. For example, accurate agriculture technologies, including smart sensors and drones, facilitate farmers to continuously check crop health, soil humidity levels, and nutrient deficiencies with unprecedented accuracy. By leveraging this data, farmers can optimize irrigation schedules, apply fertilizers more efficiently, and detect pests as well as the outbreaks of diseases early, resulting in higher yields and reduced input costs.

Furthermore, smart technologies facilitate the integration of climate-smart practices into agricultural systems, helping farmers acclimatize to varying environmental situations and mitigate the impacts of climate variation. For instance, weather forecasting models powered by artificial intelligence (AI) can provide accurate predictions of rainfall patterns and temperature fluctuations, allowing farmers to adjust planting schedules and crop selections accordingly. Additionally, innovative irrigation systems equipped with sensors and automation technology enable precise water management, conserving scarce water resources and decreasing the risk of drought-related crop failures. Moreover, smart technologies occupy a vital role in expanding market access and value chain efficiency, particularly for smallholder farmers who often struggle to access markets and obtain fair prices for their produce. Blockchain technology, for example, provides transparent and secure traceability throughout the supply chain, reducing transaction costs, eliminating intermediaries, and ensuring fair compensation for farmers. Mobile applications and e-commerce platforms connect farmers directly with buyers, enabling them to reach wider markets and negotiate better prices for their crops. The transformative impact of smart technologies extends beyond agricultural production to encompass various parts of the security of food and nutrition. For instance, mobile applications and digital platforms facilitate access to information on nutrition, health, and dietary diversity, empowering consumers to make healthier food choices. Additionally, innovations in the processing of food and preservation technologies help reduce post-harvest losses and extend the shelf life of perishable foods, guaranteeing the availability of food and reducing food waste. Likewise, smart technologies enable more effective monitoring and evaluation of the security of food interventions, allowing policymakers and development agencies to track progress, identify areas of need, and allocate resources efficiently. Advanced analytics of data as well as algorithms from machine learning can analyze large datasets to uncover patterns and trends, guiding evidence-based decision-making and enhancing the effect of food security programs.

Furthermore, to improve productivity of agriculture and food distribution, smart technologies also have the potential to enhance resilience and adaptive capacity in the presence of emerging challenges such as pandemics, conflicts, and natural disasters. For example, digital platforms for early warning and rapid response facilitate timely information sharing and coordination among stakeholders, enabling proactive measures to mitigate the effect of crises on the security of food and livelihoods. Still, the use of smart technologies can foster innovation and entrepreneurship in the agricultural sector, creating new opportunities for the growth of the economy and job creation. Startups and small businesses specializing in agri-tech solutions are emerging across the region, developing innovative commodities tailored to the necessities of smallholder farmers and rural societies. By fostering a vibrant ecosystem of innovation and entrepreneurship, smart technologies aid the overall resilience and sustainability of the agricultural sector in SSA.

The agricultural sector is witnessing a revolution driven by an array of innovative technologies aimed at addressing the diverse challenges faced by farmers globally. These solutions encompass a wide range of interventions provided by companies. For instance, biological crop treatments offered by **Alkama** which focuses on enhancing crop health while minimizing reliance on synthetic pesticides, **AMINI's** environmental data infrastructure is pivotal for informed decision-making in agriculture, facilitating better resource management and sustainability practices [16]. **Billion Carbon** pioneers' bio-fertilizer production offers sustainable alternatives to conventional chemical fertilizers, thus contributing to soil health and environmental conservation [17]. **Dvara E-Registry** introduces digital platforms to improve credit access for farmers, enhancing financial inclusion and unlocking economic opportunities in rural areas [18]. **Instacrops** employs virtual advisor models powered by AI to provide personalized recommendations for crop management, optimizing yields and resource utilization [19]. **Kheyti's modular greenhouses** extend growing seasons and improve crop quality, enabling farmers to adapt to changing climatic conditions and enhance productivity [20, 21]. **Moa Technology** leads herbicide development efforts, aiming for safer and more efficient weed management solutions to safeguard crop yields and environmental health [22]. **Spowdi pioneers solar-powered irrigation systems** reduce carbon emissions and dependence on non-renewable energy sources in agricultural operations [23]. **Unibaio** focuses on sustainable agrochemical formulations, balancing efficacy with environmental considerations to minimize ecological impact while maintaining agricultural productivity [24]. **UpTrade** facilitates efficient online trading of livestock, streamlining market access for farmers and enhancing economic opportunities in the livestock sector

[25]. **VIA (Virtual Irrigation Academy)** employs color-based tools for water and fertilizer management, enabling precision agriculture practices that optimize resource use and minimize environmental impact [26].

These technologies, supported by research and industry initiatives, are instrumental in enhancing agricultural productivity, sustainability, and resilience, thus contributing significantly to global food security and environmental conservation efforts [28-30]. Furthermore, Table 1 presents a summary of smart technologies for each target for the zero hunger goal.

Table 1: Smart technologies for the targets for zero hunger goal

Target	Indicator	Smart Technology
End hunger and ensure access to safe, nutritious, and sufficient food year-round	Prevalence of undernourishment	Precision agriculture using IoT sensors and drones for real-time monitoring and optimization of crop production
End all forms of malnutrition	Prevalence of stunting in children under 5 years	Mobile applications and wearable devices for tracking nutritional intake and promoting healthy eating
Double agricultural productivity and incomes of small-scale food producers	Volume of production per labor unit	Blockchain for transparent land ownership records, smart contracts, and digital platforms connecting farmers to markets and financial services
Ensure sustainable food production systems	Proportion of agricultural area under sustainable agriculture	Remote sensing and machine learning for optimizing farming practices and mitigating climate impact
Maintain genetic diversity in agriculture	Number of genetic resources in conservation facilities	Gene editing technologies like CRISPR for developing resilient crop varieties
Enhance international cooperation for clean energy	Percentage of renewable energy in total energy consumption	Smart grids and energy management systems for optimizing energy distribution, integrating renewable energy sources, and reducing losses
Increase investment in agriculture	Total official development assistance for agriculture	Remote sensing for monitoring land use, and digital platforms for agricultural research and extension services
Correct and prevent trade restrictions in agricultural markets	Agricultural export subsidies	Blockchain for transparent and traceable supply chains to ensure fair trade practices

Source: Authors' compilation, 2024.

5. Conclusion

Food is vital for life and well-being, but its absence causes severe physical and mental suffering, particularly in children, stopping them from reaching their full potential. Factors include inequality, climate change, conflict, disasters, and poverty exacerbate hunger. Despite efforts to attain the United Nations' SDG2 to end global hunger by 2030, the situation has worsened since 2015 as a result of the pandemic, regional conflicts, and climate change, with a notable increase in food insecurity and undernourishment in SSA. This goal aims to eradicate hunger and provide access to harmless and nutritious food for all. The rising hunger and food insecurity demand urgent, coordinated global actions. The surge in hunger perpetuates inequalities and hinders sustainable development, affecting nearly 2 billion people worldwide and significantly impacting children's health and development.

SSA countries face substantial challenges in achieving zero hunger, particularly after the global pandemic. Various studies highlight gaps in explaining the multifaceted aspects of hunger in SSA. This examination of the current state of hunger in SSA reveals significant regional disparities in the insecurity of food and undernourishment from 2015 to 2021. Countries like Nigeria, Ethiopia, Ghana, Kenya, and Zimbabwe experienced rising food insecurity due to climate variability, conflict, economic challenges, and health issues like HIV/AIDS. North African countries also faced amplified food insecurity from political instability and economic difficulties. The UN has established 14 indicators to monitor progress towards zero hunger, covering security of food, nourishment, and agricultural sustainability. Key indicators include the incidence of undernourishment, food insecurity, stunting and wasting in children, anemia in women, productivity of the agricultural sector, and government agricultural expenditures. Achieving zero hunger in SSA requires an all-inclusive approach encompassing food security, nutrition, sustainable agriculture, social equity, economic development, environmental conservation, effective governance, and community engagement. Key strategies include policies and investments for sustainable agricultural growth in Eastern Africa (Ethiopia, Kenya, Rwanda, and Uganda), consistency in genetic resource conservation in Western Africa (Ghana and Senegal), and stable agricultural policies in Southern Africa (South Africa and Botswana). Strategic interventions are also needed in countries like Angola and Zambia. SSA faces challenges such as insecurity in food production, climate variability, conflicts, economic instability, low agricultural productivity, inadequate social protection, and diseases. Solutions involve investments in agriculture, social protection programs, climate adaptation, nutrition interventions, market access, capacity building, and women's empowerment. Technologies like precision agriculture, climate-smart practices, blockchain for market access, mobile applications for nutrition, food processing innovations, and timely warning systems can significantly improve food security.

Key recommendations include increasing access to credit, technological innovation, and market access through targeted policies, enhancing international collaboration for biodiversity conservation, securing robust funding and investments for sustainable agriculture, implementing policies to address volatility of food price, and using smart technologies to boost productivity, sustainability, and resilience. Future research on achieving zero hunger in SSA should focus on enhancing the collection of data and using smart technologies like accurate agriculture and blockchain for efficient food systems. Climate-smart agricultural practices need development to address climate change impacts. Socio-economic factors, particularly gender equity and economic development, should be studied to improve the security of food.

References

1. Sachs, J., Schmidt-Traub, G., Kroll, C., Lafortune, G., Fuller, G. and Woelm, F. (2020). *The Sustainable Development Goals and COVID-19.* Sustainable Development Report 2020. Cambridge: Cambridge University Press.
2. Food and Agriculture Organization (FAO). (2021). *World Food and Agriculture – Statistical Yearbook 2021*. Rome. https://doi.org/10.4060/cb4477en
3. United Nations Children's Fund (UNICEF). (2021). Annual Report, 2021.
4. World Food Production (WFP). 2021. Annual Review, 2021.
5. World Health Organization (WHO). (2021). World Health Statistics Report, Monitoring Health for the SDGs.
6. United Nations (UN). (2024). *UN: The Main Challenges and Key Dates in 2024.* https://unric.org/en/un-the-main-challenges-and-key-dates-in-2024/.
7. United Nations. (2015). *Transforming Our World: The 2030 Agenda for Sustainable Development.*
8. World Health Statistics. (2022). *Monitoring Health for the SDGs, Sustainable Development Goals.*
9. Garnett, T., Appleby, M.C., Balmford, L.J., Benton, T.G., Bloomer, P., Burlingame, B., Dawkins, M., Dolan, L. et al. (2013). Sustainable intensification in agriculture: Premises and policies. *Science*, 341: 33–34. doi:10.1126/science.1234485.
10. Development Initiatives. (2018). *Global Nutrition Report: Shining a Light to Spur Action on Nutrition. 2018*. Bristol, UK.
11. International Food Policy Research Institute (IFPRI). (2020). *Global Food Policy Report 2020.*
12. Intergovernmental Panel on Climate Change (IPCC). (2019). *Climate Change and Land: An IPCC Special Report on Climate Change, Desertification, Land Degradation, Sustainable Land Management, Food Security, and Greenhouse Gas Fluxes in Terrestrial Ecosystems.*
13. Food and Agriculture Organization of the United Nations (FAO). (2019). *The State of Food Security and Nutrition in the World.*
14. United Nations Development Programme (UNDP). (2017). *Community Engagement and Empowerment in the Context of Sustainable Development.*
15. Food and Agriculture Organization of the United Nations (FAO). (2021). *The State of*

Food and Agriculture 2021. Making Agri-food Systems More Resilient to Shocks and Stresses. Rome, FAO. https://doi.org/10.4060/cb4476en.

16. Food and Agriculture Organization of the United Nations (FAO). (2020). *The State of Food and Agriculture 2020. Overcoming Water Challenges in Agriculture*. Rome. https://doi.org/10.4060/cb1447en.
17. Billion Carbon. (2023). *Decarbonising Food Supply Chain by Nutrient Mining from Food Waste*. https://www.billioncarbon.com/
18. International Fund for Agricultural Development (IFAD). (2019). *Rural Development Report. Creating Opportunities for Rural Youth, 2019*. https://www.ifad.org/documents/38714170/41133075/RDR_report.pdf/7282db66-2d67-b514-d004-5ec25d9729a0?ref=blog.mondato.com.
19. Instacrops (2024). *Data-Driven Agricultural Insights*. https://agtecher.com/product/instacrops/
20. Food and Agriculture Organization. (FAO). (2017). *The State of Food and Agriculture. Trends and Challenges*. www.fao.org/3/i6583e/i6583e.pdf.
21. Food and Agriculture Organization. (FAO). (2017). *The State of Food and Agriculture. Leveraging Food Systems for Inclusive Rural*. https://www.fao.org/3/i7658e/i7658e.pdf.
22. MOA. (2024). *Investigating Nature's Design for Novel Herbicide Modes of Action*. https://www.moa-technology.com/
23. Food and Agriculture Organization. (FAO). (2016). *The State of Food and Agriculture, Climate Change, Agriculture, and Food Security*. https://www.fao.org/3/i6132e/i6132e.pdf.
24. Akanmu, A.O., Akol, A.M., Ndolo, D.O., Kutu, F.R. and Babalola, O.O. (2023). Agroecological techniques: Adoption of safe and sustainable agricultural practices among the smallholder farmers in Africa. *Front. Sustain. Food Syst.*, 7: 1143061. doi: 10.3389/fsufs.2023.1143061.
25. Uplink. (2024). *Enabling Farmers to Use Livestock as Currency to Afford Key Technology*. https://uplink.weforum.org/uplink/s/uplink-contribution/a012o00001pUt7xAAC/Enabling%20farmers%20to%20use%20livestock%20as%20currency%20to%20afford%20key%20technology%E2%9C%AA.
26. Cropin. (2024). *Precision Agriculture - Harnessing Technology to Boost Productivity and Profitability*. https://www.cropin.com/precision-agriculture.
27. Our World in Data team. (2023). *End Hunger, Achieve Food Security and Improved Nutrition and Promote Sustainable Agriculture*. Published online at OurWorldInData.org. Retrieved from: https://ourworldindata.org/sdgs/zero-hunger.
28. Jose, Anand A., Nirmala Devi, K., Vijayakumar, S. and Chacko, Shiju C. (2023). Decision-making support in smart farming. *In*: K. Kalaiselvi, Jose Anand A., Poonam Tanwar and Haider Raza. *Advanced Technologies for Smart Agriculture*, 219–239. Denmark: River Publishers.
29. Augustine, C., Balaji, K., Dharanikumar, S.V. and Jose, A.A. (2023). Urban farming: Case study. *In*: K. Kalaiselvi, A.A. Jose, T. Poonam and R. Haider. *Advanced Technologies for Smart Agriculture*. 321–338. Denmark: River Publishers.
30. Althaf, S., Nafisa, B., Devadharshini, G., Geethalakshmi, M. and Jose, A. (2022). Organic farming in drainage system with advanced automation through robotics and IoT. *International Conference on Data Science, Agents, and Artificial Intelligence (ICDSAAI)*, 592–597. Chennai, India.

14 | Smart Aqua Culture Cultivation to Achieve Zero Hunger in Indonesia

Zulis Erwanto[1], Rian Mantasa Salve Prastica[2], I. Gusti Ngurah Bagus Catrawedarma[1] and Saddam Rassanjani[3*]

[1] Politeknik Negeri Banyuwangi, Banyuwangi, Indonesia
[2] Politeknik Pekerjaan Umum, Ministry of Public Works and Housing, Semarang, Indonesia
[3] Universitas Syiah Kuala, Banda Aceh, Indonesia

This research discusses the use of fog harvesting technology as an innovative solution to overcome drought and increase food security in Indonesia, by exploring its impact on community empowerment and the potential to achieve the goal of zero hunger. The integration of community-based technology in drought mitigation is expected to make a significant contribution to social, economic, and environmental transformation to support sustainable development. The technologies discussed include fog harvesting and the Biofloc method with Bubble Generator technology in increasing food production through aquaculture cultivation. This research reveals the benefits of fog-catching buildings in supporting food security by providing water for agriculture, raw water processing, and integration with aquaculture-hydroponic cultivation. This research reveals that the biofloc system and the use of bubble generators can increase the productivity of tilapia cultivation by reducing fish mortality rates, increasing growth, and optimizing feed conversion, thereby increasing harvest yields and cultivation efficiency. Policy development related to blockchain supply and demand and active community participation in maintaining fog capture technology has the potential to support the achievement of the goal of zero hunger in Indonesia through transparent, efficient, and sustainable monitoring of the availability of clean water.

*Corresponding author: saddam.rassanjani@usk.ac.id

1. Introduction

This research discusses the use of fog harvesting technology as an innovative solution to overcome drought and increase food security. Amid increasingly worrying drought challenges, people in various regions in Indonesia are struggling to maintain their food security and livelihoods. To address this problem, community-based technology has emerged as an innovative solution that not only mitigates drought but also empowers local communities to achieve zero hunger. The initial technology proposed, which is still relatively rarely used in Indonesia, is fog harvesting technology [1–6]. Its successful implementation in mountainous areas in Indonesia, which is exemplified by its implementation in Kemitir Village, Semarang Regency, Central Java, serves to offer an alternative water source for agricultural purposes, as well as encouraging community involvement and independence [7]. Installation and maintenance of fog harvesting systems require active involvement from local communities, fostering a sense of unity and shared responsibility in overcoming environmental challenges. However, implementing this technology in drought-prone areas in Indonesia is quite a big challenge, because not all areas are conducive to its application. For example, in Banyuwangi, East Java, this technology faces obstacles originating from the characteristics of windy coastal areas [8, 9].

Apart from that, fog harvesting technology also opens opportunities to increase community economic empowerment. By ensuring a more reliable water supply and increasing agricultural productivity, farmers can increase crop yields and diversify their agricultural products. However, areas that face obstacles in its implementation require additional support, such as aquaculture cultivation through biofloc [10, 11] with bubble generator technology. These steps can be a catalyst for the development of other productive sectors, including drip irrigation for agriculture and plantations [12] as well as aquaculture culture that uses irrigation canals to cultivate land fisheries in several regions in Indonesia [13–15]. Successful integration of these concepts is expected to offer a strategy to reduce hunger during the dry season in underdeveloped areas and especially dry land areas, particularly in mountainous and coastal areas. Despite limited understanding and restricted examples of implementation, the research gap lies in effectively integrating this technology as a community-based empowerment initiative to advance the sustainable development goal's (SDGs) of achieving zero hunger in Indonesia.

Community-based technology empowers communities to tackle drought while fostering sustainable change and societal transformation. Comprehensive support from stakeholders like government, NGOs, and the private sector is essential, offering education, training, financial aid, and infrastructure development. Strong partnerships and stakeholder involvement are crucial for leveraging integrated community-based technology to achieve zero hunger and sustainable development in Indonesia.

2. Fog Harvesting Tower Technology Based on Local Wisdom

This fog harvesting tower technology has dimensions of a bottom diameter of 5 m, a top diameter of 7.5 m, and a height of 7 m and is equipped with a 100 m^2 synthetic net made from bamboo material based on local wisdom. The way this technology works is by capturing mist and dew that pass-through nets on the tower, then the water goes down the piping installation to a water filter and then flows into a water reservoir. The fog harvester tower construction model implemented by the community is shown in Figure 1.

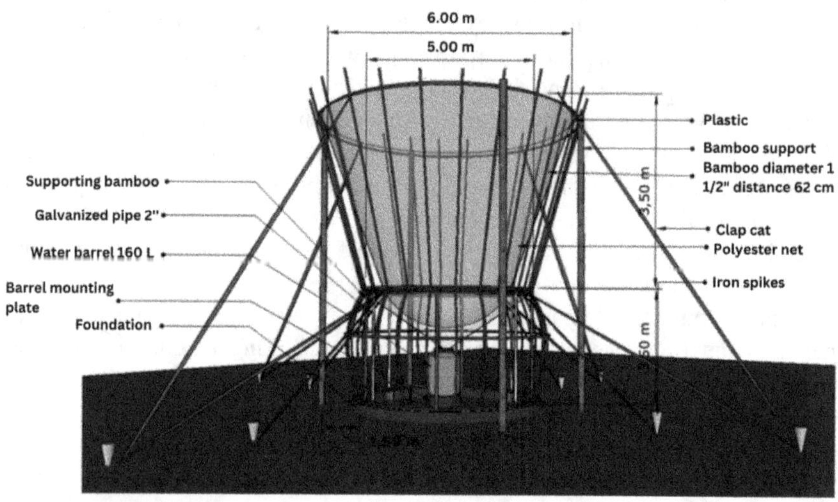

Figure 1: Fog harvesting tower building model in Alasbuluh village [8].

Fog can form when air saturated with water vapor is cooled below its freezing point. Therefore, in highland areas where temperatures are cold and humid, we often encounter fog. Fog is water vapor that condenses near the ground surface and becomes cloud-like. This usually forms because cold air causes water vapor to condense and the humidity level approaches 100% [5, 8, 16]. Fog harvesting is an innovative technology based on the collection of water collected from fog under certain climatic conditions. Several meteorological and geographical considerations are important in selecting a location: dominant wind direction (continuous wind from one direction is ideal), clouds forming below maximum ground level, sufficient space for fog collection, wind speed between 3–12 meters/second, and no major obstacles on land. In the case of coastal cloud lines, the mountains should be 5 or 10 kilometers from the coast [3].

The construction of the fog harvester building is the construction of a filter that captures small water droplets (1–40 μm) (1 micrometer = 10^{-6} m), a double-layered mesh. These nets are usually made of polypropylene or polyethylene, protected against ultraviolet, with a shading coefficient of 35%, a Raschel filter weave, and a fiber size of 1 mm. Efficiency increases as the sieve size becomes finer and the fiber width becomes smaller [17-19]. The surface area should be large enough, typical dimensions per net might be 12 m long × 4 m high (48 m^2). Typical water collection rates vary depending on the situation but appear to average between 2 and 5 liters per square meter per day, with maximum collections reaching 10 liters per square meter per day [20]. There are several alternative types of fog harvesting net models as shown in Figure 2, and the advantages and disadvantages of fog harvester buildings can be seen in Table 1.

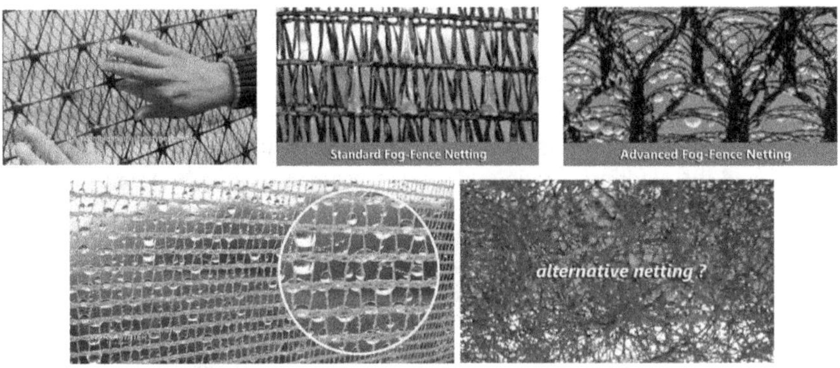

Figure 2: Types of fog harvesting nets [8, 20, 21].

Table 1: Advantages and disadvantages of fog harvesting buildings [8, 22]

Advantages	Disadvantages
1. Low construction cost 2. Simple technology and maintenance 3. The water is of good quality 4. Not affected by the dry season	1. The amount of water that can be harvested is relatively small because it depends on the weather 2. It is difficult to find polypropylene filters in some areas 3. Effects of storm damage due to location and fragility of nets if not maintained 4. Requires frequent presence of mist 5. Damage and lack of maintenance, caused by the distance between the building and populated areas

3. Biofloc Method with Bubble Generator Technology to Increase Food Production through Aquaculture

Utilizing available water resources for fish farming, like red tilapia, reduces construction costs. Training on bio floc technology (BFT) improves aquaculture efficiency, speeding up harvests. Prioritizing sustainable aquaculture methods, including climate adaptation, ensures long-term viability in meeting future food needs. The development of BFT allows increased productivity, environmentally friendly, efficient use of land and water resources. BFT is a technology that utilizes the activity of microorganisms that form flocs. The application of BFT is widely applied in industrial wastewater treatment systems and is beginning to be applied in aquaculture media water treatment systems.

Biofloc itself is composed of several components, namely: exopolysaccharides, floc-forming bacteria, and functional cycle bacteria. The following materials and methods: (i) Ingredients: Sago flour (C = 49.84%) 2.480 kg, urea 0.3925 kg, non-iodized salt 5 tablespoons, probiotics (*Bacillus* spp.) 80 ml (liquid), enough water, and yeast tape 5 grams (soaked in water warm); (ii) How to prepare it: All ingredients are mixed and stirred evenly in a tub. Then, add enough water until the mixture is like a runny paste. Wait for 3 days and stir every 4–6 hours. The final stage spreads ¼ of the mixture in the morning and can be followed up every 3 days. The process of forming organic fertilizer biofloc (as an initial starter) and the working principle of biofloc bacteria can be shown in Figure 3.

Figure 3a: Initial biofloc starter making training.

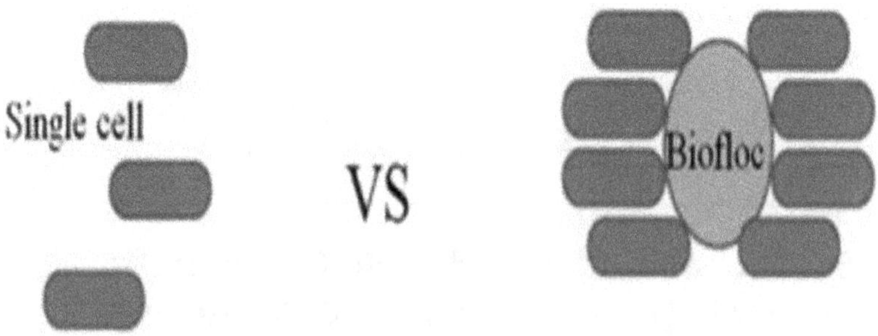

Figure 3b: The biofloc system works more effectively than ordinary bacteria, the resistance of the bacterial system is more stable, and resistance to pathogenic bacteria is more effective.

Figure 3: The process of forming organic fertilizer biofloc and the working principle of biofloc bacteria [11].

Biofloc formation involves bacteria like Bacillus circulans, Bacillus coagulans, Bacillus licheniformis, and Bacillus subtilis, working after exopolysaccharides are produced. These bacteria, especially Bacillus licheniformis, aid nutrient cycling. Biofloc systems, unlike standalone probiotics, offer stability by forming cohesive bacterial flocs, preventing disease germ entry, and maintaining stable water parameters. BFT adopts a form of environmental engineering that relies on oxygen supply and the use of microorganisms and can directly increase the digestibility of feed. The advantage of this BFT is that it can integrate with a hydroponic system, the survival rate (SR) value is up to more than 90%, and without water changes, the Feed Conversion Ratio (FCR) value reaches 1.03.

Adoption of the latest technology development for fisheries cultivation (aquaculture), includes: (1) Implementation of Integrated Multitrophic Aquaculture (IMTA) [23] through an ecosystem approach (Ecosystem Approach for Aquaculture); (2) Development of BFT which enables increased productivity, is environmentally friendly, and efficient in the use of land and water resources [10]; (3) Development of the Mina-Padi system; (4) Development of a Recirculating Aquaculture System (RAS) or closed system which can increase productivity up to 100 times and is efficient in using water and land [24]; (5) Encourages the development of local fish cultivation such as Papuyu, Belida, Tawes, Semah, Gabus, Batak fish, and other local types; (6) Development of rain and fog harvesting technology which is integrated with aquaculture and aquaponics; and (7) Ultrafine bubble oxygen technology which can increase fisheries productivity.

Transitioning from traditional to intensive fish farming methods has led to a decline in water quality [25] due to high feeding rates and accumulation of waste at the pond bottom. This waste turns into harmful ammonia, affecting fish growth [25]. Land availability and cost pose challenges for expanding aquaculture. Ideal oxygen levels for freshwater fish are 3–4 mg O^2/L [25], achievable in shallow ponds. Deeper ponds require technology beyond conventional aerators or nanobubble generators, which are costly. Microbubble generators (MBGs) offer a solution, producing tiny bubbles that dissolve slowly, enhancing oxygen levels. This supports aerobic microbial metabolism, improving water quality by breaking down organic matter. The working principle of the MBG is that pressurized water from the inlet side flows through the reduced flow cross section in the MBG. The pressure in this area of reduced cross section is less than the atmospheric pressure so that atmospheric air can be sucked in naturally in this area. The mixture of air and water forms microbubbles which come out through the outlet.

Various techniques are commonly employed to generate microbubbles, including hydrodynamic [26], acoustic or sonication [27], electrochemical [28], and mechanical agitation [27]. The hydrodynamic method is used in various fields because it consumes low energy and can be used in various environments. Included in this method are swirl flow types [29, 30], orifice [31], spherical body [32], and venturi [33–35]. Swirl flow employs centrifugal effects and strong shear forces from rapid water rotation to create microbubbles, while orifice, spherical body, and venturi methods utilize throttling effects to accelerate fluid flow, inducing turbulence and shear forces that fragment bubbles. Spherical bodies face manufacturing complexity and maintenance challenges when narrow gaps accumulate dirt, while orifice methods incur energy losses from abrupt cross-sectional reductions. Venturi systems mitigate energy losses by gradually reducing cross sections. The use of a venturi is one option to overcome the weakness in terms of energy loss due to the water flow through a gradual reduction in cross section. Modifications to the MBG venturi have been carried out by adding porous pipes [34] and several small holes [36, 37] on the air inlet side to form microbubbles.

The structure of the MBG venturi has three parts, namely converging section, throat, and diverging section as shown in Figure 4. This type has the advantages of simple structure, easy installation, no moving parts, easy maintenance, and high reliability [38]. The most important thing is low power consumption so that it is more economical to increase gas-water mass transfer [30, 39]. Another advantage is that it can produce bubbles with high density and diameters below 100 micrometers [28, 40] and is easy to regulate the diameter of the bubbles produced by adjusting the composition of the water and air discharge. Another advantage is that air can be sucked naturally into the MBG because the pressure in the throat area is smaller than atmospheric air and the air channel is open to the environment, so it is very simple and more flexible and adaptive to complex environments, without having to use a compressor to circulate air. Therefore, MBG venturi has the potential to be used in several applications such as in aquaculture and agriculture, it can increase dissolved oxygen (DO) in water to help the growth rate of fish [41] and plants [42, 43]. In wastewater processing, Mitra [44] and Kaya [45] reported a COD reduction of above 85% in water. Besides that, with its ability to supply many microbubbles, the MBG venturi can also be applied to reduce drag due to water and body friction in the movement of ships [46].

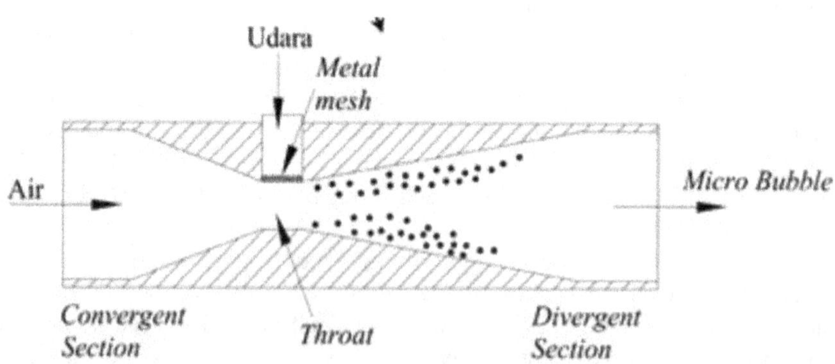

Figure 4: Venturi-metal mesh type bubble generator.

The application of the MBG venturi in fish farming can be done using several tools such as water pumps, water pipes, aerators, and air hoses with installations as in Figure 5. The greater the water flow supplied by the pump into the MBG, the smaller the bubbles that form. The required water pump specification is that the greater the pump discharge that can flow into the MBG, the smaller the bubbles that form at the same air discharge. The bubble generator installation, depicted in Figure 5, comprises several components: (1) Aerator supplies air to the bubble generator to enhance dissolved oxygen levels, crucial for fish health; (2) The fishpond serves as the farming area and must be carefully managed to prevent water turbidity caused by overfeeding or fish metabolism; (3) The bubble generator produces micro-sized bubbles by mixing air and water, contributing to oxygenation in the pond; (4) A filtered fish waste container collects debris carried by the water flow, ensuring stable water quality through periodic waste disposal; (5) Mechanical filter media, typically fishing nets, trap debris to prevent water contamination; (6) The water pump recirculates water into the pond and supplies water to the bubble generator; (7) Biological filters, made of gombong or pumice stone, foster the growth of beneficial bacteria that break down harmful substances like ammonia, maintaining water quality; and (8) Drainpipes remove excess water and solid waste from the pond, ensuring optimal conditions for fish farming.

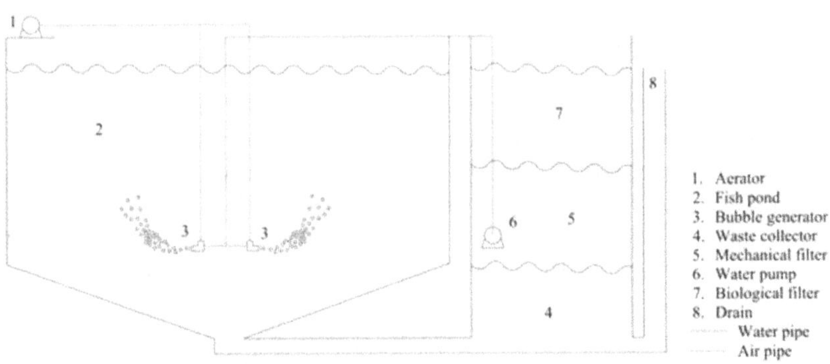

Figure 5: Bubble generator installation.

4. Application of Fog Harvesting Technology for Hydroponic Integrated Drip Irrigation Needs to Achieve the Goal of Zero Hunger

The results of the completed construction of the fog harvesting tower with tower construction dimensions of 5 meters in diameter with a height of 7 m and a building area of around ±10 m with local wisdom materials made from bamboo as in Figure 6. It is hoped that there will be further research to determine the amount of discharge. Fog water is produced by fog harvesting towers with various types of nets to obtain optimal mist water.

(a) (b)

Figure 6: (a) Construction of a fog harvesting tower, and (b) Fog harvesting tower [8].

Another simple model for a household-scale mist collector is a simple, flat, rectangular nylon net supported by poles at both ends and arranged perpendicular to the direction of the prevailing wind. Mist collecting surfaces other than plastic are usually made of fine mesh nylon or polypropylene mesh, for example, "shade cloth", which is a raschel mesh (made of flat black polypropylene filaments, 1.0 mm wide and 0.1 mm thick, in a triangular weave), installed in double layers. The results of the mist water obtained from the fog harvesting tower were not optimal because the location was not high enough and inadequate and the wind conditions were too strong. Therefore, it is necessary to have an alternative location that is much higher and shady, namely in the Teak Forest area to the west of the village. Implementing a fog harvesting program requires assessing the potential for water extraction, considering factors like fog frequency, water content, and mist collection system design. Reports from the Banyuwangi Regency Meteorology, Climatology and Geophysics Agency aid in evaluating fog occurrence. Success relies on favourable climatic conditions and strategic placement, considering wind patterns and terrain. Fog harvesters benefit by: (i) Supporting water supply for plant irrigation during dry seasons, especially for drip irrigation to boost agricultural productivity; (ii) Producing raw water for purification, ensuring clean water for daily needs; and (iii) Integrating with aquaculture and hydroponics to enhance food security in achieving zero hunger goals.

In Pondoknongko village, Banyuwangi, hydroponic cultivation and drip irrigation are integrated using polybags to maximize limited land use, as shown in Figure 7. This integration aims to optimize water utilization, particularly in mountainous and coastal areas prone to drought, by harnessing surface water potential and harvesting atmospheric water through rainwater and fog harvesting [21]. These methods are integrated with aquaculture cultivation through biofloc technology. Banyuwangi Regency's 'Sistersay' program supports food security initiatives by empowering communities through integrated livestock, fish, and vegetable cultivation systems. This program encourages residents to use home gardens for cultivating horticultural plants, such as kale, lettuce, and broccoli, supported by drip irrigation technology developed by the Center for Appropriate

Figure 7: Hydroponics with a drip irrigation system [12].

Technology Development-LIPI. Drip irrigation delivers water directly to plants through PVC pipes and plastic screw droppers, achieving up to 95% water absorption efficiency [12]. Fertigation technology, integrated with drip irrigation, is widely adopted by farmers for cultivating various vegetable crops, including tomatoes, chilies, potatoes, and mustard greens.

5. Development of Biofloc Aquaculture with Bubble Generator Technology in Banyuwangi Regency, Indonesia

Biofloc aquaculture technology offers versatility by integrating with hydroponic systems, utilizing various water sources like rainwater and mist water. In Indonesia, it's commonly used for catfish and tilapia cultivation, significantly boosting productivity. For instance, a 3-meter round pond can accommodate over 3,000 catfish, yielding more than 300 kilograms per cycle, a threefold increase [10] compared to conventional methods. The government's recent focus on tilapia cultivation using BFT further highlights its potential for enhancing fish production, particularly due to tilapia's faster growth rate as herbivores. This shift is reflected in Table 2, illustrating the productivity differences between regular ponds and the biofloc system.

Table 2: Difference in productivity results between regular pond and biofloc system

Parameters	Regular Pond System	Biofloc System
Stocking density	100 Catfish/m^3	500–1000 Catfish/m^3
	10 Tilapia fish/m^3	100–150 Tilapia fish/m^3
Harvest	120–130 days	100–110 days
Productivity	2 kg/m^3	25–30 kg/m^3
Maintenance time	4–6 months	3 months

Apart from that, tilapia fish are also able to digest floc, which is composed of various microorganisms, namely bacteria, algae, zooplankton, phytoplankton, and organic materials as part of their food source. Tilapia cultivation using the biofloc system has several advantages, such as increasing survival rate (SR) to more than 90% and without water changes. Used cultivation water is also odorless, so it does not disturb the surrounding environment and can be synergized with cultivating plants such as vegetables and fruit. This is because there are microorganisms that can break down cultivation waste into fertilizer that fertilizes plants. Another advantage is the FCR or the ratio between the weight of feed and the total weight (biomass) of fish in one cultivation cycle period, reaching 1.03. This means that 1.03 kg of feed produces 1 kg of Tilapia fish.

Figure 8 illustrates Berkah Farming Women's Group managing catfish ponds with an aquaponic system in Katulampa, Bogor City, West Java. With 40 ponds using biofloc and aquaponics, they produce 100 kilograms of catfish weekly, aiding food security during Covid-19. For tilapia raised in biofloc systems, maintaining DO is crucial as it supports fish growth and aids bacteria in breaking down waste. Tilapia require DO levels of at least 3 mg/L, necessitating Bubble Generator technology to boost oxygen levels and enhance productivity. In tilapia biofloc systems, combining a bubble generator enhances dissolved oxygen levels in ponds. Unlike aerators producing large bubbles, bubble generators efficiently boost oxygen. Research [47] shows bubble generators initially yield 7.64 mg/L of DO, dropping to 4.28 mg/L after 2 weeks. In comparison, aerators start at 6.83 mg/L but decline to 1.86 mg/L. Without these devices, DO levels decrease to 0.61 mg/L, leading to an 85% fish mortality rate due to insufficient oxygen for growth.

Figure 8: Cultivating catfish using a biofloc and aquaponics system.

Apart from increasing DO in the water, using a bubble generator can also reduce the stress level of the fish, increase the final weight of the fish which will have an impact on the specific growth rate and FCR. The stress level of the fish can be seen in the lower cortisol values when cultivating using a microbubble generator, this shows that the fish are more adaptive to salinity stress which ultimately does not affect their growth rate. The final weight of fish also increased quite significantly when using a microbubble generator compared to cultivating with a regular aerator [47]. This can be shown from Table 3 about the differences in aquaculture cultivation results using microbubble generator technology and regular aerators.

Table 3: Differences in aquaculture cultivation results between microbubble generators and regular aerators

Parameters	Microbubble Generator	Regular Aerator
The initial fish weight	114.55 g	113.33 g
The final fish weight	202.03 g	148.83 g
Increase of weight	87.48 g (76.37%)	35.5 g (31.32%)
Rate of specific growth	1.14%	0.55%
Ratio of the food conversion	1.3	1.87
The harvest time	4 times per year	3 times per year

6. Clean Water Supply and Demand Blockchain Policy in Increasing Food Productivity to Achieve Zero Hunger in Indonesia

So that the availability of clean water is maintained and able to support the achievement of SDG2 Zero Hunger, it is necessary to develop a policy related to blockchain supply and demand. Blockchain technology has provided many benefits to industry in various sectors because of its reliability in optimizing the provision of better services [48, 49]. Blockchain technology offers solutions to challenges in fog-harvesting buildings. Firstly, it enables transparent and accurate recording of harvested water, enhancing water availability monitoring. Secondly, blockchain aids in identifying and tracking polypropylene filters, simplifying their procurement and replacement. Thirdly, it facilitates tracking storm damage and net maintenance, reducing risks from adverse weather. Fourthly, real-time weather data recording and analysis aid maintenance planning. Lastly, blockchain enhances access to fog trap building information, fostering community awareness and participation in maintenance. By supporting policies related to blockchain supply and demand, Indonesia can ensure sustainable clean water availability, crucial for addressing zero hunger, especially in drought-prone mountainous and coastal regions.

However, to successfully implement this concept, active participation from the community and multi-stakeholder involvement is required through collaborative governance strategies. It is widely believed that community participation will bring benefits to the sustainability of a program. However, the reality in developing countries shows that people tend to avoid involvement in formal participation processes because they see it as a complicated official procedure [50]. In this way, stimulating the public to be proactively involved in maintaining fog capture technology requires a holistic approach involving various strategies. First, education and outreach about the importance of this technology needs to be widely disseminated, highlighting its benefits in increasing the availability of clean water and supporting food security. This can be done through outreach campaigns at the community level, seminars, and educational programs in schools. Second, governments and non-governmental organizations (NGOs) need to work together to provide training and technical assistance to communities on how to maintain and repair haze-capturing building infrastructure. Third, building strong partnerships between local communities, social activists, and other NGOs to provide moral support as well as assist in the form of human and financial resources needed for infrastructure maintenance. Fourth, providing incentives or rewards to individuals or groups who are active in maintaining technology so that it can be additional motivation for the community to be involved proactively. With these steps, it is hoped that the community will be more motivated to be involved in the construction and maintenance of this infrastructure, to ensure the sustainability of this program for the long term.

While community involvement in dialogue is supported, skepticism remains about the extent to which their views are considered. Thus, a shift towards community-led engagement is crucial [51]. Citizen participation shouldn't be merely a formality but a social innovation process enhancing public service adoption and sustainability to meet evolving societal needs [52]. Effective strategic designs can foster community ownership and responsibility for maintaining fog harvesting and blockchain technology, ensuring sustainable clean water availability and supporting Indonesia's zero hunger goal. Furthermore, a collaborative governance strategy through multi-stakeholder involvement is also key in ensuring success and sustainability in maintaining building fog harvesting and blockchain technology. In collaborative governance, stakeholders can work together to overcome complex social problems, by utilizing their diverse expertise, resources, and networks [53–55]. The government has a role as the main driver in forming policies and regulations that support the development of community-based technology. Within the collaborative governance framework, the government collaborates with NGOs, the private sector, academics, and other parties in the process of policy planning and program implementation. Through open dialogue and consultation, they can build consensus on the steps needed to increase community participation and maintain the technology.

Specifically, NGOs can play a role as important partners in increasing community participation by providing necessary technical support and training. Non-government actors can also be a bridge between government and society in facilitating dialogue and fighting for the interests of vulnerable communities

[56]. On the other hand, there is the private sector which can provide support in the form of technology, infrastructure, and financial resources to support the implementation of a program. In recent years, public–private partnerships (PPPs) have attracted public attention because they are believed to be a valuable tool for uniting public services and social needs [57, 58]. Even the involvement of the private sector in the government's agenda is believed to be an acceleration in achieving sustainable development [59–61]. Therefore, with strong multi-stakeholder involvement in a collaborative governance framework, a conducive environment that supports a sustainable agenda can be created, including efforts to build and maintain fog harvesting technology and its supporters that can support the achievement of the goal of zero hunger.

To ensure that the community empowerment program can run according to what is desired, apart from the need for commitment from all parties through action and action, the sustainability process of fog harvesting technology and its supporting technology requires an official, binding legal regulation. It cannot be denied that community-based technology adaptation regulations involve complex and sensitive decision-making because they involve various stakeholders and aspects such as security, environment, social, and economic. In the context of collaborative governance, this regulation must consider the needs and aspirations of the community and ensure protection of the public interest. The government has a central role in formulating regulations related to community-based technology adaptation. They are responsible for establishing a clear policy framework and providing direction on how the technology can be used, developed, and integrated into larger systems. These regulations should also include effective oversight and regulatory mechanisms to ensure the security and sustainability of the technology. Meanwhile, NGOs and local communities must also be involved in the process of formulating this regulation. Their participation is important to ensure that regulations reflect the needs and aspirations of society and consider the social and environmental impacts that may arise from the use of such technologies. Through open dialogue and consultation, the government can obtain diverse input and views to form participatory regulations [62]. On the other hand, although the private sector is a party that plays a big role in the success of technology development, they must also comply with existing regulations regarding the development and application of community-based technology. In this way, the security and sustainability of technology can be guaranteed, while still paying attention to the needs and aspirations of society as well as the social and environmental impacts that may occur.

Apart from that, the government can also ensure that the regulations made truly represent the various points of view and interests that exist in society. In this way, this will strengthen community involvement in the process of developing and implementing this technology. Thus, community-based technology such as fog-harvesting buildings, supported by policies related to blockchain supply and demand, has great potential to support the achievement of the goal of zero hunger in Indonesia. Only with active and collaborative involvement from various parties, can we ensure the sustainability of this program and provide a significant positive impact on communities living in mountainous and coastal areas that are vulnerable to drought problems, thereby supporting the achievement of the goal of zero hunger in Indonesia effectively and sustainably.

7. Summary

The application of community-based technologies such as fog harvesting and aquaculture with biofloc are innovative steps to overcome the challenges of drought and increase food security in Indonesia. Comprehensive support from various stakeholders is needed to maximize the potential for community empowerment through this technology, so that the goals of zero hunger and sustainable development can be achieved effectively. With its benefits in providing water supplies during dry seasons and droughts, as well as its potential in producing clean water for daily and agricultural purposes, fog catcher buildings are an important solution in increasing food security and overcoming water shortages in vulnerable areas. The integration of this technology in aquaculture and hydroponic cultivation systems also makes a significant contribution to achieving the goal of zero hunger through increasing food productivity.

Tilapia cultivation using a biofloc system provides several advantages, including increasing productivity, increasing fish survival, and reducing water use without water changes. The use of bubble generator technology in this system can further increase dissolved oxygen levels in the water, reduce stress levels in fish, and increase growth rates and feed efficiency, which ultimately can produce faster and larger harvests. The combination of tilapia cultivation with a biofloc system and the use of bubble generators promises an effective solution to increase fisheries productivity and food security. Multi-stakeholder involvement and implementation of supporting policies are the main keys in ensuring the success of community-based technologies, such as fog-catching buildings, and supporting technologies such as blockchain in achieving the goal of zero hunger in Indonesia. With strong collaboration between government, NGOs, the private sector, and society, we can ensure the continued availability of clean water and support inclusive and sustainable food security for all levels of society.

In the next stage, exploring the scalability and replicability of community-based technologies like fog harvesting and aquaculture with biofloc across different regions of Indonesia to maximize their impact on food security and water availability can be the next step of this research. Also, investigating the socio-economic implications of adopting these technologies, including their effects on livelihoods, income generation, and community resilience, to inform policymaking and implementation strategies become another crucial thing to achieve. Lastly, it is important to assess the long-term sustainability and resilience of these interventions in the face of climate change and other environmental challenges, and identify strategies to enhance their effectiveness and adaptability over time.

References

1. Kaseke, K.F. and Wang, L. (2018). Fog and dew as potable water resources: Maximizing harvesting potential and water quality concerns. *Geohealth*, 2(10): 327–332. doi: 10.1029/2018GH000171.
2. Knapczyk-Korczak, J., Szewczyk, P.K., Ura, D.P., Bailey, R.J., Bilotti, E. and Stachewicz, U. (2020). Improving water harvesting efficiency of fog collectors with electrospun random and aligned Polyvinylidene fluoride (PVDF) fibers. *Sustainable Materials and Technologies*, 25: e00191. doi: 10.1016/j.susmat.2020.e00191.
3. Morichi, G., Calixto, L.B. and Zanelli, A. (2018). Novel applications for fog water harvesting. *Journal of Geoscience and Environment Protection*, 06(03): 26–36 doi: 10.4236/gep.2018.63004.
4. Peng, Y., He, Y., Yang, S., Ben, S., Cao, M., Li, K. et al. (2015). Magnetically induced fog harvesting via flexible conical arrays. *Advanced Functional Materials*, 25(37): 5967–5971. doi: 10.1002/adfm.201502745.
5. Qadir, M., Jiménez, G., Farnum, R., Dodson, L. and Smakhtin, V. (2018). Fog water collection: Challenges beyond technology. *Water* (*Basel*), 10(4): 372. doi: 10.3390/w10040372.
6. Yu, Z., Zhu, T., Zhang, J., Ge, M., Fu, S. and Lai, Y. (2022). Fog harvesting devices inspired from single to multiple creatures: Current progress and future perspective. *Adv. Funct. Mater.*, 32(26). doi: 10.1002/adfm.202200359.
7. Taufani, A.R., Utomo, P., Maulana, T.I. and Musofa, M. (2013). Teknologi Pemanen Kabut (Fog Harvesting) sebagai Solusi Mengatasi Masalah Kekeringan pada Dataran Tinggi. *In*: *Pekan Ilmiah Mahasiswa Nasional Program Kreativitas Mahasiswa – Penelitian*, 2013. Accessed: May 03, 2024. [Online]. Available: https://media.neliti.com/media/publications/171390-ID-teknologi-pemanen-kabut-fog-harvesting-s.pdf.
8. Erwanto, Z., Ilmi, M.R. and Humadi, M.R. (2022). Penerapan Teknologi Menara Penangkap Kabut (Fog Harvesting) Untuk Kebutuhan Air Bersih. *In*: *DIKEMAS (Jurnal Pengabdian Kepada Masyarakat)*, 5(1): 32–38. 2021. Accessed: May 03, 2024. [Online] Available: https://journal.pnm.ac.id/index.php/dikemas/article/view/165/146.
9. Prastica, R.M.S. and Erwanto, Z. (2022). *Pilar Manajemen Sumber Daya Air*. Eureka Media Aksara.
10. Adharani, N., Soewardi, K., Dhamar Syakti and Hariyadi, S. (2016). Water quality management using bioflocs technology: Catfish aquaculture (*Clarias* sp.). *Jurnal Ilmu Pertanian Indonesia*, 21 (1): 35–40. doi: 10.18343/jipi.21.1.35.
11. Silviana, H., Yuniwati, I., Erwanto, Z. and Triasih, D. (2021). Pengembangan Bioflok dan Kolam Budidaya Ikan sebagai Wisata Edukasi di Desa Tulungrejo Kecamatan Glenmore Kabupaten Banyuwangi. *J-Dinamika: Jurnal Pengabdian Masyarakat*, 6(1): 100–105. doi: 10.25047/j-dinamika.v6i1.1600.
12. Jannah, R.N., Indraloka, A.B. and Mukti, M. (2023). Penerapan Teknologi Irigasi Tetes Pada Tanaman Polybag di Lahan Workshop Terpadu Sebagai Embrio Eduwisata di Desa Pondoknongko Banyuwangi. *Indonesia Berdaya*, 4(4): 1623–1632.
13. Henriksson, P.J.G., Tran, N., Mohan, C.V., Chan, C.Y., Rodriguez, U.P., Suri, S. et al. (2017). Indonesian aquaculture futures: Evaluating environmental and socioeconomic potentials and limitations, *Journal of Cleaner Production*, 162: 1482–1490. doi: 10.1016/j.jclepro.2017.06.133.

14. Senff, P., Partelow, S., Indriana, L.F., Buhari, N. and Kunzmann, A. (2018). Improving pond aquaculture production on Lombok, Indonesia. *Aquaculture*, 497: 64–73. doi: 10.1016/j.aquaculture.2018.07.027.
15. Collier, W.L. (2019). Aquaculture and artisanal fisheries. *In*: Gary E. Hansen (Ed.), *Agricultural and Rural Development in Indonesia*, 275–294. Routledge. doi: 10.4324/9780429048364-23.
16. Toth, G. (2010). *The Environment Canada Handbook on Fog and Fog Forecasting*. Canada. Environment Canada.
17. Azeem, M., Noman, M.T., Wiener, J., Petru, M. and Louda, P. (2020). Structural design of efficient fog collectors: A review. *Environ. Technol. Innov.*, 20: 101169. doi: 10.1016/j.eti.2020.101169.
18. Kowalski, N.G., Shi, W., Kennedy, B.S. and Boreyko, J.B. (2021). Optimizing fog harps. *ACS Appl. Mater. Interfaces*, 13(32): 38826–38834. doi: 10.1021/acsami.1c08995.
19. Shi, W., Anderson, M.J., Tulkoff, J.B., Kennedy, B.S. and Boreyko, J.B. (2018). Fog harvesting with harps. *ACS Appl. Mater. Interfaces*, 10(14): 11979–11986. doi: 10.1021/acsami.7b17488.
20. Mileta, M., Beysens, D., Nikolayev, V., Milimouk, I., Clus, O. and Muselli, M. (2007). Fog and dew collection projects in Croatia. *In*: *Proc. International Conference on Water Observation and Information System for Decision Support*. Ohrid, Republic of Macedonia.
21. Kowaas, C.D.M.D., Kanafani, F., Fathaya, F.A., Irfan, M., Pratiwi, V.A. and Arbaningrum, R. (2019). Kajian Bangunan Penangkap Kabut dan Penampung Air Hujan untuk Daerah Sentul sebagai Ganti Air Bersih. *Widyakala Journal*, 6: 21–28. doi: 10.36262/widyakala.v6i0.167.
22. Akvopedia. (2016). *Panen Air Hujan/Pengumpulan Embun dan Kabut/Pengumpulan dan penyimpanan kabut*. Accessed: May 03, 2024. [Online]. Available: https://akvopedia.org/s_wiki/index.php?title=Panen_Air_Hujan_/_Pengumpulan_Embun_dan_Kabut_%20/_Pengumpulan_dan_penyimpanan_kabut&oldid=35327.
23. Chopin, T. (2013). Aquaculture, Integrated Multi-trophic (IMTA). *In*: Paul Christou, Roxana Savin, Barry A. Costa-Pierce, Ignacy Misztal and C. Bruce A. Whitelaw (Eds.), *Sustainable Food Production*, 184–205. New York, NY: Springer New York. doi: 10.1007/978-1-4614-5797-8_173.
24. Helfrich, L.A. and Libey, G.S. (1990). *Fish Farming in Recirculating Aquaculture Systems* (*RAS*). Accessed: May 04, 2024. [Online]. Available: https://koiorganisationinternational.org/sites/default/files/Fish-Farming-in-Recirculating-Aquaculture-Systems-RAS.pdf.
25. Samsundari, S. and Wirawan, G.A. (2013). Analisis Penerapan Biofilter dalam Sistem Resirkulasi Terhadap Mutu Kualitas Air Budidaya Ikan Sidat (Anguilla bicolor). *Jurnal Gamma*, 8(2): 86–97. [Online]. Available: http://ejournal.umm.ac.id/index.php/gamma/article/view/2410.
26. Agarwal, A., Ng, W.J. and Liu, Y. (2011). Principle and applications of microbubble and nanobubble technology for water treatment. *Chemosphere*, 84(9): 1175–1180. doi: 10.1016/j.chemosphere.2011.05.054.
27. Xu, Q., Nakajima, M., Ichikawa, S., Nakamura, N. and Shiina, T. (2008). A comparative study of microbubble generation by mechanical agitation and sonication. *Innovative Food Science and Emerging Technologies*, 9(4): 489–494. doi: 10.1016/j.ifset.2008.03.003.

28. Uesawa, S., Kaneko, A. and Abe, Y. (2012). Measurement of void fraction in dispersed bubbly flow containing micro-bubbles with the constant electric current method. *Flow Measurement and Instrumentation*, 24: 50–62. doi: 10.1016/j.flowmeasinst.2012.03.010.
29. Tabei, K., Haruyama, S., Yamaguchi, S., Shirai, H. and Takakusagi, F. (2007). Study of micro bubble generation by a swirl jet (Measurement of Bubble Distribution by Light Transmission and Characteristics of Generation Bubbles). *Journal of Environment and Engineering*, 2(1): 172–182. doi: 10.1299/jee.2.172.
30. Terasaka, K., Hirabayashi, A., Nishino, T., Fujioka, S. and Kobayashi, D. (2011). Development of microbubble aerator for waste water treatment using aerobic activated sludge. *Chem. Eng. Sci.*, 66(14): 3172–3179. doi: 10.1016/j.ces.2011.02.043.
31. Sadatomi, M., Kawahara, A., Matsuura, H. and Shikatani, S. (2012). Micro-bubble generation rate and bubble dissolution rate into water by a simple multi-fluid mixer with orifice and porous tube. *Exp. Therm. Fluid Sci.*, 41: 23–30. doi: 10.1016/j.expthermflusci.2012.03.002.
32. Sadatomi, M., Kawahara, A., Kano, K. and Ohtomo, A. (2005). Performance of a new micro-bubble generator with a spherical body in a flowing water tube. *Exp. Therm. Fluid Sci.*, 29(5): 615–623. doi: 10.1016/j.expthermflusci.2004.08.006.
33. Gordiychuk, A., Svanera, M., Benini, S. and Poesio, P. (2016). Size distribution and sauter mean diameter of microbubbles for a venturi type bubble generator. *Exp. Therm. Fluid Sci.*, 70: 51–60. doi: 10.1016/j.expthermflusci.2015.08.014.
34. Afisna, L.P., Juwana, W.E., Indarto, I., Deendarlianto, D. and Nugroho, F.M. (2017). Performance of porous-venturi microbubble generator for aeration process. *Journal of Energy, Mechanical, Material, and Manufacturing Engineering*, 2(2), 73–80. doi: 10.22219/jemmme.v2i2.5054.
35. Yoon, R.-H. (1993). Microbubble flotation. *Miner. Eng.*, 6(6): 619–630. doi: 10.1016/0892-6875(93)90116-5.
36. Sun, L., Mo, Z., Zhao, L., Liu, H., Guo, X., Ju, X. and Bao, J. (2017). Characteristics and mechanism of bubble breakup in a bubble generator developed for a small TMSR. *Annals of Nuclear Energy*, 109: 69–81. doi: 10.1016/j.anucene.2017.05.015.
37. Zhao, L., Mo, Z., Sun, L., Xie, G., Liu, H., Du, M. and Tang, J. (2017). A visualized study of the motion of individual bubbles in a venturi-type bubble generator. *Progress in Nuclear Energy*, 97: 74–89. doi: 10.1016/j.pnucene.2017.01.004.
38. Huang, J., Sun, L., Liu, H., Mo, Z., Tang, J., Xie, G. and Du, M. (2020). A review on bubble generation and transportation in venturi-type bubble generators. *Experimental and Computational Multiphase Flow*, 2(3): 123–134. doi: 10.1007/s42757-019-0049-3.
39. Basso, A., Hamad, F.A. and Ganesan, P. (2018). Effects of the geometrical configuration of air–water mixer on the size and distribution of microbubbles in aeration systems. *Asia-Pacific Journal of Chemical Engineering*, 13(6): doi: 10.1002/apj.2259.
40. Yoshida, A., Takahashi, O., Ishii, Y., Sekimoto, Y. and Kurata, Y. (2008). Water purification using the adsorption characteristics of microbubbles. *Jpn. J. Appl. Phys.*, 47(8R): 6574. doi: 10.1143/JJAP.47.6574.
41. Akhtar, M.S., Rajesh, M., Ciji, A., Sharma, P., Kamalam, B.S., Patiyal, R.S. et al. (2018). Photo-thermal manipulations induce captive maturation and spawning in endangered golden mahseer (Tor putitora): A silver-lining in the strangled conservation efforts of decades. *Aquaculture*, 497: 336–347. doi: 10.1016/j.aquaculture.2018.08.003.

42. Dahrazma, B., Naghedinia, A., Ghasemian Gorji, H. and Saghravani, S.F. (2019). Morphological and physiological responses of *Cucumis sativus* L. to water with micro-nanobubbles. *Journal of Agricultural Science and Technology*, 21(1). [Online]. Available: http://jast.modares.ac.ir/article-23-15964-en.html.
43. Bagatur, T. (2014). Evaluation of plant growth with aerated irrigation water using venturi pipe part. *Arab J. Sci. Eng.*, 39(4): 2525–2533. doi: 10.1007/s13369-013-0895-4.
44. Mitra, S., Daltrophe, N.C. and Gilron, J. (2016). A novel eductor-based MBR for the treatment of domestic wastewater. *Water Res.*, 100: 65–79. doi: 10.1016/j.watres.2016.04.057.
45. Kaya, Y., Bacaksiz, A.M., Bayrak, H., Gönder, Z.B., Vergili, I., Hasar, H. and Yilmaz, G. (2017). Treatment of chemical synthesis-based pharmaceutical wastewater in an ozonation-anaerobic membrane bioreactor (AnMBR) system. *Chemical Engineering Journal*, 322: 293–301. doi: 10.1016/j.cej.2017.03.154.
46. Hashim, A., Yaakob, O.B., Koh, K.K., Ismail, N. and Ahmed, Y.M. (2015). Review of micro-bubble ship resistance reduction methods and the mechanisms that affect the skin friction on drag reduction from 1999 to 2015. *J. Teknol.*, 74(5). doi: 10.11113/jt.v74.4650.
47. Heriyati, E., Rustadi, R., Isnansetyo, A., Triyatmo, B., Istiqomah, I., Deendarlianto, D. and Budhijanto, W. (2022). Microbubble aeration in a recirculating aquaculture system (RAS) increased dissolved oxygen, fish culture performance, and stress resistance of Red Tilapia (*Oreochromis* sp.). *Trends in Sciences*, 19(20): 6251. doi: 10.48048/tis.2022.6251.
48. Ali, O., Jaradat, A., Kulakli, A. and Abuhalimeh, A. (2021). A comparative study: Blockchain technology utilization benefits, challenges, and functionalities. *IEEE Access*, 9: 12730–12749. doi: 10.1109/ACCESS.2021.3050241.
49. Anand, J., Shobika, M. and Sneha, T. (2022). Programmable stickers to monitor perishable goods using blockchain. *In*: *2022 Second International Conference on Computer Science, Engineering and Applications* (*ICCSEA*), 1–5. IEEE. doi: 10.1109/ICCSEA54677.2022.9936484.
50. Swapan, M.S.H. (2016). Who participates and who doesn't? Adapting community participation model for developing countries. *Cities*, 53: 70–77. doi: 10.1016/j.cities.2016.01.013.
51. King, C. and Cruickshank, M. (2012). Building capacity to engage: Community engagement or government engagement? *Community Dev. J.*, 47(1): 5–28. doi: 10.1093/cdj/bsq018.
52. Castelnovo, W., Misuraca, G. and Savoldelli, A. (2016). Smart cities governance. *Soc. Sci. Comput. Rev.*, 34(6): 724–739. doi: 10.1177/0894439315611103.
53. Lang, A. and Brüesch, C. (2020). Collaborative governance in program implementation: The development of E-relocation notification in the Swiss Canton of Zurich. *International Journal of Public Administration*, 43(12): 1083–1095. doi: 10.1080/01900692.2019.1665681.
54. Schoon, M., York, A., Sullivan, A. and Baggio, J. (2017). The emergence of an environmental governance network: The case of the Arizona borderlands. *Reg. Environ. Change*, 17(3): 677–689. doi: 10.1007/s10113-016-1060-x.
55. Lee, E.W. (2012). Civil society organizations and local governance in Hong Kong. *In*: Stephen Wing-kai Chiu and Siu-lun Wong (Eds.), *Repositioning the Hong Kong Government*, 147–164. Hong Kong University Press. doi: 10.5790/hongkong/9789888083497.003.0007.

56. Banks, N., Hulme, D. and Edwards, M. (2015). NGOs, states, and donors revisited: Still too close for comfort? *World Dev.*, 66: 707–718. doi: 10.1016/j.worlddev.2014.09.028.
57. Berrone, P., Ricart, J.E., Duch, A.I., Bernardo, V., Salvador, J., Piedra Peña, J. and Rodríguez Planas, M. (2019). EASIER: An evaluation model for public–private partnerships contributing to the sustainable development goals. *Sustainability*, 11(8): 2339. doi: 10.3390/su11082339.
58. Wang, H., Xiong, W., Wu, G. and Zhu, D. (2018). Public–private partnership in *Public Administration* discipline: A literature review. *Public Management Review*, 20(2): 293–316. doi: 10.1080/14719037.2017.1313445.
59. Castelblanco, G. and Guevara, J. (2022). Building bridges: Unraveling the missing links between public–private partnerships and sustainable development. *Project Leadership and Society*, 3: 100059. doi: 10.1016/j.plas.2022.100059.
60. Haque, Md. N., Saroar, M., Fattah, Md. A. and Morshed, S.R. (2020). Public–private partnership for achieving sustainable development goals: A case study of Khulna, Bangladesh. *Public Administration and Policy*, 23(3): 283–298. doi: 10.1108/PAP-04-2020-0023.
61. Ma, M., Wang, N., Mu, W. and Zhang, L. (2022). The instrumentality of public–private partnerships for achieving sustainable development goals. *Sustainability*, 14(21): 13756. doi: 10.3390/su142113756.
62. Eriksson, E., Fredriksson, A. and Syssner, J. (2022). Opening the black box of participatory planning: A study of how planners handle citizens' input. *European Planning Studies*, 30(6): 994–1012. doi: 10.1080/09654313.2021.1895974.

15 | Leveraging Information and Communication Technology (ICT) Solutions for Food Security as Panacea for Hunger Alleviation in Nigeria

Torruam, Japheth Terande
Department of Computer Education, College of Education Oju, Benue State, Nigeria

This chapter explores the potential of Leveraging Information and Computer Technology (ICT) Solutions for Food Security as Panacea for Hunger Alleviation in Nigeria. With a growing population, insecurity, fuel subsidy removal which puts increasing pressure on food production, ICT offers a way to improve productivity, efficiency, and sustainability in the agricultural sector in Nigeria. By harnessing ICT tools such as drones, sensors, Global Positioning System (GPS) technology, and data analytics, farmers can make informed decisions about crop management, resource allocation, and pest control. Nigeria, as a country facing food insecurity challenges, stands to benefit significantly from the adoption of precision agriculture practices. ICT solutions enable real-time monitoring of crop conditions, allowing for timely interventions to prevent crop losses due to diseases or pests. Nigeria, as a developing country, can enhance food production, improve food security, and ultimately alleviate hunger in the country by embracing ICT innovations in agriculture. However, challenges such as the high cost of ICT accessories, limited access to technology, inadequate infrastructure, electricity, insecurity, and low digital literacy among farmers must be addressed to fully harness the potential of precision agriculture. Government support, the private sector, and other relevant stakeholders in the agriculture ecosystem involvement, and capacity-building programs are essential to promote the widespread adoption of ICT in agriculture and achieve sustainable food security in Nigeria.

Email: torruam@gmail.com

1. Introduction

Nigeria faces the challenge of food insecurity and hunger due to various factors such as growing population, insecurity, fuel subsidy removal which puts increasing pressure on food production, inefficient agricultural practices, limited access to resources, and unpredictable weather conditions like many other developing countries. The increased hunger in the country has resulted in several protests in the states of the country such as Kano, Niger, etc.

According to [18], desperate Nigerians protested against soaring food prices as an economic crisis forced people to skip meals and eat poor-grade rice used as fish food. Women in northern Nigeria have even resorted to digging up anthills in search of grain stored by the insects to feed their children. This can be attributed to a large extent to the insecurity due to which farmers hardly go to their farms for fear of the unknown, unpredictable weather conditions at the little places in the country where farmers attend to their farms, and lack of adequate rainfall causes inadequate yield and sometimes excess rainfall that causes terrible flood, fuel subsidy removal where the cost of transporting goods and services has skyrocketed causing terrible hunger in the country. A hunger condition that sparked protests in several parts of the country which necessitated the influential traditional emir of Kano Aminu Ado Bayero to warn that Nigerians faced "economic hardships, hunger, and starvation" and called on the president to take urgent action.

The advent of Information and Communication Technology (ICT) innovations in recent years has emerged as a promising solution to improve agricultural productivity, ensure food security, and alleviate hunger in the country. ICT has been defined as a broad-based technology (including its methods, management, and application) that supports the creation, storage, manipulation, and communication of information [6]. With ICT, precision agriculture (PA) involves the use of advanced ICTs such as drones, sensors, GPS systems, and data analytics to optimize agricultural practices. By integrating ICT innovations into agriculture, farmers can increase crop yields, reduce input costs, and minimize environmental impact. This will enable farmers to collect real-time data on soil conditions, crop health, and weather patterns, allowing for precise decision-making in resource management and crop production. According to [1] it is pertinent to note that the agricultural sector in Nigeria has really come of age and the application of ICT to its operation is absolutely necessary for the growth and development of the industry. ICT must be taken as a priority. In this regard, we look up to the policymakers and stakeholders in the agricultural sector in Nigeria to put their heads together and work towards making this happen.

About 90% of the Nigerian farmers are smallholders, owning less than two hectares of land for agricultural activities. Farming is habitually the primary source of revenue for those smallholder farmers in the country, therefore making them vulnerable to price fluctuations and extreme weather events that bring about reduction in their agricultural productivity. The lack of access to resources such as financing, land, and water by smallholder farmers are not only what they suffer from but also from a variety of asymmetries in markets. This includes power structures and information access due to lack of ICT innovations adoption in the country. The adoption of precision agriculture (PA) in the country has the capability of improving the productivity and livelihoods of smallholder farmers and increasing their productivity for hunger alleviation in the country. This is consequently vital for achieving United Nations Sustainable Development Goals (SDGs) 2 on or before 2030. Government investment in ICT innovations for smallholder farmers in their small farm size will eliminate the challenges small farm size poses on several farmers which hinder from achieving maximum agriculture productivity to alleviate hunger in the country. Technological investments enhance productivity and yields and eliminate the barrier to benefiting from economies of scale in the country. PA has the capacity to radically transform the hunger situation in the century. The recent revolutions in PA have brought in farm digitalization and mechanization, reducing the need for physical monitoring, labor, etc. The revolution in ICT innovations increased crop yields substantially and ensured the supply of enough food for the rapidly growing population of Nigeria. However, the under adoption of PA revolution techniques and technologies led to several adverse impacts, including loss of soil fertility, depletion and pollution of water resources, and increased incidence of livestock and human diseases.

PA is now on the cusp of another revolution, the digital revolution has help address several issues faced by the agricultural sector. PA, which is a form of agricultural digitalization, is impacting all stages of agri-food supply chains ranging from planning, growing, and harvesting to marketing, consumption, and waste disposal. The use of digital technologies at the growing stage also does not pose the kinds of environmental harms that resulted from the Green Revolution. These ICT innovations or digital revolution is a key foundation for PA, a farm management approach that uses data and technology to make farming simpler, more efficient, and more productive for alleviation of hunger in Nigeria. ICT innovations enable farmers to apply agricultural inputs like water, fertilizers, and pesticides to be applied in precise amounts to get increased average yields for alleviation of hunger in the country compared to traditional cultivation techniques in the country. The birth of PA has often been linked to the introduction of Global Positioning System (GPS) guidance for tractors yet its adoption in Nigeria is largely limited owing to high costs and technological limitations.

The advances in ICT innovations with its potential to transform the agricultural sector, its growing non-affordability are now making PA unavailable to smallholder farmers in Nigeria. PA has evolved beyond data-driven farm management that focuses on managing crop production inputs such as seeds, water, fertilizers, and pesticides to increase production, reduce waste, and protect the environment. Although data is also essential for PA since various kinds of data relevant for farming such as soil parameters, weather data, crop health, and geographical location can be collected through various means of data collection including on-site sensors, remote sensing using satellites and drones, and weather stations. This data can be integrated and analyzed to generate customized, actionable information and insights that meet the needs of individual farms and farmers and delivered to them via accessible technologies like mobile phones which will enhance their farm productivity for hunger alleviation in the country. PA, solutions which are powered by ICT innovations can improve productivity and crop yields to alleviate hunger. It helps to protect the environment by reducing chemical inputs, foster gender inclusion, and help make farming profitable and attractive for smallholder farmers. PA is transforming all stages of agri-food supply chains ranging from production and processing to distribution and consumption, thereby making food available to people to alleviate hunger in the country. It provides access to the farm-level data to be used beyond the objectives of PA. It can be used to leverage supply-chain actors other than farmers such as input suppliers, credit and insurance providers, and retailers to improve their service offerings to farmers. Farm-level data availability in Nigeria can also enable informed policymaking and help governments address various agricultural issues, including threats from climate change and pests, farmer livelihoods, food security, and food safety to alleviate hunger in the country.

Against this background, this chapter explores how leveraging ICT innovations PA can serve as a panacea for hunger alleviation in Nigeria. This is with the hope of digging out ways for achieving the United Nations (UN) SDG2. This goal is aimed at creating a world free of hunger by 2030. The global issue of hunger and food insecurity has shown an alarming increase since 2015, a trend exacerbated by a combination of factors in Nigeria including subsidy removal, corruption, pandemic, conflict, climate change, and deepening inequalities.

2. Importance/Impact of ICT on Food Security

The use of ICT in precision agricultural practices is inevitable since it provides very important tools and platforms for improving the fortunes of farmers in Nigeria. It furnishes information to the farmers which help them in decision-making and also improves on their farming techniques for an improved productivity and alleviating hunger in the country. A well-informed farmer has great potential to become a successful farmer. This is because the farmer knows when and how best to farm, where and when to sell his products, where and when to obtain loans or grants to improve his farm and productivity. The following are considered to be some of the benefits of ICT in food security to alleviate hunger in Nigeria.

1. **Radio/Television:** With these ICT gadgets, farmers can connect to radio stations to get current information that can help them for farm improvement, farming methods, and productivity for alleviating hunger in Nigeria. The radio can also give a guide and directions on current seed and farm product prices, thereby creating more awareness and attracting more people to venture into precision agriculture. Due to the fuel subsidy removal in Nigeria, radio

receivers are cheaper compared to other ICT innovations in agriculture to afford and maintain and so the majority of the rural farmers can afford it and prefer to rely on it for information. Television can also be used because it affords the farmers opportunity to visualize the demonstration of experts in the field of agriculture. By doing so the farmer can understand better and also become better equipped in the current trends in agriculture.

2. **Mobile Phones:** The rapidly increasing use of mobile phones in Nigeria as a source of information has become a major means of disseminating information to the urban and rural farmers through the short message service (SMS) and various social media platforms. The farmers receive information directly on their phones, and sometimes in their local language. Through this means, farmers can be offered services that will help them achieve better yields and secure better prices to alleviate the hunger situation in the country. Through this, a platform where farmers and traders across the country can share market information via mobile networks and the Internet can be created. This service can help to avoid food being wasted and increase food supply, thereby alleviating hunger in the country. Reaching out to farmers directly through the mobile phones will create awareness among the youth about the profitability in farming. This will, therefore, greatly encourage them to take farming as an occupation for hunger alleviation in the country.
3. **Improvement in Food Supply Chain:** In a country like Nigeria, where there is a prevailing food crisis and hunger, food wastage cannot be condoned. Therefore, wastage of food can be minimized if there is good communication amongst the farmers, marketers, and consumers in the country. The use of ICT innovations such as radio frequency identification (RFID) tags can be attached to farm produce to ensure that they are kept in good condition and also to track inventories to ensure optimal supply of farm produce for the alleviation of hunger in the country. Also, GPS can be used for fleet management for agricultural product distribution.
4. **Monitoring and Early Warning:** ICT innovations in PA has the potential of creating a systematic monitoring and creation of a database of farm outputs and its consumption rate which will go a long way to enhance food security and thereby alleviating hunger in the country. Computer networks and databases can be used to collect and analyze information from the farm's future use. The use of mobile phones and the internet can go a long way to ensure that information reaches the farmers about impending disasters which can help in averting the forthcoming danger. Also, high resolution radiometers and satellites can be deployed for remote sensing of agricultural resources. The deployment of ICTs in the agriculture sector can help to improve food yields and enable farmers to better forecast crop yields and production to alleviate hunger in the country. The use of ICTs to share data increases the number of farmers profiting from the information for better productivity.
5. **Automation of Farm Production:** ICT innovations bring new innovations in PA where new farming methods are introduced and novel farming methods

are automated for improved efficiency and productivity for alleviating hunger. E-record keeping through appropriate software is deployed for proper farm inventory. Statistical tools and computer applications can be used for proper farm budgeting (e-budgeting) in the country. Automation, apart from ensuring that wastage is reduced, also ensures that farmers are adequately informed about the prospects and pitfalls ahead. For example, the following ICT for development (ICT4D) innovation can be found in Nigeria. A rice germination electronic cooker, a robotic tubing-grafting system for fruit-bearing vegetable seedlings, and an air bubble machine and multi-functional, ultrasonic machine for fruit cleaning. Other examples of the importance of ICT in food security include: e-banking, use of internet social media for promotion and marketing, online farmers' market for e-ordering and delivery and e-learning/training for farmers through the internet.

3. Challenges Related to Food Insecurity and Hunger in Nigeria

In Nigeria, there is persistent hunger prevalent in the country resulting in protests and threats of protests from various groups including the Nigerian Labour Congress (NLC) which is the hub of other unions in the country. A large number of people are hungry in Nigeria and as a matter of fact, Nigeria is not seen to be working towards meeting SDG2 which is eventually slowing down the progress towards all other millennium development goals (MDGs). While the removal of fuel subsidies increases national revenue and production, it does not translate into more food generation and does not deliver food for everyone.

The protests and threats of protests from various parts of Nigeria concerning insufficient availability of food causing terrible hunger in the country is a verification that there is no hesitation that food is the most basic of all human survival needs. Although many efforts have been made by the Nigerian government to improve the quality as well as production of food supplies in the country, food insecurity remains prevalent. At the moment, Nigeria is currently facing significant challenges related to food insecurity and hunger. The situation is perceived to be caused by a variety of factors including a growing population, corruption (it is disheartening that even inputs supplied for farmers by the government are sold away by top government appointees for personal interest, thereby making it out of reach of the actual farmers), insecurity (Herders-Farmers conflict, Banditry, etc.), fuel subsidy removal, inefficient and lack of ICT integration of modern agricultural practices, limited access to resources, and unpredictable weather conditions. This has resulted in protests and unrest across the country, with many Nigerians struggling to access sufficient food to feed their families. With the removal of fuel subsidy, there is an increase in revenue generation but the more money is pumped into agriculture and other supporting ministries and agencies to boost food production, the more it is looted away by corrupt Nigerians causing pain to the general populace of the nation.

The emergence of ICT innovations which have been practiced in precision agriculture techniques in recent years offers hope as a potential solution to these challenges. By leveraging precision agriculture techniques that utilize advanced ICT tools like drones, sensors, GPS systems, and data analytics, farmers can optimize their practices to increase food. With the integration of ICT innovations into agriculture, the government can deal directly with the actual farmers and monitor their activities. On the other hand, farmers can collect real-time data on soil conditions, crop health, and weather patterns, allowing them to make more informed decisions about resource management and crop production for increased food production in the country. On the whole, this has the potential of helping to address many of the underlying causes of food insecurity and hunger in the country. Due to Nigeria's alarming increase in hunger and food insecurity since 2015, it is essential for the country to make possible efforts to integrate ICT innovations and adopt ICT solutions like PA for increased food production. Embracing ICT innovations for PA with the United Nations UN's SDG2, which aims to create a world free of hunger by 2030, will be instrumental in helping Nigeria to achieve the SDG2 by 2030.

Besides the above challenges, the high cost of ICT accessories, limited access to technology, inadequate infrastructure, and low digital literacy among farmers need to be addressed. This will enable the Nigerian farmers to harness the potential benefits of PA as a significant tool for increased food production. Taking advantage of ICT innovations to improve agricultural productivity and ensuring food security in Nigeria will in no measure ultimately alleviate hunger in the country. However, this could be achieved only by government support, private sector involvement, and capacity-building programs to enable Nigerians to harness the power of ICT innovations in the country. ICT incorporation in agriculture for better productivity is a promising solution to the rapidly growing population, fuel subsidy removal, and insecurity which has led the agricultural sector under increasing pressure to produce more food to tackle the effect of hunger in the country. Food insecurity is also strongly linked with other global issues, such as population growth, surge in energy demand, as well as competition for land and water, and issues of climate change [3].

According to the World Bank [16], there are three pillars identified for effective food security. These pillars are food availability, food accessibility, and food utilization. The problem of food insecurity, mainly among farming households in Nigeria, is long-standing. The food crisis is chiefly a result of the header-farmers' crisis, corruption, and many other factors which has hampered the Nigerian economy to the realities on the ground for sustainable development causing terrible hunger in the country. This is so because the main goal of food security is for individuals to be able to obtain adequate food needed at all times, and to be able to utilize the food to meet the body's needs. However, 100,000 hunger-related deaths are reported every year which testify to Nigeria's failure to achieve national food security. The recent escalation of food prices has transformed food insecurity in Nigeria from a difficult development problem into an emergency. Nigerians' hunger situation is best defined as being in an intensive care unit. The growing incident of food shortages as manifested in skyrocketing food prices and the low growth rate of agriculture underlines the problem in the food sector. The problem posed by hunger and malnutrition, arising from acute shortages of food has become critical in Nigeria. The country is now classified as one of the countries experiencing hunger [16].

4. Benefits of Precision Agriculture (PA) in Improving Crop Yields, Resource Efficiency, and Sustainability

PA is the science of improving crop yields and assisting management decisions using high-technology sensor and analysis tools. It is a new concept adopted throughout the world to increase production, reduce labor time, and ensure the effective management of fertilizers and irrigation processes. It uses a large amount of data and information to improve the use of agricultural resources, yields, and the quality of crops [9]. With the emergence of ICT innovations, an area of land within a field may be managed with different levels of input depending on the yield potential of the crop in that particular area of land.

PA is a sure way and a promising one to benefit both farmers and society for increasing food production effectiveness while improving stewardship of the environment [15]. These principles are fundamental to the current resurgence of interest in eco-efficiency that is driven by food price spikes in Nigeria. Therefore, this is overlying the progressive concern about the degradation of the agroecosystems in Nigeria and the world at large. For Nigeria to achieve the UN sustainable goal which aimed at putting food on the table of every Nigerian, these drivers have to refocus the country's concern on the dual aims of improving food security while protecting the environment. As such, it makes it a particularly appropriate time to take stock of the relevance of the principles of PA for agricultural improvement in Nigeria for sustainability and food security.

McBratney and Whelan [11] considered industrial agriculture, based on inorganic fertilizers, pesticides and other inputs, generally uses the risk-averse premise that given uncertainty in space and time, uniform within-field treatment is the best strategy. But on the contrary, according to Vanlauwe et al. [17], PA recognizes the fine-scale heterogeneity of agricultural fields, as many subsistence farmers have traditionally done. Whereas subsistence farmers often have to concentrate inputs in fertile microsites as a risk-minimizing strategy, in industrialized agriculture PA focuses on optimizing farm inputs by translating site-specific crop demands into variable management practices [10, 15]. According to Wheeler and Blackmore [5], who have highlighted the other benefits of PA in twofold as the cost of producing the crop in an area of land can be reduced, and the risk of environmental pollution from agrochemicals applied at levels greater than those required by the crop can be reduced. PA has immense benefits for farmers in the following ways:

1. Efficient use of precision agricultural equipment informs farmers on soil characteristics and weather which helps them to plan and improve scheduling of operations, which can increase machinery utilization rates and lower per-acre cost to avoid unnecessary wastage. The use of GPS-based guidance systems can allow farm machinery operators to achieve greater field efficiency under difficult conditions. They can reduce overlap and missed applications of inputs (e.g. spraying), helping fatigued operators maintain higher field efficiency and effective productivity of food to alleviate hunger in the country.
2. Risk reduction of the field level is another area where precision farming provides site-specific management that can point out problems with growing conditions which reduce variability in net returns. At the farm level, adopting this form of farming known as precision farming, information can be used to improve variety choice, crop rotation, and other agronomic practices that reduce the risk to increase productivity. This information on crop growth during the season can help farmers make more informed market decisions that can enhance their out from the farm.
3. Management of differentiated agricultural products is an area that takes care of the future. Precision technology has the capability to help farmers differentiate their production within a particular field. For instance, farmers might segregate higher protein crops for marketing in more rewarding channels. In addition, precision farming technology allows the supplementary control that is required when farmers are managing the production of differentiated agricultural products as opposed to the production of regular bulk crops in Nigeria. This enables farmers for proper documentation of crops' conditions and control of inputs to meet the very specific requirements of these crops for more agricultural productivity.
4. Environmental stewardship is also an area where precision technology helps in the application of chemical and fertilizer that will better match crop requirements. This helps for the prevention of the overapplication of chemicals and fertilizers on the farm. It is important to note that over-application of the chemical and fertilizer on the farm can be non-beneficial to the environment. The management practices generated by precision farming technologies will promote good land stewardship

In agriculture, precision farming techniques leverage sensors, data analytics, and automation to optimize resource use, minimizing water, fertilizer, and pesticide applications. These ICT-driven approaches enhance efficiency, reduce environmental impact, and contribute to sustainable development and better productivity.

5. Challenges Faced by Nigerian Farmers to Embrace Precision Agriculture to Improve Crop Yields, Resource Efficiency, and Sustainability

The adoption of modern technologies in the production and distribution of agricultural products is very low in Nigeria. Therefore, the sector depended more on manual labor for farm activities. This is a result of the lack of adoption of modern technological innovations in Nigeria, particularly as it relates to the mechanization of agriculture to improve productivity which has the potential of reducing hunger in the country.

Furthermore, local farmers can barely afford modern technologies and lack maintenance capacity [13]. Those local farmers have limited access to technology, inadequate infrastructure, low digital literacy, etc. They lack the technical knowhow and even those with little knowledge of it lack the facilities to preserve food items such as yam, beans, corn, etc., which results in wastage (in Benue State which is the food basket of the nation for instance, the amount foodstuff wastage is alarming) thereby, further deepening the insecurity level of food. Also, the lack of food processing apparatus sometimes leaves farmers with no choice but to consume a significant fraction of their harvest within a short period. Food processing is an important aspect of agriculture that prevents wastage of food items that cannot be easily stored in their original form by transforming them into another form that can enable their preservation [14]. For instance, in Benue State, mangos, citrus, etc. can easily be preserved when converted into fruit juice and avoid it being wasted away.

Modern agriculture has become so highly industrialized and dependent on electricity which has eluded Nigeria and has become a mirage for successive governments to achieve, making it impossible to practice mechanized farming. Also, most of the agricultural farm products are produced in rural areas of the country and need to be transported from one location to the other via air, water, or land. This whole process requires a lot of power and fuel to keep food prices low and affordable for the common man. However, with escalating prices of fuel as a result of the removal of fuel subsidy, the cost of transporting farm produce has also skyrocketed taking food outreach of the common and low-income earners, thereby creating hunger in the country.

It has been identified that using ICT has a lot of gains in curbing the food crisis in Nigeria, but these will depend on deploying the rapidly growing range of ICT applications in PA. Unfortunately, despite the bright prospect of ICT in curbing the food crisis in Nigeria, it is grossly lacking in Nigeria, in particular; let alone its circulation across various states in the country. A lot of problems are confronting the use of ICT in curbing the food crisis in Nigeria. These are the problems associated with ICT usage in curbing food crisis in the country:

1. ***Professional knowledge and technical expertise***: ICT is a fairly new area of importance especially in developing communities like Nigeria, etc. It

is a highly technical field, and to understand how it curbs the food crisis in Nigeria, one first has to understand the operational functionality of the materials. Materials differ in terms of technical components, design and set-up, and also in terms of functionality. Same are multidimensional capable of various functions such as giving logical outcomes, manipulating information, etc. Without a user who is knowledgeable enough, ICT cannot create the desired change and progress aimed at curbing the food crisis in Nigeria to alleviate hunger in the country. User's knowledge has a great impact on the effective application of ICT innovations to curb the national food crisis.

2. ***Lack of computers***: Computers are very essential for an effective application of ICT innovations to curb the national food crisis in Nigeria, but they are still very expensive and despite spirited efforts by the government agencies, nongovernmental organizations (NGOs), corporate organizations, and individuals to donate computers to as many agricultural institutions as possible, there still remains a large percentage of agricultural institutions unable to purchase computers for use to disseminate information to farmers and training purposes.
3. ***Lack of electricity***: Nigeria, as one among the developing countries, has not been able to connect all parts of the country to the national electricity grid. This has affected rural training of farmers in the modern innovations in ICT. Furthermore, there is no stable electricity and infrastructure to attract foreign investment which Nigeria is lacking. Consequently, the agricultural sector is left handicapped and may not be able to use ICT facilities to achieve maximum gains for curbing the food crisis in Nigeria and alleviating the hunger condition in the country.
4. ***Broken-down computers***: While a good number of agricultural institutions have benefited from donated used computers, they have not been adequately equipped with the same regarding maintenance and repair; hence, it's very common to see agricultural institutions full of broken-down computers, some repairable and some not. This has actually been a major problem, and therefore, many governments have now put strict measures on any person, NGOs, or corporate bodies willing to donate second-hand computers (it is seen as a dumping ground); e-waste management.
5. ***Burglary***: The fact that ICT innovation facilities are still very expensive makes them a target for thieves who usually have ready markets at a much less price. This has made many agricultural institutions incur extra expenses trying to burglar proof the precision agricultural materials' rooms. This extra expense makes some agricultural institutions shy away from purchasing these materials.
6. ***Lack of internet or slow connectivity***: Most agricultural institutions are not able to connect to the World Wide Web, due to the high costs involved in the connectivity. This has made it extremely difficult to train farmers on modern

technology because many rural areas have no access to the internet and even GSM (global system for mobile communication).

7. ***Increased moral degradation***: Internet pornography, cyber bullying, and other antisocial behaviors are worrisome emerging problems that prevent many agricultural institutions and farmers who have GSM services and internet from using ICT despite its numerous advantages in curbing the food crisis in Nigeria to alleviate hunger.

6. The Potential of ICT Innovations in Addressing Agricultural Challenges in Nigeria

ICT plays many important roles in enhancing agricultural extension administration such as helping to produce knowledgeable and well-informed farming communities, individually and collectively through diagnosing problems, interpreting data, and applying their meanings [8]. According to Chadwick [4] ICT helps in providing knowledgeable and new technology, up-to-date information and services for increased agricultural production, improving market access, capacity-building and empowerment, and information for improvement, management of newly developed agricultural practices and methods emanated from extension education.

The livelihoods of small-scale farmers can be enhanced through ICT since it has the capacity and developmental tools. New ICTs represent the greatest package to date for self-education, distance learning, sustainable development, and women empowerment [12]. Agvu and Chan [2] reaffirm that ICTs have become the most effective method of training, informing, and disseminating proven technologies to rural farmers; hence, an extension worker can learn new technologies, rainfall forecasts, and commodity prices among others, and use that information to advise farmers in rural villages. ICT has considerable instrumental value for other related innovations; the internet which is a functional member of the ICT family is very beneficial in providing opportunities for distance education and training, thus overcoming some of the problems of location and lack of time in family-run small business [7]. Moreover, publications from the internet are produced in downloadable election form instead of paper format, thereby allowing access to information unlike how one had to go to the library to look for books by professionals.

ICT is one of the promising areas of PA and aids in the sharing of knowledge. Mobile phones, in combination with radio, enable messages to be given to a large number of listeners. In Nigeria, cassava growers receive market information through a new initiative known as the integrated cassava project. Its information is based on phones, the internet, and online market prices. For example, ICT services obtained by the cassava growers include prices, demand volumes and offers, trade assistance, training, SMS alerts, and technical messages.

7. Strategies to Promote the Adoption of ICT in Agriculture

The Nigerian government and private sector, as a matter of priority, make all efforts to promote and support the adoption of ICT for improved agricultural productivity through capacity-building programs to tackle the current trend of hunger in the country. The integration of ICT in agriculture can be utilized for providing accurate, timely, and relevant information and services to the farmers, thereby facilitating an environment for more remunerative agriculture. With ICT facilities, farmers can be updated on temperature, humidity, and rainfall with additional parameters such as atmospheric pressure, solar radiation, wind speed, and soil moisture, etc.

The Nigerian government at this point needs to create an agricultural website and encourage farmers to use it for dissemination and assimilation of vital agriculture information on the part of the government and the farmers such as online detailed contents, crops, crop management techniques, fertilizers and pesticides, and many other agriculture-related materials. Most of the small-scale farmers sell their agricultural products to middlemen who now determine the prices by hoarding the products to the detriment of the farmers.

The hoarding of agricultural products makes the products insufficient in the markets which are instrumental in the skyrocketing of prices beyond the reach of low-income earners, thereby causing unnecessary hunger in the country. However, with the provision of commodity prices and market information on a real-time basis available on the internet, the farming community can be provided with choices they lack today. This will ensure better price realization and stimulate a drive towards better productivity. Also, with e-commerce farmers can sell their products online and sell their products right inside his farm. What the farmer needs to do is to register his location and products, to ensure that products ordered online can be traced to a particular farmer which can also widen the market for the farmers. At this point, it is pertinent to say that one does not claim to have exhausted what ICT can do for agriculture with the above discussion.

The situation and the importance of ICT innovations integration for sustainable food productivity in Nigeria has called for a need for the government to create an enabling environment towards improved rural ICT infrastructure and accessibility to the internet. And, other mobile communication technologies for poor rural farmers which would improve the overall use of ICT facilities for a robust agricultural productivity and in turn provide more food, thereby alleviating hunger in the country and achieving SDG2 on or before 2030. Additionally, the government at various levels should establish agricultural-based ICT training centers for farmers in urban and rural areas to acquire the new skills that are essential to use the modern ICT innovations in agriculture for better productivity. Furthermore, the post University graduate, especially Corp members who are on a one-year National Youth Service and agricultural extension workers can be trained on the application of the ICT innovations and employed to train farmers and manage the facilities. Also, the government and private sectors should intensify the use of ICT facilities such as the radio and television programs, mobile technology, and social media platforms as a means to reach more farmers at the same time. Government and NGOs advocacy to farmers should be encouraged to educate stakeholders on the benefits of using ICTs and integrating the technologies for a high agricultural productivity for the alleviation of hunger. Other strategies to promote the adoption of ICT in agriculture which is known as PA are as follows:

1. ***Investment in ICT infrastructure***: The Nigerian government has to prioritize investment in ICT infrastructure to cater for PA needs such as internet connectivity, mobile networks, and data centers to ensure reliable access to ICT tools for farmers to practice PA across the country for easy access of information, thereby improving agricultural productivity and alleviating country-wide hunger.
2. ***Subsidies and incentives***: The government should create an enabling environment to farmers by offering subsidies, grants, tax incentives, or other financial support to farmers and agri-businesses to facilitate the adoption of ICT innovations in PA and make the technology more accessible and affordable to farmers for more agricultural productivity and thereby alleviate hunger.
3. ***Capacity-building and training***: The Nigerian government needs to develop training programs and workshops to create awareness on ICT innovations in agriculture for PA practice and educate farmers on how to use these ICT tools effectively, and improve their technical skills and knowledge of modern farming practices for an increased productivity and hunger alleviation nation-wide.
4. ***Research and development funding***: The government needs to allocate resources to support research and development in the field of ICT innovations for precision agricultural practice. Also encouraging ICT innovations and the development of tailored technologies that address the specific needs of Nigerian farmers for increased productivity and alleviating hunger.
5. ***Public-private partnerships***: Private sector inclusiveness fosters collaborations between government agencies, private sector companies, research institutions, and international organizations to leverage each other's expertise, resources, and networks in promoting the adoption of ICT innovations in PA, thereby increasing farmers' productivity and hunger alleviation across the country.
6. ***Data security and privacy regulations***: Implement policies and regulations to ensure the protection of farmers' data collected through ICT tools, safeguarding their privacy and preventing misuse of sensitive information gives farmers confidence for ICT innovations adoption for an enhanced agricultural productivity and hunger alleviation.

7. ***Market access and information systems***: Creating an enabling environment for the widespread adoption of ICT innovations in PA by establishing information sharing platforms and market data systems that provide farmers with real-time market prices, weather forecasts, and other relevant data to improve decision-making and enhance market access for agricultural products and alleviation of national hunger and food insecurity.
8. ***Sustainable agriculture practices***: Creating an enabling environment for the widespread adoption of ICT innovations in PA also encourage the integration of ICT-based PA technologies with sustainable farming practices to promote environmental conservation, reduce resource wastage, and enhance the pliability of agricultural systems to climate change, thereby ultimately contributing to the alleviation of hunger and food insecurity in the country. The role of ICT in agriculture in the alleviation of hunger and achieving SDG2 in Nigeria is enormous.

8. How PA and ICT Innovations can Contribute to Alleviating Hunger in Nigeria

PA and ICT innovations create an enabling environment for farmers to be more productive and profitable. Nigerian farmers need to be internationally competitive and produce agricultural products of high quality while meeting sustainability goals. When their agricultural products meet the international standards, they can export and get more money to invest in their farm thereby producing more food to alleviate the hunger in the country. To remain competitive, their agricultural producers need rapid access to emerging technologies.

Technological change has been the basis for increasing agricultural productivity and promoting agricultural development. Research affects the productivity of farming systems by generating new technologies that, if appropriate to farmers' circumstances, will be rapidly adopted. Historically, researchers and extension workers have been primarily responsible for identifying and injecting economic and environmental factors into the process of developing and introducing agricultural innovation. This is typically characterized as a top-down process, whereby researchers develop the innovation, extension workers promote its use, and farmers either adopt or reject the innovation based on the features important to them. These technological changes may include those that more precisely target pests and diseases, administer water more efficiently, reduce wastage following harvesting, etc.

Another ICT innovation area that has the potential to help end the hunger situation in Nigeria by enabling more efficient and sustainable agriculture practices, predicting and mitigating the effects of climate change, improving food distribution, and enhancing food security is Artificial intelligence (AI). PA can be considered as one of the key ways that AI can help to increase food production in Nigeria to alleviate the hunger in the country. AI systems have the potential to analyze data from sensors and drones. These help farmers to determine the best time for planting and harvesting crops, optimize the use of fertilizer in their farms, and monitor the health condition of the crop, thereby creating an opportunity for higher food production to alleviate the hunger situation in Nigeria. It has the capacity to improve farmers' crop yields and reduce food waste. This is crucial in Nigeria where food scarcity is a major issue which has led to several protests because of the shortage of food in the country.

Farmers can prepare for natural disasters and mitigate the effects of climate change are other important areas where AI can help Nigerian farmers to increase their food production and alleviate hunger in the country. AI systems can predict the climate patterns and weather events, which can provide farmers with information on how to best manage their crops during droughts and floods, especially the states of Adamawa, Benue, etc., which are always affected by floods. This can help prevent crop failures and ensure food security in the country. AI can also help end hunger in Nigeria by improving food distribution and optimize delivery routes for food aid and track food supply chains to reduce food waste as a target to alleviate hunger and ensure that food reaches those who need it most in every part of the country. This is particularly important in Nigeria where access to food is currently a major challenge facing the country. This form of ICT innovation has the potential of improving food security in Nigeria by monitoring the quality of food in the country and detecting contaminated food supplies in the country. This can be achieved by analyzing data from sensors and cameras.

AI systems can identify potential risks and provide early warnings that can be useful for the prevention of food-borne related sickness and reduce wastage of food in Nigeria where food is already insufficient. Though AI has the potential to be a powerful tool in the fight against hunger in Nigeria to alleviate the country from hunger, it is important to note that it is not a silver bullet. AI must be combined with other efforts, such as policy changes, education, and investment in ICT infrastructures, to create a comprehensive approach and enabling environment to address food insecurity in the country. For example, the government can invest in the education and training of farmers to help them understand how to use AI facilities that can improve their farming practices and productivity. Financial support can also be given to existing farmers and a startup package for upcoming and smallholder farmers who lack access to the technology and resources needed to take advantage of AI to boost their food production.

As a country in an intensive care unit food wise, Nigerian governments need to invest in infrastructure to improve food distribution and storage. These infrastructures could be building roads and transportation networks that can facilitate the movement of food from farms to markets, and as well as investing in storage facilities that can help prevent food waste in Nigeria. Policymakers must work relentlessly to address the root causes of hunger in Nigeria and implement policies that ensure the integration of ICT innovations which have the potential to be a game-changer in the fight against hunger in Nigeria. Thus, in creating enabling environment for an efficient and sustainable agriculture practices, predicting and mitigating the effects of climate change, improving food distribution, and enhancing food security, ICT innovations can help ensure that everyone has access to the food they need to live healthy, thereby achieving the UN SDG2 and alleviate hunger in Nigeria. It is important to note that ICT innovations alone cannot solve the problem of hunger in Nigeria. These must be combined with other efforts to create a comprehensive approach to addressing food insecurity in Nigeria and alleviating the country of hunger on or before 2030.

9. Conclusion

The increasing food prices in Nigeria and enormous productivity challenges in the country are clear indications that Nigeria is not food secure. In this regard, the Government resolved to implement the National Agricultural Technology and Innovation Policy (NATIP). This policy aimed at unlocking the country's untapped potential through the massive deployment of ICT innovations in PA knowledge for more agricultural productivity to alleviate hunger in the country. The technology and innovation deployment in the agricultural sector is to make Nigeria food and nutrition secure. The policy replaces the Agricultural Promotion Policy which ended in December 2020. The policy is expected to leverage the quick achievements in ICT by building on agricultural gains that sustained food supply and nutrition for about 200 million Nigerians to alleviate hunger in the country. The integration of ICT in agriculture in Nigeria can alleviate the current hunger situation in the country. ICT can be utilized to provide accurate, timely, relevant information and services to the farmers. This can facilitate an environment for more remunerative agriculture in Nigeria for a greater food production. The role of ICT in agriculture in the Alleviation of Hunger and achieving SDG2 in Nigeria is enormous.

Although the challenge of food insecurity is a global phenomenon currently, presently it is more prevalent in Nigeria despite the fact that Nigeria is a rich country blessed with abundant fertile land for agriculture and enormous human and mineral resources. This is a country that has a rich history of agricultural practice. Unfortunately, narratives have all soon changed due to the discovery of oil in the country making it to focus more attention on oil, and forgetting agriculture which has landed the country into a hunger situation where the country cannot even produce enough food for her citizens. The huge financial prospect of the oil resources has distracted the nation from its agricultural blessing as all attention has been diverted from agriculture which could have been produced for the nation for alleviation of hunger. Sadly, even the so-called financial returns from the oil resources are consistently manipulated and diverted by corrupt government officials leaving the common man to suffer the consequences, such as hunger, starvation, and malnutrition. Food security can still be restored in the country but some factors are still militating against the restoration, such as insufficient production, inefficient policies and corruption, hoarders and farmers conflicts and banditry, climate change and natural disasters, low technology for processing and storage, etc. It is therefore important for the Nigerian government to surmount these challenges by going back to the drawing board to provide an enabling environment and promote the agricultural sector by providing credit facilities to farmers to enhance their agricultural productivity. Adoption of PA should be encouraged and promoted by all means and at all levels for social networking and cooperation among farmers. It is also necessary for the government to provide the basic infrastructure such as accessible roads and electricity, and make education more attainable to build the capacity of farmers, thereby making them more productive.

Recommendations

The Nigerian government and the private sector, as a matter of priority, should make all efforts to promote and support the adoption of ICT through the relevant bodies such National Orientation Agency (NOA), Television, Radio, etc. They should organize capacity-building programs on the usage of ICT to tackle the current trend of hunger in the country.

References

1. Agbata, C.F. (2019). *Transforming Nigeria's Agriculture Sector with Technology*. [Online] Available at: https://punchng.com/transforming-nigerias-agriculture-sector-with- technology/. [Accessed 10 January, 2024].
2. Agwu, A.E. and Chah, J.N. (2007). Access and utilization of modern information communication technologies among extension personnel in Benue State of Nigeria. *Proceedings of the 12th Annual Conference of Agricultural Extension Society of Nigeria (AESON)* 7–21. Maiduguri 4–7 July 2007.
3. Behnassi, M., Pollmann, O. and Kissinger, G. (2013). *Sustainable Food Security in the Era of Local and Global Environmental Change*. New York: Springer Dordrecht Heidelberg. https://doi.org/10.1007/978-94-007-6719-5.
4. Chadwick, B.P. (2003). *Knowledge Technologies for Sustainable Development. Future Direction in Agriculture and ICTs*. USAID, Winrock International, USA.
5. Earl, Wheeler and Blackmore. (1997). Precision farming: The management of variability. *The Journal of the Institution of Agricultural Engineers, Landwards*, 18–23.
6. French, C.S. (1996). *Data Processing and Information Technology*. London: Thomson Learning.
7. Fillip, B. (2000). *Distance Education in Africa: New Technologies and New Opportunities*. Japan International Cooperation Agency. USA.
8. Gurstein, M. (2000). *Rural Development Security a "Community Informatics". Based Conceptual Framework*. CAB International: Africa Regional Centre.
9. Mulla, (2013). *Encyclopedia of Soils in the Environment* (2nd Edn.), 4: 710–725. https://doi.org/10.1016/B978-0-12-822974-3.00261-5. [Accessed 10 January, 2024].

10. Mueller, N.D., Gerber, J.S., Johnston, M., Ray, D.K., Ramankutty, N. and Foley, J.A. (2012). Closing yield gaps through nutrient and water management. *Nature*, 490: 254–257.
11. McBratney, A.B. and Whelan, B.M. (1999). The "null hypothesis" of precision agriculture. *In*: J.V. Stafford (Ed.), *Precision Agriculture '99*, 947–957. Sheffield, UK: Sheffield Academic Press.
12. Mohan (2001). Bridging the digital divide. *In*: P. Pinstrup Anderson and P. Pandya-Lorch (Eds), *Unfinished Agenda: Perspectives on Overcoming Hunger, Poverty and Environmental Degradation*, 257–265. Washington, D.C: International Food Policy Research Institute.
13. Nwajiuba, C. (2013). Nigeria's agriculture and food security challenges. *Agriculture and Food Security*, 45–53.
14. Saheed, M. and Isa, E. (2017). The challenges of food security in Nigeria. *Open Access Library Journal*, 4: 1–22.
15. Srinivasan, A. (Ed.). (2006). *Handbook of Precision Agriculture*. New York: Food Products Press.
16. World Bank (2007). *Income Poverty: The Latest Global Numbers*. Retrieved from www. worldbank.org.
17. Vanlauwe, B., Bationo, A., Chianu, J., Giller, K.E., Merckx, R., Mokwunye, U., Ohiokpehai, O., Pypers, P. et al. (2010). Integrated soil fertility management: Operational definition and consequences for implementation and dissemination. *Outlook on Agriculture*, 39: 17–24.
18. The Guardian Newspaper. (2024). Food crisis grips northern Nigeria sparking protests. https://guardian.ng/news/food-crisis-grips-northern-nigeria-sparking-protests/.

16 | Augmented Reality Based Food Wastage Reductions in Oman Food Restaurants

Saleem Basha[1*], Rajbunisa[2] and Sunil M.R.[3]

[1] RIEU, University of Buraimi, Oman
[2] Sult Bin Malik, Muscat, Oman
[3] Department of Computing, De Montfort University Kazakhstan, Almaty, Kazakhstan

Augmented Reality (AR) has been applied to numerous fields and is still gaining impetus to be further infused in diversified territories. This work suggests the importance of AR being pervaded into the restaurant and fast-food industry. Businesses need to retain their current customers and expand the customer's base significantly. Thus, businesses must find ways to keep the customer engaged and fully satisfied with the facilities they are offering. This led to a study that focuses on the integration of AR in the restaurant and fast-food industry which can help in improving the marketing of these businesses and elevate businesses. In addition, this research also emphasizes the potential contribution of AR to Sustainable Development Goal (SDG) 2: Zero Hunger. The app designed with the help of AR technology, helped the restaurants to provide top of the range dining experience to the customers by displaying high-definition videos and pictures of food items using a mobile or tablet. This assisted the customers in better selection of food as the visuals are more appealing than the text. By leveraging AR to improve customer interactions and streamline business processes, the sector can play a role in reducing food wastage and backing efforts to combat hunger. This app not only provides the businesses with a high profit margin and increased ROI (return on investment), but also aligns with global efforts towards achieving SDG2: Zero Hunger, making it an ideal scenario for businesses and society at large.

*Corresponding author: saleem.b@uob.edu.om

1. Introduction

Food is life! Mostly we would say that people eat food to live but nowadays it is the other way round, they live to eat. In today's times, it has become sort of a trend for people to go out to eat. It may be due to various factors like a busy workplace routine, no basic cooking skills, and expansion of the food industry allowing you with different options to choose from. Historically, it has been observed that it was considered as a woman's responsibility to cook food. But with the modern age where women are getting educated and liberated, they are not only confined to the household chores but are successfully stepping up to different fields and possibilities. This may also be counted as one of the reasons to crave for outside food when people get so tired fulfilling their workplace duties. So, restaurant dining has become a common experience of everyday life, but that wasn't always the case. Let us go back in time and try to visualize the history of the restaurant industry and see how things have evolved from the early days in the 18th century to today.

1.1. History of the Restaurant Industry

- **The Early Days**
  In the late 18th century, France was the first among many countries to have started the trend of restaurants. But back then, things were quite different. People only used to order soups or restoratives. It was the first place to offer a menu with multiple choice. Before such concepts came into existence, there used to be roadside inns from where people use to buy food but usually whatever the cook had made and whenever it was ready. There was no food made to order [9].
- **Nineteenth- and Twentieth-Century Restaurants**
  Due to the growth in transportation services in America during the 19th century, people could travel longer distances without any hassles. Long-distance travel meant that there was a need for refreshment and food services during the period of travel. Thus, the restaurants became the need of the hour. The first American restaurants tried to follow their French counterparts, but they used local American cuisines. By the time they stepped into the 20th century, their restaurants started to reflect the American food, flavors, and lifestyles. In the mid-20th century, fast food restaurants like McDonalds came into existence due to the fast lifestyle of the people who didn't have time to sit and eat [9].
- **Modern Restaurants**
  In contrast to the early days, modern restaurants are more meticulous about the ambience and flawless dining experience. They focus more on food styling, labeling, branding, marketing, and décor. They invest more on the training and grooming of the staff to make them look presentable. They can choose how they want to present their restaurant to the customers. They also provide customers with a wide variety of options to choose from. There have been

many changes in the history of the restaurants. The industry has transformed from what it was in the past [9].

It has not been a long time since Augmented Reality (AR) and Virtual Reality (VR) technologies seemed far-fetched ideas to be incorporated in daily lives. But they have been gaining importance and have been developing and getting expanded into new areas of businesses as well as the lives of consumers.

AR is a direct or indirect live view of a real-world environment whose elements are augmented by computer-generated sensory input like video, sound, GPS data or graphics. With this ability, AR is greatly helping businesses to expand considerably by providing real-life experiences to the customers.

There has been a considerable amount of study regarding the implementation of AR in educational [7, 8] and other fields. So, moving on to other business ventures, such as restaurant, fast food, and beverage industry, seems to be a fair field for the application of AR. Restaurants are a booming business venture and with the passage of time restaurant owners have revolutionized their way of marketing by adopting modern trends to pace up the sales. A significant change in the way of serving, the interior of a restaurant, and even the staff uniform styles have been observed in order to make the visitors comfortable within that environment.

Technology is no doubt one of the best ways to engage your visitors. AR is one such innovation that can help in increasing your restaurant sales and popularity. It allows you to overlay digital content on the real world using your own phone, engaging the user and customer more quickly and in a new way. Just imagine the visitors scanning the restaurant logo with their mobile devices and seeing the menu the restaurant is offering! Or, they scan a menu and watch the video of how the meals are cooked, or see a head chef appearing on a phone screen with a welcome address!

AR can be used in these industries as tools for improving staff training which can prove cost-effective for the business and create better consumer experience by creating an interactive app that can change the restaurant menu by including visually appealing pictures and videos of food and drinks to attract customers. AR can thus help the food industry to pace up sales and popularity in a competitive environment. It allows us to overlay digital content on the real-world using a phone, engaging the user and customer more quickly and in a new way.

Due to the Covid-19 pandemic, the restaurant and hotel industries are facing a lot of losses and there is a significant drop in revenue. Customers are eager to buy food but are unsure about the conditions in which they are prepared due to the fear of the virus. While placing the order and receiving the packaged food from the restaurant staff, there will be inevitable human contact which should be avoided as much as possible to alleviate the chances of infection. In this case, the AR app can help to avoid contact with restaurant staff by ensuring availability of the menu on your phone and at your fingertips so that without even contacting the staff you can select the food items and place the order. This not only addresses health concerns but also aligns with Sustainable Development Goal (SDG) 2: Zero Hunger, as the restaurant industry plays a role in minimizing food wastage and promoting sustainable sourcing methods.

There are hordes of restaurants across Oman offering delectable varieties of food and beverages. In this competitive environment, each restaurant is trying its level best to survive and flourish. AR can provide a platform for them to prosper. With online food ordering systems gaining momentum, AR can act as a cherry on the cake by providing a means for a classic dining experience with a contemporary outlook.

There are several reasons as to why restaurants should invest in creating an AR app. Some of the main reasons are as follows:

(a) Offering AR food menus
(b) Better food selection and interactive dining experience
(c) Improved customer engagement and experience
(d) Know customer preferences
(e) Search niche customers
(f) Enhance marketing executions and elevate business.

1.2. Food Wastage in Restaurants

The United Nations' Food and Agriculture Organization (FAO) estimates that restaurants and other food service providers account for around 12% of the total food waste generated globally [12]. Going a bit deeper, the FAO suggests that about a quarter of food wastage across the entire supply chain occurs at the consumption stage, so a lot of that food waste is happening in places like restaurants when food is prepared but not fully consumed [12].

A report by the UN Environment Programme and the World Resources Institute backs this up, finding that the hospitality sector, encompassing restaurants, hotels, and catering, is responsible for approximately 12% of all food waste globally [13].

The World Business Council for Sustainable Development also points to the food service sector, including restaurants, generating around 12% of global food waste [14]. So, it seems this 12% figure is a pretty consistent estimate from multiple reputable international organizations.

Clearly, restaurants and the wider food service industry have a significant role to play in addressing the global food waste challenge. Improving inventory management, portion control, and finding ways to donate or repurpose excess food could go a long way. But there's still a lot of work to be done in this area.

1.3. Sustainable Development Goals (SDGs): Zero Hunger

The 2030 Sustainable Development Goals (SDGs) [10] are a globally project initiative at addressing concerns that would enable a more equitable and sustainable future. These goals, which were adopted by all United Nations

Member States in 2015, have 17 interrelated objectives that address concerns such as poverty, healthcare, education, gender equality, and environmental protection. Sustainable Development Goal 2: Zero Hunger aims to end hunger, enhance food security, improve nutrition, and boost agriculture by 2030.

SDG2 aspires to ensure that every person, particularly those from economically disadvantaged backgrounds, has access to enough, safe, and nutritious food throughout the year. Undernourishment and Sustainable Agriculture are crucial indicators.

(a) **Undernourishment:** By 2030, the goal is to eliminate hunger and provide safe, nutritious food for all, particularly vulnerable groups.
(b) **Sustainable Agriculture:** Invest in rural infrastructure development, research, extension services, and technology advancements to boost agricultural productivity [11].

Looking at the advantages AR can offer to the restaurant and fast-food industry, this study attempts to focus on the food industries in Oman by providing them with an interactive app to improve sales and customer satisfaction, and contribute to SDG2: Zero Hunger. The integration of AR in this context not only aligns with global sustainability goals but also addresses current challenges faced by the industry, making it a promising avenue for growth and positive societal impact.

2. Literature Survey

There have been many related works by various authors where AR and machine learning [15–17] have been applied in different fields. Emphasizing on the different authors' works relating to the use of AR in the restaurant and fast-food industry, presented are some research work relevant to the topic.

Sabir R. et al. [1] in their paper have discussed customer satisfaction in the restaurant industry. They have devised a model of customer satisfaction based on different variables such as responsiveness, product quality, physical design, and price. The physical design or layout of the organization has been considered to have an influence on the clients as well as workers. The physical design pertains to an artificial environment which can be improved for customer interaction so as to capture their interests. But they also added that the environment alone is not the basic criteria for customer satisfaction. Other factors like service quality and the primary reason for the customers to go to a restaurant should also be taken into consideration.

Raut K. et al. [2] have talked about building an application for a restaurant where the user can see the 3D model of the menu and food options for better choices and food preferences. Their research was basically focusing on developing an AR Android app which would help in providing better services to the customers. Real-time tagging, an interactive menu, and food identifier were some of the features provided by this app. Their proposed system was able to help consumers browse through the menu and select food based on the pictures, videos, ratings by other customers, and ingredients related to the dish. As a result, the consumer's interest in the app was built which allowed them to have a view of the food on their fingertips. This helped the restaurant to improve their sales. They also stated that some restaurants who do not pay attention to the food presentation can use such apps for better impact on customers and eventually their growth. The problems with language and pronouncing the odd names of the dishes was also removed.

Nayyar A. et al. [3] have discussed the application of AR and VR technology in the tourism and hospitality industry. They have described in detail the concept of AR and VR as well as the display technologies used for them. They have talked about the application areas of AR and VR and how they can influence the tourism and hospitality industry by providing better planning and management, marketing as well as an effective entertainment tool for the consumers. They have also highlighted the impact of AR and VR in the tourism and hospitality industry pertaining to education, navigation, interactivity, exploration, marketing, and the real-time experience provided by them. From their study, it is evident that many institutions have invested in the VR/AR apps which will help their customers to make decisions before purchase. They have shed light on the cost-effective nature of these technologies as well as the power to attract potential customers which can be beneficial for the food industry business.

Ivkov M. et al. [4] have researched about the current innovations in the restaurant industry. This study relies on information obtained through managers of restaurants about various factors which can impact the restaurant industry and its growth. The managers had to fill questionnaires wherein they ranked different innovation areas by allocating points to each area. It was observed that the criterion relating to the design and atmosphere of the restaurant was ranked third and the criterion relating to the technology application was ranked fourth. Basically, the study concluded that innovations are necessary as the restaurant sector is expected to bloom in the next ten years where technology application will be occupying 66% of the market following innovation trends.

Al Khamisi A. et al. [5] have discussed in their paper the rise of AR and its current and future application areas starting from medical learning, education, and advertising to entertainment, design, etc., they have also shed light on how AR can be used in manufacturing, industrial, and commercial fields in the future. AR can be used to revolutionize the customer sector by being introduced in social engagement, entertainment, and marketing.

Basha M.S. et al. [6] in their study have designed an app for improving the educational system in Oman by using AR technology across some courses in a college for enhancing the comprehension and understanding capabilities of students. The app has also been helpful for the teachers for providing much clearer demonstration of topics to the students. Their paper shows a comparative analysis in the results of a particular course studied by students across different semesters with and without using AR. The results have shown a tremendous rise in the grades of the students as well as the mark average of the class.

Basha M.S. et al. [7] in their research article have investigated the past to find references to the use of AR and its applications during those times that can be attributed to its current usage in educational as well as other sectors. They did an extensive study starting with the traditional educational method till the modern educational practices. They concluded that modern research in innovation and development focuses more on AR applications. Through the background study, they observed that AR has many benefits over the traditional educational model. It influences the behavior and learning abilities of the students.

Basha M.S. et al. [8] in their study have compared the academic performance of the language learning students with and without the use of AR. The research implementation and results of this study divulge how the teaching and learning process was enriched with the integration of AR and how the academic achievements were influenced. This work also assessed the acceptance level of the students pertaining to the AR technology and the hurdles faced during each phase of development. The benefits and problems of adopting AR in the educational system were also revealed. However, the researchers also identified the lack of tools and cost of development as bottlenecks during this project. But the readiness and enthusiasm with which the app was used made the project worthy and successful.

3. Methodology

The restaurant and food industries are compelled to enhance their interaction with the customer. To showcase the recipes, an AR application was developed and implemented in two of the busiest restaurants in Muscat. The methodology starts with identifying a suitable restaurant. A suitable restaurant was identified, and several meetings were conducted with the management and the chief chef to decide upon the criteria suitable for augmentation. The following points were noted:

Selection criteria for augmented reality application:

1. Selection of starters
2. Selection of main course
3. Selection of juices and soft drink
4. Special dishes
5. Rarely ordered dishes
6. Expensive dishes
7. Unique dishes
8. Children-centric dishes
9. Colorful dishes.

Upon the discussion with management, the recipes and dishes were selected based on the above criteria. The management had decided to augment starters, main course, juices, and special dishes for augmentation. In the above said categories, some dishes were selected for augmentation based on the selection criteria. On every off day the chef is directed to prepare the dish. Professional cameraman was employed to shoot the dishes in high-definition photos. For every dish around 200 photos were taken and a short video clip of the chef describing the dish. The photos were then converted as 3D models by the authors.

4. Augmented Reality Application Development

The application development has two folds: initial analysis, and interface design.

4.1. Initial Analysis

The Initial analysis is the preliminary step to start building the AR application. It comprised of a planning and research approach. The planning and research approach went hand-in-hand to ensure the research methodology which supports the result. Part of the initial analysis is prediction of risk factors too. The environmental setup for the development of the proposed system will be in identifying a full set of Game Engine and AR framework.

4.2. Interface Design

The interface design is the crucial part of the research, because the application is going to be used by diverse customers which includes all age groups and epicure and non-epicure. A small miss in the design may deflate the customer's enthusiasm to use the application. The research completely depends on the customer survey for their satisfaction. Enough care is taken to design the user interface. The user interface must be user friendly and simple enough to convey all the necessary information to the customer.

The interface design of the AR application was carefully crafted to ensure a seamless and intuitive user experience. The main objectives were to make the application accessible to customers of all ages and culinary backgrounds, while providing them with comprehensive information about the dishes in an engaging and visually appealing manner.

The development of the interface design was an iterative process, with extensive user testing and feedback being incorporated to refine the layout, navigation, and overall usability of the application. This attention to detail has resulted in an interface that is both visually appealing and highly functional, effectively catering to the diverse needs and preferences of the restaurant's customers. Figures 1 and 2 depict the AR framework and interface design.

The AR application development deals with integrating target image, video, audio and 3D models.

AR Application

The development of AR applications is more expensive than any other application development. Figure 3 is the block diagram of AR application. The application

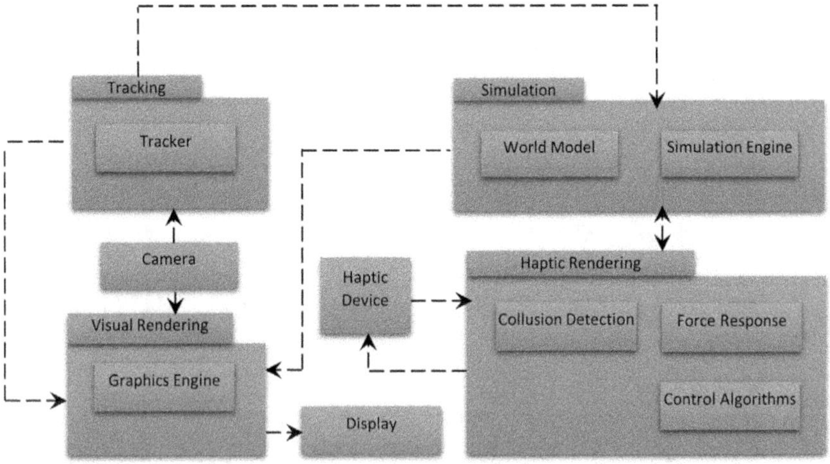

Figure 1: AR framework.

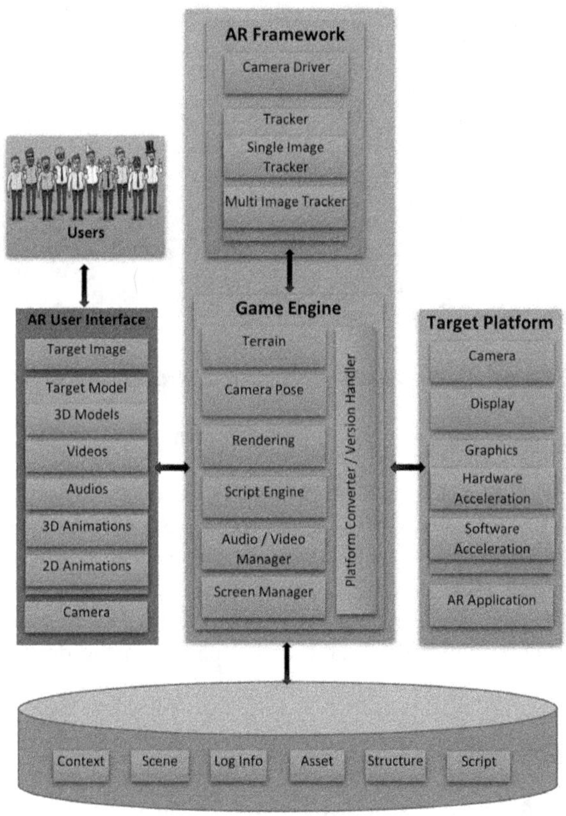

Figure 2: User interface.

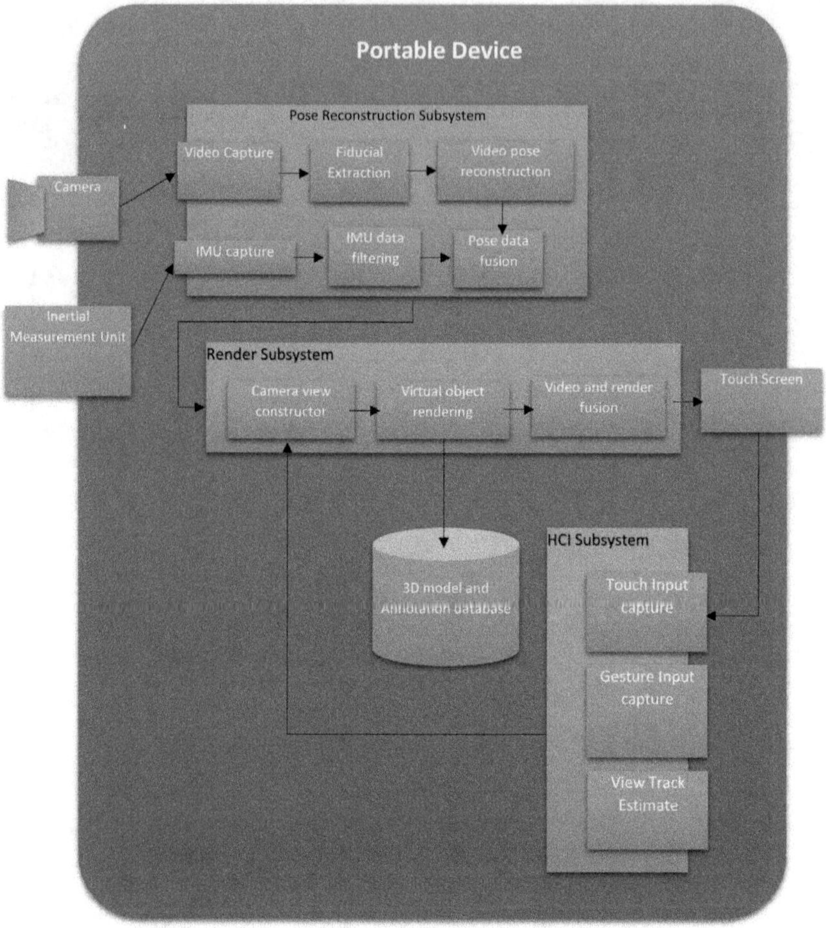

Figure 3: Block diagram of AR application.

has two screens, a welcome or landing screen as shown in Figure 4 and the AR screen as shown in Figure 5. The home screen has the hotel logo and general information, and a button to enter the next screen. The AR screen opens the device camera and starts scanning the object in front of it. Once the device camera finds the target image (the image already known to the application), then the necessary 3D model or the image will appear over the device camera as a floating layer.

5. Implementation

The AR application was developed for two restaurants. The application was developed in Unity 3D game engine with Vuforia AR framework for target

Figure 4: Augmented reality home page.

Figure 5: Augmented reality page.

image. A Canon DSLR camera is used for 3D model generation. The Blender software is used to generate 3D models and for video editing.

Camera setting for photography: ISO 400, aperture f/2.8, shutter speed 1/200 sec.

The application was under study for 15 days with the two restaurants, every customer of the restaurants was given the tablet in which the AR application runs. The customers were allowed free access to utilize and explore the application.

The application gives the real 3D model of the food that they were going to order which gives a rich experience to the customers, to the children and customers non-native to the menu language. Finally, feedback was elicited for the analysis purpose. The result of the analysis is shown in Table 1.

Table 1: Results of the customer feedback analysis

No.	Statement	4 Strongly Agree	3 Agree	2 Disagree	1 Strongly Disagree	Average
1	Service objectives were stated clearly and met.	11	3			3.79
2	The interface was well organized.	9	5			3.64
3	The information presented was relevant and useful.	9	5			3.64
4	Adequate time was stated to find the dishes.	11	3			3.79
5	The application helped me to find the dishes and preview them.	12	2			3.86
6	There was sufficient opportunity for me to participate in the discussions.	11	3			3.79
7	The visual aids were clear and easy to follow.	11	3			3.79
8	The method used in the application was appropriate for the outcome.	10	4			3.71
9	Gives a new food ordering experience.	10	4			3.71
10	The application promotes sustainable food production.	11	3			3.79
11	The app supports reducing food waste in the restaurant industry.	12	2			3.86
12	The food options available align with healthy and nutritious choices.	10	4			3.71

6. Contribution to Sustainable Development Goal: End Hunger

The AR application contributed to achieving the SDG of eliminating hunger in the following ways:

1. ***Reducing Food Waste***: The application provides real-time information on available food options in restaurants, including serving sizes. This enables clients to make better informed selections, reducing the likelihood of over ordering or buying more than they want. By encouraging customers to order only what they need, the app helps minimize food waste at the restaurant level, contributing to the goal of reducing overall food waste. Reducing food waste not only conserves resources but also helps ensure that available food reaches those in need, supporting the objective of ending hunger.
2. ***Nutritional Information and Meal Planning***: The AR app provides nutritional information for menu items, promoting healthier and more sustainable food choices. The app can additionally assist users plan well-balanced meals, which improves overall health and reduces the financial strain on healthcare systems. Healthy and sustainable food choices not only benefit the individual but also contribute to the larger goal of ending hunger by promoting a more equitable and efficient food system.
3. ***Promoting Sustainable Food Practices***: The app provides information on sustainable and locally produced food options. This encourages customers to choose items with a reduced environmental impact, helping to promote sustainable agriculture and reduce the carbon footprint connected with food production and transportation. Sustainable food practices, such as supporting local producers and minimizing the environmental impact of food supply chains, contribute to the overall resilience and stability of the food system. This, in turn, helps address the root causes of hunger by ensuring a more reliable and accessible food supply, particularly for vulnerable populations.

To further improve, the restaurants were advised to integrate the following elements in the AR app to encourage sustainability and contribute to the larger objective of eradicating hunger.

1. ***Educating Consumers***: Using an AR app to provide awareness about the global issue of hunger, the need of sustainable food practices, and how individual choices affect the wider ecosystem of food. Increased awareness might result in better consumption practices. Raising awareness on these critical issues can motivate customers to make more sustainable and hunger-mitigating choices, ultimately contributing to the achievement of the SDG of ending hunger.
2. ***Donation Campaigns and Initiatives***: AR app to promote charitable donations and campaigns for hunger relief. Restaurants can partner with local

food banks or non-governmental organizations (NGOs), and the app can urge customers to donate a portion of their bill to such causes.

3. ***Customer Loyalty Rewards for Sustainable Choices***: Implement a loyalty program to reward customers for sustainable decisions, such as choosing locally sourced food, using recyclable containers, or participating in food donation programs.
4. ***Food Rescue and Redistribution***: The AR app could integrate features that connect restaurants with local food rescue organizations, enabling the donation of surplus or unsold food items. This would be a tremendous step towards reducing food waste by ensuring that edible food reaches those in need, rather than being discarded. By facilitating the rescue and redistribution of food, the app can directly contribute to the goal of ending hunger by providing access to nutritious meals for food-insecure individuals and communities.
5. ***Supply Chain Transparency***: The app could provide information about the sourcing and production methods of the restaurant's ingredients through the AR app can empower customers to make informed choices about the environmental and social impact of their food selections. Transparency around the supply chain can encourage customers to support restaurants and food providers that prioritize sustainable and ethical practices. This increased awareness and ability to make informed choices can drive the adoption of more sustainable food production and distribution systems, further contributing to the goal of ending hunger.
6. ***Advocacy and Policy Engagement***: The AR platform could be leveraged to raise awareness about hunger-related policies and encourage customers to engage with local and national governments to support initiatives that address the root causes of hunger. By providing information and tools for civic engagement, the app can empower customers to become active advocates for policies and programs that address food insecurity, malnutrition, and the systemic issues underlying hunger. Advocacy and policy engagement, facilitated through the AR app, can help catalyze broader societal change and drive progress towards the SDG of ending hunger.

7. Conclusion and Future Directions

AR has a wide range of possibilities to explore but the apprehension of setting foot into a new territory was a challenge which bore fruitful results. Applying the AR technology into the restaurant and fast-food industries in Oman has paved the way for a new revolution in the dining business. The analysis and results show a high degree of customer satisfaction and interest in this technology. Visual aids for food selection in restaurants gave a new experience to customers which was interactive as well as captivating. The AR application proved to be a successful step in promoting the restaurant and its food offerings in the most appealing way. It has not only helped the restaurants to provide better customer service, increase revenue generation, but also holds the potential to contribute significantly to the

SDG of ending hunger. By incorporating features that highlight sustainable and locally sourced food options, providing nutritional information and spreading awareness about responsible food choices, the AR application goes beyond enhancing the dining experience. It aligns with global efforts to address food insecurity, reduce environmental impact, and foster sustainable practices within the food industry.

In future, this technology can be propagated into the beauty, personal care, and cosmetics markets of Oman, which have a huge customer base, in anticipation of better customer experience and achieving global standards in the said domains.

References

1. Raja Irfan Sabir, Muhammad Irfan, Naeem Akhtar, Muhammad Abbas Pervez and Asad ur Rehman. (2014). Customer satisfaction in the restaurant industry: Examining the model in local industry perspective. *Journal of Asian Business Strategy*, 4(1): 18–31.
2. Kunal Raut, Priyanka Khare, Aishwarya Kamble and Sachin Deshpande. (2018). Augmented reality in restaurant. *International Journal of Innovative Research in Computer and Communication Engineering*, 6(2): 1199–1204.
3. Anand Nayyar, Bandana Mahapatra, DacNhuong Le and G. Suseendran. (2018). Virtual Reality (VR) and Augmented Reality (AR) technologies for tourism and hospitality industry. *International Journal of Engineering and Technology*, 7(2.21): 156–160.
4. Milan Ivkov, Ivana Blešić, Karolina Simat, Dunja Demirović, Sanja Božić and Vidoje Stefanović. (2016). Innovations in the restaurant industry: An exploratory study. *Economics of Agriculture*, 63(4): 1169–1186.
5. Abrar Omar Alkhamisi and Muhammad Mostafa Monowar. (2013). Rise of augmented reality: Current and future application areas. *International Journal of Internet and Distributed Systems*, 1: 25–34.
6. Saleem Basha, Mohamed Abbas, Gazala Yusufi and Rajbunisa. (2019). Augmented reality based education for the improvement for sustainable learning ability in Oman educational system. *International Journal of Advanced Trends in Computer Science and Engineering*, 8(1) Sp 2: 110–115.
7. Saleem Basha, Mohamed Abbas, Khalfan Al Masruri, Said Al Saadi, Rashid Al Azri, Adnan and Gazala Yusufi. (2020). Augmented reality brings the fifth dimensional experience in educational system – An intensive breakdown. *International Journal of Innovative Technology and Exploring Engineering*, 9(5): 1911–1919.
8. Saleem Basha, Mohamed Abbas, Khalfan Al Masruri, Said AL Saadi, Rashid Al Azri, Adnan and Gazala Yusufi. (2020). Impact of augmented reality assisted language learning on students academic achievements at tertiary level in Sultanate of Oman. *Journal of Critical Reviews*, 7(17): 2821–2828.
9. The Camden Chronicle. (2019). *History of the Restaurant Industry*. Online Features, found in https://thecamdenchronicle.com/history-of-the-restaurant-industry/.
10. United Nations General Assembly. (2015). *Transforming Our World: The 2030 Agenda for Sustainable Development, A/RES/70*. United Nations: New York, NY, USA.

11. unstats.un.org. (2023). *Goal 2: End Hunger, Achieve Food Security and Improved Nutrition, and Promote Sustainable Agriculture: SDG Indicators*. Found in https://unstats.un.org/sdgs/report/2023/Goal-02/ .
12. FAO. (2011). *Global Food Losses and Food Waste*. http://www.fao.org/3/a-i2697e.pdf.
13. UNEP and WRI. (2021). *Food Waste Index Report 2021*. Found in https://www.unep.org/resources/report/unep-food-waste-index-report-2021.
14. WBCSD. (2018). *CEO Guide to Food System Transformation*. Found in https://www.wbcsd.org/Programs/Food-and-Nature/Food-Land-Use/Resources/CEO-Guide-to-Food-System-Transformation.
15. Seedha Devi, V., Sumathi, K., Mahalakshmi, M., Jose Anand, A., Anita Titus and Naga Saranya, N. (2024). Machine learning based efficient human activity recognition system. *International Journal of Intelligent Systems and Applications in Engineering*, 12(5s), 338–346.
16. Kavitha Balamurugan, Jose Anand, A., Rammya, B., Dinesh Kumar, R., Kalanithy, K. and Mathan, J.N. (2023). Implementation of effective rainfall forecast model using machine learning. *In*: *4th International Conference on Smart Electronics and Communication* (*ICOSEC*), 1655–1660. Trichy, India.
17. Ponmalar, A., Jose Anand, A., Muthamizhan, T., Shobana, T. and Vishwath, J.H. (2022). Automatic forensic analysis of criminal navigation system using machine learning. *In*: *IEEE First International Conference on Computational Science and Technology* (*ICCST*), 1–5.

Index

www.ingramcontent.com/pod-product-compliance
Lightning Source LLC
LaVergne TN
LVHW021725021026
849161LV00004BA/730
* 9 7 8 1 0 3 2 8 8 2 6 7 3 *